D1319850

Breaking Up Bell

A CERA Research Study

Breaking Up Bell
Essays on Industrial Organization and Regulation

Edited by
David S. Evans
CERA Economic Consultants, Inc., Evanston, Illinois

With Contributions by

Robert Bornholz • William A. Brock • David S. Evans
Sanford J. Grossman • James J. Heckman
Michael Rothschild and José A. Scheinkman

North-Holland
New York • Amsterdam • Oxford

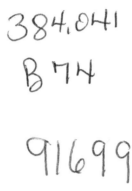

384.041

B 74

91699

Elsevier Science Publishing Co., Inc.
52 Vanderbilt Avenue, New York, New York 10017

Sole distributors outside the United States and Canada:

Elsevier Science Publishers B.V.
P.O. Box 211, 1000 AE Amsterdam. The Netherlands

Library of Congress Cataloging in Publication Data

Main entry under title:

Breaking up Bell.
    (A CERA research study)
    "October 1, 1982."

    Includes index.
    1. American Telephone and Telegraph Company—Addresses, essays, lectures.
    2. Telephone—United States—Addresses, essays, lectures.
    3. Telecommunication—United States—Addresses, essays, lectures.
    4. Industrial organization (Economic theory)—Addresses, essays, lectures.
    5. Monopolies—Addresses, essays, lectures. 6. Trade regulation—
    United States—Addresses, essays, lectures. I. Evans, David Sparks.
    II. Brock, William Allen. III. Series.
HE8846.A55B73 1983   384'.041   82-21122
ISBN 0-444-00734-2

Manufactured in the United States of America

Contents

Preface

The U.S. Department of Justice filed an antitrust suit against the American Telephone and Telegraph Company on November 20, 1974. They charged that AT&T had used its dominant position in the telecommunications markets to suppress competition and enhance its monopoly power. They sought the divestiture from AT&T of the Bell operating companies and the divestiture and dissolution of Western Electric, AT&T's manufacturing subsidiary.

Several years of legal maneuvering followed this filing. AT&T argued that pervasive state and federal regulation of its activities made it immune from the antitrust charges lodged by the Justice Department. It asked the Court to dismiss the suit on these grounds. In September 1978, the Court declined and ordered the parties to prepare for trial. Several more years of pretrial activity followed. During this time, the Justice Department decided to seek only the divestiture from AT&T of the local exchange facilities held by the Bell operating companies.

The trial began on January 15, 1981. But, a few days later, the parties announced they had reached a tentative settlement. The trial stopped. The Court pressed the parties to either formalize the settlement or resume trial. The parties were unable to conclude the settlement under the Court's timetable. The trial resumed on March 4, 1981. The Government rested its case in early August. AT&T moved for dismissal. The Court declined, finding that, "The testimony and documentary evidence adduced by the government demonstrate that the Bell System has violated the antitrust laws in a number of ways over a lengthy period of time." AT&T began pleading its case in mid-August. Despite recommendations by several Cabinet members and a presidential commission that the Justice Department drop the case, the trial continued.

On January 7, 1982, less than two weeks before AT&T was scheduled to complete its case, the parties reached a settlement. AT&T agreed to divest the local exchange facilities held by the Bell operating companies, thereby losing better than half of its assets. The Justice Department agreed to release AT&T from a 1956 Consent Decree which prohibited AT&T from serving unregulated markets. Unfettered by this decree, AT&T will be able to offer various computer and information services. The Court approved the settlement, with some minor modifications, on August 24, 1982. AT&T is scheduled to complete divestiture by September 1984.

The litigation that led to the break up of the Bell System raised numerous economic issues. The Justice Department argued that AT&T tried to crush the competition by *pricing without regard to cost,* a novel version of predatory pricing, and preventing competitive interconnection with the local exchange *bottlenecks.* AT&T replied that it relied on long-run incremental cost pricing, a respected economic concept, and that its competitors were trying to creamskim its lucrative markets. It claimed that divestiture would entail enormous social costs because telecommunications is a natural monopoly, and can therefore be provided most efficiently by a single firm, and because it realized enormous savings from its horizontally and vertically integrated structure. The following essays address these arguments, which have obvious relevance to other regulated and unregulated industries.

The authors of these essays were consultants to the Justice Department on the AT&T litigation between June 1980 and January 1982. The essays evolved from research we began during this litigation, but do not necessarily reflect the views of the Justice Department or members of the trial staff who prosecuted the case. The authors of one essay do not necessarily share the views expressed by the authors of any other essay. The authors of each essay are listed in alphabetical order.

We would like to thank James Denvir and Thomas Spavins of the U.S. Department of Justice for much encouragement and many helpful comments. John Bender, Robert Bornholz, Alan Brazil, Thomas Coleman, Christopher Flinn, John Moen, Imran Mudassar, Brook Payner, S. Ramachandran, and George Yates provided valuable assistance at various stages in our research. Nan Roche prepared the graphs. Kathryn Evans, with the assistance of Janice Gilmore and Katherine Glover, typed the manuscript under impossible deadlines with unfailing good cheer. None of these individuals are responsible for any shortcomings this volume may have.

David S. Evans

Evanston, Illinois
September 1, 1982

List of Figures

Chapter 10

List of Tables

List of Contributors

Robert Bornholz
Senior Research Associate,
CERA Economic Consultants, Inc.

William A. Brock
Professor of Economics,
University of Wisconsin at Madison

David S. Evans
President,
CERA Economic Consultants, Inc.

Sanford J. Grossman
Professor of Economics,
University of Chicago

James J. Heckman
Professor of Economics,
University of Chicago

Michael Rothschild
Professor of Economics,
University of Wisconsin at Madison

José A. Scheinkman
Professor of Economics,
University of Chicago

Breaking Up Bell

Chapter 1
Introduction

David S. Evans

The introduction of competition into the telecommunications industry and the breakup of the Bell System mark the end of an era in which AT&T* controlled most every aspect of the telephone business in this country. During this era, AT&T walked a thin line with remarkable agility. In 1875, Alexander Graham Bell filed his patent on the telephone only hours before Elisha Gray filed a competing claim. This stroke of luck ultimately prevented Western Union, which purchased Gray's patents, from controlling the telephone system as well as the telegraph system. The Supreme Court rejected a rival patent claim by Daniel Drawbaugh, a village mechanic, by a one-vote margin.

AT&T's basic patents expired in 1894. Within 14 years, independent telephone companies had garnered more than half of the telephone subscribers in this country. AT&T regained control over the telephone industry after 1907 by aggressively acquiring independent telephone companies. This practice soon ran into the antitrust laws. In 1913, AT&T avoided an antitrust suit, which probably would have sought to dismantle the company, by agreeing not to acquire any more competing telephone companies. AT&T staved off attempts to nationalize the telephone industry—most of the European systems had been nationalized by 1912—until 1918, when the telephone and telegraph systems were "postalized" for national defense reasons. But government control had such disappointing results that the systems were returned to private control a year later.

AT&T entered the 1920s as a regulated monopoly with firm control over the telephone business. Of the almost 14 million telephones in this country in 1921,

---

*In this book, AT&T always refers to the American Telephone and Telegraph Company and *US v. AT&T* always refers to *United States of America v. American Telephone and Telegraph Company, Western Electric Company, Inc., and Bell Telephone Laboratories, Inc.*, CA No-74-1698, filed in the U.S. District Court for the District of Columbia.

64% were controlled by the Bell System and 32%, although owned by independent telephone companies, were connected with the Bell System. The Willis–Graham Act of 1921 permitted AT&T to purchase competing independent telephone companies and thereby consolidate its controls over the telephone business. Most of the states had public utility commissions that regulated telephone rates and encouraged the cessation of telephone competition. The Mann–Elkins Act of 1910 empowered the Interstate Commerce Commission (ICC) to regulate interstate telephone service. By 1934, when the Federal Communications Commission (FCC) began regulating interstate telephone service, AT&T looked much as it did in 1982. It operated 80% of all telephones and embraced almost all telephones in its nationwide system. Western Electric, its manufacturing arm since 1882, supplied most of the telecommunications equipment used by the Bell System. Bell Laboratories, established jointly by AT&T and Western Electric in 1925, performed most of the research and development work for the Bell System.

Despite several threats to its control over the telephone business, AT&T lived the quiet life of regulated monopoly from the 1920s through the 1960s. A massive FCC investigation in the 1930s attacked Western Electric's role as monopoly supplier to the Bell System. But AT&T rebuffed the investigation's proposals for modifying AT&T's relationship with Western Electric. In early 1949, partly through the urging of an attorney who had worked on the FCC investigation prior to joining the Antitrust Division, the Justice Department charged AT&T with monopolizing the market for telecommunications equipment and sought the divestiture and dismemberment of Western Electric. Seven years later, by agreeing to limit its business to the provision of common carrier services and to limit Western Electric's business to the provision of equipment to the Bell System, AT&T persuaded the Eisenhower Administration, whose lack of enthusiasm for the litigation is well documented, to drop the case.[1]

During the 1950s and 1960s, several tenacious entrepreneurs nibbled at AT&T's monopoly markets. The Hush-a-Phone Company tried to sell a plastic telephone attachment that reduced background noise. The Carter Electronics Corporation tried to sell a device for interconnecting two-way radios with the telephone system. Microwave Communications, Inc. (MCI) wanted to establish private line service between Chicago and St. Louis. AT&T responded to these threatened competitive incursions aggressively. It forebade interconnection of competitive terminal equipment with the Bell System and vigorously opposed entry into private line service.

Several lengthy FCC inquiries ensued. The 1968 *Carterfone Decision* ruled that AT&T could not unreasonably prohibit the connection of terminal equipment manufactured by AT&T's competitors to the telephone system. The 1971 *Specialized Common Carriers Decision* permitted entry into private line service. Thus, in the space of four years, the FCC unleashed competition into the markets for terminal equipment and intercity private line service. AT&T faced its most serious competitive challenge in 50 years.

AT&T's reaction to the businesses that entered the newly competitive telecommunications markets prompted the Justice Department, in late 1974, to file an antitrust suit. The suit, as later modified, sought the divestiture from AT&T of the local exchange facilities held by the Bell operating companies. In early 1982, a few months after the Reagan Administration almost dropped the case and a few weeks before the end of a trial that many observers believed AT&T would have lost, AT&T agreed to divest its local exchanges by September 1984.[2] The local exchanges, which the Justice Department argued were the source of AT&T's power to stifle competition, comprised more than half of AT&T's assets. The Justice Department agreed to release AT&T from the 1956 Consent Decree which precluded AT&T from competing in unregulated markets. In late August 1982, the Court approved the settlement.

The introduction of competition into the telecommunications industry and the breakup of the Bell System controvert arguments that AT&T has made with remarkable consistency over the last hundred years and that few economists or regulators have challenged. Theodore Vail, AT&T's president from 1907 to 1919, railed against competition which merely duplicates facilities available from the Bell System and which tries to share the profitable parts of the telephone business "without assuming the burden for the unprofitable parts."[3] AT&T opposed MCI's 1963 application to establish private line service on the grounds that "MCI's proposal was an attempt to attract only the more profitable segment of the total communications market—a 'creamskimming' arrangement—[and] that MCI had made no showing of a need for the proposed services."[4] In the litigation that led to the breakup of the Bell System, AT&T's "fundamental defense [was] that [its] behavior reflects not a continuing course of conduct to monopolize any market but instead a good faith effort to . . . compete fairly and thereby mitigate the creamskimming effects of new entry. . . ."[5]

Theodore Vail also argued that " 'Interdependence,' 'intercommunication,' 'universality,' cannot be had with isolated systems under independent control, however well connected. They require the standardization of operating methods, plant facilities and equipment, and that complete harmony and co-operation of operating forces that can only come through centralized or ommon control."[6] AT&T opposed interconnection of competitive terminal equipment on the grounds that this equipment would threaten the integrity of the system. AT&T opposed competition in intercity private line service on the grounds that it could provide all telephone service more cheaply than its competitors. In opposing the breakup of the Bell System, AT&T claimed that the telecommunications network is a natural monopoly which "can be planned, constructed, and managed most efficiently by an integrated enterprise that owns the major piece-parts of the facilities network and maintains research, development, manufacturing and systems engineering capabilities."[7]

The profound change in one of this country's most important industries and in the constitution of the world's largest corporation obviously raises numerous economic issues. The following nine essays address some of these issues. The first six essays require only a modest background in economics. Chapter 2 examines the early history of the telephone industry in order to determine whether previous public policies which discouraged competition were desirable and whether the telephone industry was a natural monopoly. Chapter 3 critiques arguments that AT&T used predatory tactics to squelch competition in intercity telephone service. Chapter 4 discusses economic aspects of creamskimming entry and examines whether AT&T faced a serious creamskimming problem. Chapter 5 analyzes whether common ownership is necessary for the efficient provision of telephone service. Chapter 6 reviews the econometric evidence that the telephone industry is a natural monopoly. Chapter 7 examines the impact of divestiture on the cost of capital to the companies that comprise the Bell System. The last three essays require a technical background in economics. Chapter 8 analyzes pricing and predation in regulated industries. Chapter 9 examines the sustainability of natural monopoly against uninnovative entry under alternative assumptions concerning the strategies followed by the entrant and the incumbent. Chapter 10 reports multiproduct cost function estimates and natural monopoly tests for the Bell System.

Although these essays are largely independent of one another, they serve three common purposes. First, they analyze many of the key arguments presented in the antitrust litigation that led to the breakup of the Bell System. Chapter 3 analyzes the Justice Department's case that AT&T engaged in anticompetitive behavior. Chapters 4, 5, 6, and 7 analyze AT&T's case that the conduct alleged by the Justice Department was a reasonable response to socially undesirable creamskimming entry and that divestiture would wreak havoc in the telephone industry. Second, they address whether competition in the telephone industry is desirable or workable. Chapters 2, 5, 6, 7, and 10 suggest that breaking up AT&T's monopoly and introducing intercity competition into the telephone industry will not make this industry less efficient. Chapters 3, 4, 8, and 9 examine whether the participants in the telephone industry can use predatory tactics to prevent socially desirable competition. These chapters find that the major opportunity for predation, and the major obstacle to socially desirable competition, involves using the regulatory process to impose artificial burdens on competitors. Both AT&T and its competitors will undoubtedly avail themselves of this opportunity. Third, they make several theoretical and empirical contributions to the literature on industrial organization and regulation. Chapter 5 argues against the transactions cost theory of vertical integration. Chapter 8 analyzes Ramsey pricing by dominant firms facing a competitive fringe; the incentives held by regulated and unregulated businesses to engage in predatory strategies; and the incentives of a regulated firm with increasing returns to scale to invest in barriers to entry. Chapter 9 analyzes the sustainability of natural monopoly against uninnovative entrants when the entrants assume that the incumbents will maintain output rather than maintain price, as assumed in the existing sustainability literature. Chapter 10 reports a procedure for testing for the existence of natural monopoly and uses Bell System data to demonstrate this procedure.

Any beginning student of the telecommunications industry will be struck by how little basic economic research has been published on this important industry during the last century.[8] The long lack of competition in this industry has prevented the kinds of market experiments from which economists reap rich data and the regulatory process has scrambled the data reported by AT&T and other telephone companies.[9] The following essays consequently rely on empirical research that is skimpier and much less rigorous and reach conclusions that are far more tentative than we would have liked. The introduction of competition into the telephone industry and the breakup of the Bell System will, we hope, provide researchers with the necessary data for testing some of the hypotheses we advance in the following pages.

NOTES

1. Report of the Antitrust Committee of the House Committee on the Judiciary on the Consent Decree Program of the Department of Justice, 86th Congress, 1st Session, January 30, 1959.

2. Judge Greene's *Opinions* on AT&T's motion for dismissal and on the settlement proposal gave rather clear indication that he would have found that AT&T had violated the Sherman Act and would have provided structural relief at least as strong as that contained in the settlement.

3. AT&T, *Annual Report*, 1910, p. 33.

4. AT&T, *Defendants' Third Statement of Contentions and Proof*, p. 583.

5. *Ibid.*, p. 2084.

6. AT&T, *Annual Report*, 1909, p. 23.

7. AT&T, *Defendants' Third Statement of Contentions and Proof*, p. 35.

8. The first major economic treatise on the telephone industry was J. Warren Stehman's *The Financial History of the American Telephone and Telegraph Company* (Boston: Houghton-Mifflin, 1925), which was his Ph.D. thesis, submitted to the Department of Economics at the University of Chicago. The second major treatise was Gerald Brock's *Telecommunications Industry* (Cambridge: Harvard University Press, 1981). In the hundred-odd years since Bell filed his patent, scarcely any articles on the industry appeared in professional economic journals. Unlike other public utilities, little empirical work on the telephone industry has been conducted. James Bonbright, for example, in his classic *Principles of Public Utility Rates* (New York: Columbia University Press, 1961), p. 17, complained that "the telephone industry, despite its vast facilities for statistical and economic research, has never seen fit to publish elaborate studies of its cost functions."

9. See Appendix A to Chapter 10 for a discussion of the problems with the existing data on the telephone industry.

Chapter 2

The Early History of Competition in the Telephone Industry

Robert Bornholz and David S. Evans

The FCC'S decision to permit entry into intercity telecommunications and AT&T's divestiture of its local exchanges promise an era of vigorous competition in the provision of intercity telecommunications services. Already, numerous companies have established or have announced plans to establish intercity telecommunications systems. Local service remains a monopoly. But many industry observers believe that cable television systems and other new technologies may, in the foreseeable future, force local service competition.

The introduction of competition reverses long-standing public policies towards the telecommunications industry. Many students of public utility regulation have argued that the telephone industry is a natural monopoly and that competition necessarily leads to waste and public inconvenience. J. Warren Stehman, an early economic historian of AT&T, argued that "the telephone industry is, perhaps to a greater degree than any other public utility, essentially monopolistic in character."[1] He claimed that telephone competition leads to the inefficient duplication of plants and inhibits communication by subscribers on competing systems. Alfred Kahn observed "it seems clear that [local] service is a natural monopoly: if there were two telephone systems serving a community, each subscriber would have to have two instruments, two lines into his home, two bills if he wanted to be able to call everyone else."[2] He continued, "That the provision of local telephone service is a natural monopoly is generally conceded. The Bell System makes a powerful argument that the same is true of the entire national telecommunications network."[3] By and large, public policies towards the telecommunications industry have been based on similar beliefs that competition in this industry is undesirable.

In order to shed light on whether public policies which discouraged competition were prudent, whether the telephone industry was a natural monopoly, and whether renewed competition will promote efficient telephone service, this chapter reviews the early history of the telephone industry. It has two sections. The first describes the evolution of the Bell System from 1876 to 1921, and shows how the Bell System established its control over the telephone industry.[4] The second reviews the history of competition between the Bell System and the independent telephone movement.[5]

The Evolution of the Bell System[6]

Alexander Graham Bell, a Scottish immigrant and Boston speech professor, filed, on February 14, 1876, the patent application that became the cornerstone of the Bell System. He had, since 1872, experimented with methods for transmitting sound by electrical current. His early interest in transmitting musical tones quickly shifted to transmitting speech. Following several promising but ultimately unsuccessful attempts to transmit sound, he filed a patent application that described two methods for electrically transmitting "vocal or other sounds."[7] In the first method, speech vibrated a membrane placed near an electromagnet. In the second method, speech caused variations in the electrical current between a transmitter and receiver. Using the second method, on the evening of March 10, 1876, he and his assistant, Thomas Watson, had the first telephone conversation.

The previous year, Bell had formed the Bell Patent Association with financial backing from the wealthy fathers of two deaf students he had been tutoring: Thomas Sanders, a Salem leather merchant, and Gardiner Greene Hubbard, a Boston lawyer. After he obtained additional telephone patents and popularized the telephone through numerous lectures and demonstrations, Bell and his partners began the commercial exploitation of their patents. On July 9, 1877, they replaced the Bell Patent Association with the Bell Telephone Company.

The Bell Telephone Company, largely under Hubbard's direction, appointed regional agents to promote the telephone. It leased telephones to customers who were responsible for constructing their own lines, often with assistance from the Bell agent's construction company. The agent received a rental commission as well as any profits from his construction business. Early telephone lines connected two points. By 1878, Hubbard and his associates recognized the importance of a telephone exchange which, by connecting telephone lines to a central switchboard, would enable subscribers to communicate with each other. They formed the New England Telephone Company to promote the construction, but not to engage in the actual operation, of exchanges in New England. The new company licensed local operating companies in Boston and Lowell while the parent company licensed local operating companies in New York and Chicago.

Half interest in the New England Telephone Company was sold to a group of investors—they are usually referred to as Boston capitalists—connected with G. L. Bradley. Additional interest in the Bell Telephone Company was sold to Sanders. Bradley and Sanders thus obtained operating control over both companies. By 1879, these companies were starved for capital and involved in a patent dispute with Western Union, the largest telegraph company. Western Union had purchased a telephone patent from Elisha Gray, a Chicago inventor who had filed a caveat on the variable-resistance method of transmitting speech on the same day Bell filed his first patent, but who had not actually invented a working telephone. While Bell sued Western Union for patent infringement, Western Union rapidly developed its telephone business. Colonel William Forbes, described as a "swashbuckling and aristocratic financier fully in the tradition of Raleigh and Lord Nelson," replaced Sanders as the principal source of financing and became a Bell Telephone Company board member.[8] At his urging, the two Bell companies were consolidated into the National Bell Telephone Company. The larger stockholders in the new company agreed to sell stock and voting proxies only to each other, in an attempt to thwart Western Union's efforts to absorb its competitor. Forbes became president of the new company and Theodore Vail, the superintendent of the Railway Mail Service and later president of AT&T after the Morgan banking interests took control of the Bell System, became its general manager.

In November 1879, the Bell–Western Union patent infringement suit was settled out of court. Western Union withdrew from the telephone business and sold its telephone system—56,000 phones in 55 cities—to Bell. Bell agreed to stay out of the telegraph business and to pay Western Union a 20% royalty on all Bell telephones. The agreement remained in force for 17 years. Western Union thereby denied Jay Gould, who had been trying to establish a telephone–telegraph combine between the Bell licensees and the second largest telegraph company, a critical weapon. The Bell interests thereby avoided protracted litigation with the financially stronger telegraph company.

By 1880, Bell had eliminated its main rival and had obtained financing for establishing its control over the nascent telephone industry. Its president, William Forbes, and its general manager, Theodore Vail, began creating an organization which, in most respects, was to endure until 1982. Over the next 5 years, it would consolidate its control over the local operating companies, establish a long-distance company to weld the local exchanges into a unified system, and consolidate its manufacturing licensees into a wholly owned manufacturing subsidiary. In order to obtain additional capital for the consolidation, the National Bell Telephone Company was replaced by the American Bell Telephone Company on April 17, 1880.

Lacking capital and anxious to promote the telephone, Bell had licensed numerous operating companies between 1877 and 1881. Under the five- to ten-year contracts signed by these companies, Bell received $20 per telephone per year and could purchase the licensee's property at fair valuation when the control expired. Beginning in 1881, Bell encouraged the operating companies to replace their temporary contracts with permanent contracts which continued the licensee fee, rescinded Bell's right to take over the operating company's property, and provided Bell with 30–50% ownership in the operating companies.

The local operating companies were territorial monopolies under the new contracts. They could construct long-distance lines to connect the exchanges within their territories. They were prohibited from connecting their exchanges with those of another operating company: this task was reserved for the parent company. They were also prohibited from connecting with independent telephone companies. If they violated these or any other term of their contracts, Bell could purchase their property for "a reasonable price not exceeding the actual cost."[9]

The new contracts served several purposes. First, they provided Bell with considerable control over the operating companies. Second, they enabled Bell to increase its control over the operating companies by purchasing additional stock in them. Bell made a concerted effort after 1900 to attain majority interest in the operating companies. By 1934, Bell owned 100% of 18 operating companies, majority interest in 3 operating companies, and minority interest in 2 other operating companies. Third, Bell levered its managerial resource by relying on local entrepreneurs familiar with local demand for telephone service and with connections to local political interests. Fourth, Bell shifted much of the risk of developing the telephone business from its stockholders to stockholders, usually local businessmen, in the local operating companies.

Telephone lines could carry conversations no farther than 20 miles in 1877. They were quickly improved. Using metallic circuit systems, they could carry conversations several hundred miles when, in 1884, Bell established an experimental line between Boston and New York. With this line operating successfully, Bell installed a line between Philadelphia and New York the following year. In order to finance, construct, and operate its long-distance system, Bell created the American Telephone and Telegraph Company in 1885. By 1894, AT&T had interconnected most major cities in the Northeast and Midwest, including New York, Chicago, Boston, Cleveland, Detroit, Toledo, Milwaukee, Cincinnati, Dayton, Indianapolis, and most of the larger cities in between.[10]

Over this period, the operating companies consolidated themselves into larger operating companies and created a tighter web of toll lines within their territories. In 1883, Bell formed the New England Telephone and Telegraph Company by consolidating eight operating companies in Maine, New Hampshire, Vermont, and most of Massachusetts. Forbes said this merger was "the outcome of a strong conviction on the part of those interested in telephone matters that it was for the best interests of all concerned, to bring large territory under one management." According to Stehman, "Officials of the companies consolidated defended it as a result of the demand for toll service through the district."[11] By 1894, this large operating company had established a highly interconnected telephone system in New England. Stehman says that, "None of the other companies which were to emerge later had in this period reached anything approaching a similar state of development."[12]

In 1882, Bell established Western Electric as its manufacturing arm. Western Electric, which was founded by Elisha Gray and Enos Barton in 1869, had become the major supplier to telegraphic equipment to Western Union and the supplier of telephones, using Gray's patents, to Western Union's short-lived telephone subsidiary. Financial interests associated with Western Union had acquired majority ownership in Western Electric. Bell bought these interests out and thus took control over Western Electric. Bell consolidated several other manufacturers that had been licensed to make telephones into Western Electric. Bell then made Western Electric its exclusive manufacturing licensee. Under the terms of a contract signed on February 6, 1882, Western Electric could manufacture equipment under any of Bell's patents, Bell could acquire Western Electric's patents at cost, Western Electric was required to supply any equipment required by the operating companies for cost plus a 20% profit, and Western Electric was prohibited from supplying equipment to independent telephone companies in the U.S. and Canada.

By 1885, Bell firmly controlled the telephone business. It had entered into permanent relationships with the operating companies. Although it did not have controlling interests in them, it had considerable leverage over them. It controlled the long-distance lines which provided them with access to other operating companies. It had the right to take over their property if they violated their contracts with Bell by, for example, connecting with an independent company. It also controlled their major supplier, Western Electric. It had 156,000 telephones carrying 756,000 average daily conversations.[13]

With the expiration of Bell's two basic patents in 1893 and 1894, entrepreneurs surged into the telephone industry, establishing exchanges in rural areas that Bell had not yet penetrated and in areas where Bell operating companies offered unsatisfactory service. Eighty-seven independent telephone systems were established in 1894. More than 4000 independent telephone systems had been established by 1902.[14] To combat this competitive encroachment, Bell vigorously expanded its plants by establishing new exchanges, enlarging old exchanges, and extending its long-distance system. This expansion was financed by numerous public stock offerings. By 1899, Bell's Board of Directors, still controlled by the Boston capitalists—Forbes, Bradley, and Cochrane—owned less than 1% of the outstanding stock.[15]

Massachusetts laws inhibited the expansion and consolidation of the Bell System during the 1890s. These laws prevented the American Bell Telephone Co. from obtaining more than 30% control over its licensees, prevented it from paying dividends on its stock, and inhibited it from raising needed capital. Consequently, during 1899, it transferred its stocks and bonds to the American Telephone and Telegraph Co., its New York subsidiary, which would control the Bell System for the next 85 years.

Between 1900 and 1907, the Boston capitalists gradually lost control over AT&T to the Morgan banking interests. Two groups of financiers—Widener–Elkins from Philadelphia and Rockefeller–Stillman from New York—joined forces in 1899 to develop an independent long-distance system with independent exchanges in Bell-controlled Boston and New York. They formed a company that acquired a controlling interest in the Erie Telegraph and Telephone Co., which owned roughly 115,000 phones in the Midwest, as well as a company that was to develop independent systems in New York and Boston. Widener and Elkins, allegedly influenced by J.P. Morgan, withdrew from the venture. The Erie Company, which had incurred a large debt for expansion of its facilities, was unable to find alternative stock financing. As a result, according to a leading financial newspaper [16]

> When the Erie people found themselves burdened with a debt of $7,500,000, and the Philadelphia people unable to finance it, arrangements were made to return the company to Boston, and to the care of the Old Colony Trust Co., the chairman of whose board of directors is a director of the American Bell Telephone Co., and one of whose Vice-Presidents is upon the Erie Board.

The chairman of Old Colony, T. Jefferson Coolidge, was also associated with J.P. Morgan. Meanwhile, AT&T, then lead by F.P. Fish, had purchased controlling interest in the Telephone, Telegraph and Cable Company, thereby obtaining a 20% interest in Erie. In 1902, Coolidge and Fish agreed on a reorganization plan for Erie whereby AT&T assumed control over Erie's properties. Within two months, AT&T, then in urgent need of cash, sold 50,000 shares of stock to a group backed by Morgan under the condition that George Baker and John Waterbury, two Morgan allies, would become directors of AT&T. Danielian, a member of the FCC staff that investigated AT&T during the late 1930s and author of a book on AT&T, implies that the consolidation of Erie into Bell was a *quid pro quo* for the subsequent stock sale and that Morgan was behind Coolidge as well as Baker and Waterbury. [17]

By 1902, the independent telephone companies operated 44% of all telephones. [18] In 1901, Vail, then a stockholder but not yet a director of AT&T, recognized that "the worst of the opposition has come from a lack of facilities afforded by our companies,—that is, either no service, or poor service" and that, to "meet these increasing demands, increasing amounts of money will be needed each year." He estimated that at least $200 million would be required and urged a "broad financial policy covering a period of no less than five years." [19] The Baker–Morgan interests on the board of directors, between 1902 and 1907, pressed various methods for financing AT&T's huge capital requirements. In 1906, AT&T sold $150 million of convertible bonds to a group headed by J.P. Morgan, with $100 million of purchases staggered between April 1906 and January 1908 and with an option for $50 million more. The Morgan syndicate had considerable difficulty selling the bonds. By June 1908, only $10 million of bonds had been placed, leaving the Morgan syndicate with $90 million.

This situation gave Morgan considerable leverage with AT&T. The Boston capitalists lost control over AT&T. Morgan representatives took control over the executive committee, and Theodore Vail replaced F.B. Fish as president. This changing of the guard took place in 1907. Since 1907, there has been no major shake-up in the control over AT&T. Many of the policies followed by AT&T in the 1970s were promulgated by Vail during his presidency between 1907 and 1919. Much of the rhetoric used by AT&T to argue against competition and to enhance regulation was used by Vail early in his presidency. The board of directors has perpetuated itself. The president of AT&T has passed powers, in every instance, to a subordinate who has spent most of his career at AT&T.

Prior to 1907, Bell used several methods to impede the development of an independent telephone system. Its operating companies were prohibited from connecting with the independents. Its manufacturer was prohibited from supplying the independents. Its patent arsenal permitted numerous patent suits against the independents. It reduced its prices for telephone service. It pursued a policy of rapid expansion of telephone service. These methods were remarkably ineffective. By 1907, the independent telephone companies operated 51% of all telephones.

The Walker Report found that[20]

> With the advent of Baker–Morgan control of the Bell System in 1907, there was an abrupt change in the Bell System policy from one of meeting competition through rapid expansion to one of financial competition through absorption and purchase of independents. This change in policy is evidenced by correspondence among company officials and banking interests and by such external manifestations as the curtailment of expansion shown by construction records, a decrease in the rate of growth in telephone stations, by reversal of the former policies of refusal to interconnect with independents, and refusal to sell them Bell System telephone apparatus and equipment.

Bell thus revised its strategy from competing directly with the independents to absorbing the independents into its own system. It tried to acquire independent exchanges that competed directly with its own exchanges. It tried to acquire independent exchanges that were strategic links in the regional independent systems. Finally, it tried to persuade independent exchanges that did not compete directly with its own exchange to interconnect with its system. As a result of these policies, the percentage of all telephones operated by independent telephone companies declined from 51% in 1907 to 45% in 1912 and the percentage of all telephones operated by independent telephone companies that did not connect with the Bell System declined from 37% in 1907 to 17% in 1912.[21]

The leaders of the independent telephone movement resisted the Bell System's advances. In 1912, Clarence Mackay, a major force behind the development of competing independent systems, and other backers of the independents complained to the Justice Department that AT&T was violating the Sherman Antitrust Act. The Attorney General threatened AT&T with an antitrust suit to dismantle the Bell System. At the same time, the Interstate Commerce Commission, which had assumed jurisdiction over the telephone industry in 1910, launched an investigation of AT&T's practices. AT&T sought a compromise in order to fend off this two-pronged threat.[22] Under the Kingsbury Commitment, AT&T agreed to purchase only noncompeting independents and to connect noncompeting independents with the Bell System.

The Kingsbury Commitment remained in force until 1921, when the Willis–Graham Act permitted the consolidation of competing telephone exchanges subject to regulatory approval. Between 1912 and 1921, the percentage of all telephones operated by independent telephone companies declined from 45% to 36% and the percentage of all telephones operated by independent telephone companies not connected to the Bell System declined from 17% to 4%.[23] Thus, the Kingsbury Commitment failed to preserve the independents as serious competitors. Between 1921 and 1934, the percentage of all telephones operated by independent telephone companies declined from 36% to 21%.[24] By 1934, virtually all telephones were connected with the Bell System and there was no direct competition with the Bell System.

There are two alternative interpretations of this history of competition and monopolization in the telephone industry. First, the fact that the independent telephone companies were unable to provide a competitive nationwide telephone system, the fact that so many independents apparently found it profitable to join the Bell System through acquisition or interconnection, and the fact that the Bell System succeeded in establishing a single, universal, interconnected telephone system suggest that a single firm with common ownership over the pieces of the network can provide telephone service more efficiently than multiple firms. The competitive process, in effect, revealed that the telephone industry was a natural monopoly. Second, the fact that AT&T was unable to impede the independent telephone movement through direct competition in price and quality and the fact that AT&T had to resort to the same kind of merger tactics that created monopolies like Standard Oil suggest that the telephone industry was not a natural monopoly. AT&T circumvented the competitive process in order to establish its unnatural monopoly. After reviewing the competitive strategies adopted by the Bell companies and by the independent telephone companies, the next section examines these alternative interpretations in more detail.

Telephone Competition[25]

The first Bell patent, which covered methods for transmitting vocal sounds, expired in March 1893. The second Bell patent, which covered the telephone handset, expired in January 1894. Nine hundred patents, covering improvements made since 1877, remained in force. The most important of these—the Berliner patent—covered the telephone transmitter. In 1877, Bell acquired the rights to and filed a patent on a variable-resistance transmitter developed by Emile Berliner. The Patent Office did not approve the patent until late 1891, thereby extending Bell's patent monopoly until 1908. The Berliner patent became the subject of much contentious litigation. It was rescinded by the Federal Court of Massachusetts in 1894 and reinstated by the Supreme Court in 1897. But a subsequent Supreme Court decision construed the patent so narrowly as to destroy its effectiveness in extending Bell's monopoly.[26] With Bell's patent wall quickly disintegrating and with the formation of independent telephone manufacturers who supplied all the equipment necessary for building telephone exchanges, independent telephone companies sprouted up across the country. Between 1894 and 1902, the number of independent exchanges increased from 154 to 4017[27]; the number of independent telephones increased from 15,000 to 970,000; and the percentage of telephones operated by the independents increased from 6% to 44%.[28]

This section has three parts. The first part reviews the development of the independent telephone system. It describes competition between the local exchanges and the development of the independent toll system. The second part analyzes the pricing and interconnection strategies adopted by competing telephone companies. The third part evaluates the historical evidence that the telephone industry was a natural monopoly during the early part of this century.

System Building

The independent exchanges were started mainly by local entrepreneurs responding to local profit opportunities. They faced common problems and, because they seldom competed against one another, a common adversary: the Bell System. Recognizing these facts, representatives from the independent exchanges met in 1897 and formed the National Association of Independent Telephone Exchanges. While remaining separately owned, they agreed on a five-part strategy for competing with the Bell System.[29]

Cooperation between the independents for mutual protection and development.

Cooperation between the independents for court battles with the Bell System.

Development of long-distance toll lines between the independents.

Establishment of an independent telephone system in Chicago.

Development of an independent long-distance company connecting the major commercial centers.

The first three parts of this strategy proved quite successful. The independent association provided a convenient forum for exchanging information, discussing competitive strategies, and negotiating with the Bell System. Beginning in 1897, Bell began losing major patent infringement suits.[30] Toll lines connecting nearly all independent exchanges were quickly established.

The second two parts of this strategy failed miserably. Four ventures were organized to develop long-distance service between the major commercial centers.[31] The first was to establish independent exchanges in New York and Boston and to establish long-distance service between these cities. The second was to connect independent exchanges in Baltimore and Pittsburgh. The third was to establish independent exchanges in Columbus and Cleveland and to establish long-distance service between these cities. The fourth group planned to connect Minneapolis, St. Paul, and other cities along the Mississippi. The independents failed to secure subway privileges, necessary for laying underground cable, in New York and Boston, thereby leaving these Eastern commercial centers under the absolute control of the Bell System. The independents also failed to obtain a franchise in Chicago, thereby leaving the major Midwestern commercial center under Bell System control. The independents succeeded in developing a system that encompassed Baltimore, Washington, and Pittsburgh (this was controlled by the United States Telephone and Telegraph Company) and a somewhat less successful system that included Cleveland and Columbus. Given that there was probably a considerable demand for long-distance communication with the major commercial centers—Boston, New York, and Chicago—the independent's inability to establish exchanges in these cities or to connect with Bell System exchanges in these cities was a major handicap.

Despite this failure, the independents made deep competitive inroads into the Bell System. They controlled 51% of all telephones by 1907. Although most of the independent exchanges served predominantly rural Midwestern areas, many independent exchanges competed directly with Bell exchanges in the larger cities. After the turn of the century, "over half of all Bell exchanges in cities over 5000 faced competition."[32] In 1916, well after the heyday of independent telephony, 20% of all Bell exchanges still faced direct competition from independent exchanges.[33] Moreover, the independents provided a considerable portion of toll service as well as local service. By 1907, when they operated 51% of all telephones, the independents carried 20% of all toll traffic.[34]

The independents, as the strategy detailed above suggests, hoped to develop an independent system capable of competing with the Bell System on a national, systemwide basis. In 1905, James B. Hoge, a major leader of the independent telephone movement, stated[35]

> The Independent officers and stockholders have for years been making the claim that there would be two great telephone companies the same as there are two telegraph companies. Did it ever occur to you that we are fast approaching the realization of that prediction[sic]? Today we have almost a continuous system from the eastern slope of the Rocky Mountains to the Atlantic Coast.

He said the independents' aim was to "build up a complete competitive system."[36] The independents developed a standard logo that would differentiate members of the independent telephone system from the Bell companies.[37] According to Hoge, "A number of state associations have adopted the shield as their toll sign, and a large number of independent toll line companies also have adopted the shield, for the reason that they see the importance of having some uniform emblem for long-distance business, which everyone that travels will be able to recognize."[38]

The independents tried to eliminate competition among themselves through territorial consolidations and divisions. Hoge noted[39]

> The subject that has just been discussed with reference to protecting Independent operating companies against competition, from an Independent standpoint, was taken up some days ago at the Kansas meeting, which adopted a resolution that

in cases of this kind the association secretary should notify all manufacturers of these members and the territory they were operating in, and if any manufacturer sold any material to compete with these companies he would be immediately notified by the association or their members that they would refuse to buy from that manufacturer. I believe that substantial manufacturers are willing to cooperate with us, but the only way to get that cooperation is through organization. It's necessary for us all to go into the association and work as a unit.

Nevertheless, as late as 1916, there was competition between independents in a third of the cities in which there were competitive exchanges.[40]

Several large, regional, independent telephone companies emerged from the consolidation of independent exchanges. There was little effort during the heyday of independent telephones, however, to form a commonly owned independent telephone system.[41] Many independents adopted the so-called "Ohio plan" for developing a national independent organization which could encourage standardization of facilities, develop an independent toll system, and stand firm against the Bell System. Hoge, for one, had several role models in mind.[42]

For the last few years the steel companies of the United States have been operating as a community of interests, the same as the great railroad corporations, and it has not only proved satisfactory to the patrons of the companies, but has also proven eminently satisfactory from an operating and security-holding viewpoint. At this time there is a meeting in Washington City of the International Railway Congress, made up of representatives of the various steam railways in the world. They hold their conventions every five years, this being the first one ever held in the United States. Their plan is that of presenting subjects of special importance, which are afterwards taken up in a logical way and discussed, then submitted to a committee, which brings in what seems to it the consensus of the opinions of the delegates present. That report is then discussed, and if it receives the approval of a majority of the delegates present, it is ratified as the report of the convention. This does not in any way bind any of the railway systems to adopt it, but it is quite likely to be very carefully considered by all successful railway operators.

It is clear that telephone competition took an extremely narrow form in the early part of this century. There were seldom more than two competing exchanges in any one city. Exchanges usually had access to only one system of toll lines. Competition took place between two competing telephone systems: the Bell System on the one hand and the Independent Telephone System on the other hand. Before examining the strategies used by these systems, the remainder of this part reviews the history of competition at the exchange level and the development of independent toll systems.

By the turn of the century, direct competition between local exchanges was common. Of 1002 cities that had telephone service and populations greater than 4000, 41% were served exclusively by Bell, 14% were served exclusively by an independent, and 45% were served by two or more competing telephone systems. Assuming a third of the competition involved competing independents, more than two-fifths of the Bell exchange in larger cities faced direct competition from an independent telephone company. In 1916, after the Morgan–Vail strategy of financial acquisition had eliminated direct competition between Bell and many independent exchanges, only one-fifth of all Bell exchanges faced direct competition from an independent telephone company.[43]

The independents generally developed their systems in districts where the Bell System had not provided service. Frederick Dickson argued[44]

> It was the policy of the Bell company to develop only in our larger cities and towns, and to develop there only in the thickest built up centers. In every case with which I am acquainted the Bell company waited until the Independent company began laying its conduits and stretching its cables before offering accommodations to the residence section of the city.

Consequently, competing exchanges tended to divide the cities geographically, with some overlap, usually in the business districts.[45]

Between 8% and 13% of telephone subscribers in cities with two competing systems had service from both systems.[46] There was greater duplication in larger cities. Most duplicate subscribers were businesses. For example, 4900 of 5200 duplicate subscribers in Pittsburgh were businesses.[47] Bell cited these duplications as evidence that competition was inefficient. In the first of many yearly broadsides against the independent telephone movement, President Vail argued[48]

> Duplication of plants is a waste to the investor. Duplication of charges is a waste to the user. . . . Given the same management, the public must pay double rates for service, to meet double charges, on double capital, double operating expenses and double maintenance.

The independents made two counterarguments. First, the cost of duplication was not substantial. According to a rebuttal prepared by the independents[49]

> [Duplicate investments] are mostly in the business districts nearest the exchange where the cable circuits by reason of short lengths and the most economical sizes, are cheapest. . . . Switchboards, if not connected, are cheaper separated than combined. . . . Two pole lines may represent waste when they are parallel with no more of a load than could be borne on one. They may have no element of waste with a greater load, or when shared with other wire-using companies. . . . Of the subway and conduct system, only that smaller portion is waste which is represented by the costs of opening and repaving the streets. . . . The cost of interior wiring and instruments is duplicated only in proportion to the duplication of telephones.[50]

Second, telephone competition reduced rates and increased the number of subscribers. Duplicate subscribers reached more subscribers for a lower total rate under competition than they did under monopoly. The situation in Columbus was often cited as an example.[51]

> The present Columbus subscriber, who finds it either convenient or necessary to keep two such services, pays $54.00 for one and $40.00 for the other, or a total of $94.00 per year, $2.00 less annually for two than one cost him before competition, and what does he get? Connection with 22,000 telephones instead of 2,000, or, in other words, competition has brought to the alleged burdened businessman who has to keep two telephones, 20,000 more telephones to talk with and has handed him a $2.00 yearly rebate in the bargain.

Thus, competition improved social welfare by reducing price and stimulating output.[52]

Competition between more than two systems was uncommon. One observer found[53]

> [T]he experience of more than fifty of the principal cities and towns have demonstrated that two systems can be profitably maintained to a great advantage to all concerned. Telephone competition is always limited to two systems, and there can never be general competition.

He also noted

> There is a place for two telephones in about every community of 10,000 population and up, occupied by a Bell company, because it generally fails to develop a territory. Competition is desirable unless there is a development by one company of one telephone to five or six persons in the community. With full development there is very little room or demand for competition.

As shown in the last part of this section, competition in this limited sense was possible because (a) the average cost per subscriber of providing telephone service increased with the number of subscribers and (b) there was little demand for universal service within any given city at prevailing prices.[54]

The independents, like the Bell System, recognized the importance of long-distance connections. The strategy adopted at the first meeting of the independents had the development of long-distance companies with connection to the major commercial centers as its centerpiece. The president of the National Association of Independent Telephone Exchanges urged the development of an independent toll organization in 1901.[55] Z.G. Houck, writing in 1905, noted that "every one of us will acknowledge that our success or failure depends in a great measure upon whether or not we heed the many pleas made for better long-distance lines."[56]

Although the independents failed to establish long-distance connections to the major commercial centers, they succeeded in developing a fairly extensive network of toll lines. According to Hoge, by 1905 the independents had a continuous system from the Rockies to the Atlantic Coast.[57] By 1910, independent toll lines stretched from coast to coast.[58] The independents carried 20% of all long-distance messages in 1907 and 28% in 1912.[59]

The independent telephone system used three organizational methods for establishing toll service between independent exchanges. First, independent exchanges were merged into large regional independent operating companies which also provided toll service between these exchanges. Several large systems served parts of the Midwest and Southwest.[60] Second, independent exchanges formed toll associations which divided up the revenues from toll calls carried over toll lines provided by the individual members of the toll. For example, an independent toll association in New York reported an elaborate arrangement for reimbursing members for toll calls carried over their lines.[61] Clearinghouses for dividing revenue became quite common.[62] Third, separate toll companies were formed to provide long-distance service to independent exchanges.[63] The United States Telephone Co., for example, developed a long-distance system in Ohio, Indiana, and Illinois. Interestingly, this company prohibited its member exchanges from connecting with Bell toll lines for a period of 99 years. The courts subsequently struck these contracts down.[64] As with exchanges, there were seldom more than two competing lines regardless of the type of organization used to coordinate toll service.

Strategies

In order to meet the competition head-on, Bell lowered the rental rates on its telephones thereby encouraging the operating companies to expand their systems and lower their rates.[65] Between 1885 and 1893, its average rental rate ranged from a low of $10.76 to a high of $11.68 per telephone. Its average rental rate declined precipitously to $7.78 in 1894, $4.36 in 1895, $3.74 in 1896, $3.26 in 1897, and $2.90 in 1898. Its operating companies incurred an average expense per telephone of $58 in 1893.[66] The decrease in the rental rate of $3.70 in 1894 (from $11.48 to $7.78) therefore reduced operating expenses by 6%. Its operating companies incurred an average expense per telephone of $47 in 1898, a decrease of $11 from the 1893 level. The rental rate per telephone decreased by $8.58 between 1893 and 1898. Therefore, 80% of the decrease in operating expenses can be ascribed to the decrease in the rental rates. The operating companies apparently reduced their rates to subscribers in response to the lower rates. The average revenue per telephone decreased from $90 in 1894 to $70 in 1898. These rate reductions greatly accelerated the installation of Bell telephones. Between 1885 and 1894, the number of Bell telephones increased by 6.26% per year. Between 1895 and 1906, the number of Bell telephones increased by 21.54% per year. The yearly percentage of increase in the number of Bell telephones rose dramatically from 1.48% in 1894 to 14.47% in 1895, and to a high of 34.48% in 1898.[67]

Average operating revenue per telephone and average expense per telephone declined continually during the competitive period. Between 1895 and 1909, average revenue declined by 55%, from $70.00 to $31.50, and average expenses declined by 45%, from $57.70 to $30.00. Theodore Vail, citing these figures, argued that reduction in rates followed closely on reduction in expenses, and that "reduction in expenses was the result of the broad policy of development and improvement, the policy of the Bell system from the beginning, and not forced upon it by competition."[68] There is undoubtedly some truth to his claim that "development and improvement" in the art of telephony decreased expenses. Although there were no major technological advances during this period, the telephone industry probably became increasingly efficient at building and operating telephone exchanges.[69]

A crude calculation shows, however, that reduced rental rates accounted for about two-fifths of the reduced operating expenses. Between 1893 and 1901, the telephone rental rate decreased by about $11.00 from $11.48 to $.50.[70] In 1902, AT&T replaced the rental charge with a charge of 4.5% of operating income. Operating income per telephone declined from roughly $15 per telephone in 1902 to roughly $10 per telephone in 1913. Therefore, the effective charge per telephone declined from $.68 in 1902 to $.45 in 1913. Between 1893 and 1913, average expense per telephone decreased by about $26 from $58 to $32. Therefore, the reduced rental rate accounted for 42% ($11/$26 = .42) of the cost reductions experienced by the operating companies during the competitive period.[71] This figure suggests that competition did, indeed, force substantial cuts in telephone rates by the Bell System.

When confronted by competition, Bell exchanges slashed rates deeply. In Indianapolis and Toledo, the Bell exchange reduced its rate for business service from $72 to $54, and for residential service from $72 to $24, after competitive entry by independents which charged $40 for business service and between $24 (Indianapolis) and $30 (Toledo) for residential service.[72] In Cleveland, the Bell exchange reduced its business rate from $120 to $84 when an independent offered business service for $72.[73] In Columbus, the Bell exchange reduced its business rate from $96 to $54 when an independent offered business service for $40.[74]

The Bell exchanges and the independent exchanges accused each other of predatory pricing. AT&T claimed that the independents forced its exchanges to "meet competitive rates that are not based upon a proper recognition of the cost of doing the business or an adequate appreciation of the amount that should be set aside from earnings for maintenance, reconstruction and depreciation. . . ." and that "the competing companies, having an erroneous idea of the cost of giving service, for the most part undertook to secure subscribers by offering telephone service at a price that was unremunerative."[75] The independents claimed that "In most places the Bell Companies give away their service after an independent company has been started. . . ."[76] They cited instances where Bell exchanges offered service for $.50 per annum at the same time that AT&T was charging the exchange $.50 per annum for renting Bell telephones.[77] The charge by AT&T that its competitors underestimated their own costs and the charge by its competitors that it engaged in predatory pricing were remarkably similar to—and as difficult to judge as—charges made more than 70 years later.

AT&T also reduced its rate dramatically in cities where the independents had not established exchanges. In 1894, the average revenue per telephone in exchanges which faced no competition was $76.41. In 1899, the average revenue per telephone in exchanges that faced no competition was $65.51, and $57.02 in exchanges that did face competition. In 1904, the average revenue per telephone in exchanges that faced no competition was $40.05, and $36.89 in exchanges that did face competition. In 1909, the average revenue per telephone in exchanges that faced no competition was $35.71 and $31.32 in exchanges that did face competition.[78] Vail used these figures to argue that competition had little effect on reducing rates and that rate reductions and the expansion of service resulted from technological change. He said that "As methods, plant and apparatus became more fixed and permanent, methods of operating improved, operating expenses declined, and reductions in rates followed—not because of competition."[79] The sudden reduction in rates in response to competition, the lack of technological improvements in telephony, and the abrupt reductions in rental rates for Bell telephones in response to competition make his conclusion a bit farfetched. Nevertheless, it is not entirely obvious why AT&T lowered rates prior to actual entry by an independent.

There are three possible explanations.

1. City governments, through their power to grant and revoke franchises, had considerable leverage over the Bell exchanges. They could threaten to replace the Bell exchanges with an independent exchange or allow an independent exchange to compete with the Bell exchange. These threats were believable because (a) independent exchanges had demonstrated that they could provide service comparable to that provided by Bell exchanges, (b) some cities and towns had actually expelled Bell exchanges, and (c) many cities had permitted exchange competition. Some anecdotal evidence supports this hypothesis. The Bell franchise in Chicago came up for renewal in 1907.[80] At the same time, the independents lobbied heavily for a competing frachise. In return for effectively granting a monopoly franchise to the Bell exchange, the City Council of Chicago required the Bell exchange to reduce residential rates. The availability of a competitive alternative undoubtedly aided the city in obtaining this and other concessions from the Bell exchange.[81]

2. Bell may have followed a precommitment strategy for deterring entry by independents.[82] A simple albeit unrealistic example illustrates this strategy. It costs telephone company I F_I dollars to construct a telephone exchange and c_I dollars per subscriber to provide telephone service over this exchange. It is easy to show that a telephone company will establish an exchange if it is guaranteed annual revenues per subscriber p_I greater than p_I^* where

$$p_I^* = \frac{rF_I}{q_I} + q_I \tag{2.1}$$

where q_I is the number of subscribers and r is the real interest rate.[83] A price equal to p_I^* enables the telephone company to cover its operating expenses c_I

and earn a competitive return on its investment F. A price greater than p_I^* enables the telephone company to make a supracompetitive return on its investment F. The resale value of the exchange is zero, by assumption, so that F is a sunk cost. After the telephone company has established the exchange and incurred the sunk cost F, it will provide service if its annual revenue per subscriber p_I exceeds its annual cost per subscriber c_I.

By building the exchange first, telephone company I can deter entry by even more efficient telephone companies. Let us explain. Telephone company E has costs F_E and c_E. It will establish an exchange if it can obtain annual revenues per subscriber p_E greater than p_E^* where

$$p_E^* = \frac{rF_E}{q_E} + c_E \qquad (2.2)$$

Suppose telephone company E is more efficient than telephone company I—F_E is less than F_I and c_E is less than c_I—so that p_E^* is less than p_I^*. Then telephone company E could undercut telephone company I's price and make a profit. But telephone company I would be willing to lower its price to as low as c_I in order to remain in business. Recognizing that this will be telephone company I's response to entry, telephone company E will enter only if p_E^* is less than c_I. Therefore, telephone company I's sunk investment of F_I dollars in its telephone exchange deters entry by more efficient telephone companies with F_E and c_E such that $p_I^* > p_E^* > c_I$.

If the sunk cost of establishing telephone service were large, the company that initially established service would realize a first mover advantage.[84] The race to establish service first may have encouraged the Bell exchanges to charge lower prices than they would have charged in the absence of competition. There are two major problems with this explanation. First, as discussed below, the sunk costs of establishing telephone service were not particularly large relative to the variable costs of providing telephone service and, therefore, were not particularly effective deterrents to entry by more efficient telephone companies. Second, the prices Bell charged between 1885 and 1906 are not consistent with the hypothesis that Bell followed a precommitment strategy. If Bell followed a precommitment strategy, it would have had incentives to accelerate its investment in deterrent capacity during the years prior to 1894 when its patents expired. In fact, the yearly percentage increase in Bell telephones fluctuated from a low of 5.04 to a high of 8.97 between 1885 and 1892, dropped to 2.16 in 1893 and 1.48 in 1894, leaped to 14.47 in 1895, and fluctuated between 14.47 and 34.48 between 1895 and 1906.[85] It is possible, however, that Bell failed to follow a precommitment strategy prior to 1894 because (a) it expected that its remaining patents would deter entry or (b) it failed to recognize the value of a precommitment strategy.[86]

3. The threat of entry may have forced Bell to reduce prices. The theory of contestable markets sheds light on this possibility. A market is contestable to the extent that the threat of entry attentuates the ability of incumbent firms to

charge supracompetitive prices. A market is perfectly contestable if entrants can enter costlessly when prices exceed the competitive level and exit costlessly when prices fall below the competitive level. Hit-and-run entry thereby prevents incumbent firms from maintaining supracompetitive prices. A market is perfectly noncontestable when entry is impossible so that incumbent firms can charge supracompetitive prices without fear of entry. A market is more contestable the smaller sunk costs are and the faster incumbents can respond to competitive price cuts. As shown above, large sunk costs deter entry. The more quickly the incumbent can change its prices, the smaller are the profits earned by hit-and-run entrants.

The sunk costs of establishing telephone exchanges were not so large as to deter entry by slightly more efficient telephone companies. A proposal by an independent telephone company to establish an exchange in Chicago estimated that it would cost approximately $15 million to establish an exchange. Of this $15 million, probably no more than $10 million would have been unrecoverable in the event of bankruptcy.[87] The annual operating expense was estimated to be approximately $5 million. The system was to service approximately 100,000 subscribers. Assume that an incumbent telephone company had these costs. Thus, $F_I = \$10$ million, $q_I = 100{,}000$, and $c_I = \$50$. Assume that the real interest rate was 4%. Then, substituting into equation (2.1)

$$p_I^* = \frac{(\$10 \text{ million})(.04)}{100{,}000} + \$50 \tag{2.3}$$

$$= \$4 + \$50 = \$54$$

Such an incumbent would be vulnerable to entrants with $p_E^* < c_I$. An entrant whose operating costs were 8% lower than the incumbent but whose costs of establishing an exchange were the same as the incumbent could, therefore, drive the incumbent from the market. Consequently, the sunk costs of establishing an exchange provided incumbent telephone companies with, at best, a modest margin of comfort.[88] This conclusion is supported by Stehman's finding that[89]

> . . . the establishment of a new company requires relatively small initial financial outlay. . . . The extreme case is represented by the rural lines which in many cases were built almost entirely by the farmers who were to use them. They used old wire, strung on fence posts and trees and secondhand instruments. The whole required a very small expenditure.

To the extent that entrants were no more efficient than incumbents, incumbents could arguably deter entry and maintain supracompetitive prices by lowering prices in response to entry, bankrupting the entrant, and raising prices thereafter. There is considerable evidence that prices were sufficiently flexible for incumbents to pursue this strategy.[90] Long-term contracts between entrants and telephone subscribers, however, limited the effectiveness of this strategy. The entrant could offer to provide subscribers with cheaper telephone service for a number of years. This offer would force the incumbent to make the same kind of offer and thereby prevent the incumbent from maintaining supracompetitive prices. The franchise process provided one method for enforcing such contracts. Subscribers could achieve the same result by forming their own company.[91] The fact that Bell exchanges in some localities raised prices after eliminating competition would have encouraged Bell subscribers in other locations to make long-term contracts with an independent telephone company or to establish their own independent telephone company.

Although the local telephone markets obviously were not perfectly contestable, the fact that sunk costs provided only a modest barrier to entry and that subscribers could make long-term contracts for telephone service suggests that the threat of entry may have encouraged the Bell exchanges to charge competitive prices.

As explanations 1 and 3 suggest, competition limited Bell's ability to earn supracompetitive profits. Between 1876 and 1894, Bell earned an average yearly return on investment of 46%. During this period nominal interest rates were approximately 5% and the rate of *deflation* was approximately 1% per year. Between 1900 and 1906, Bell earned an average yearly return on investment of 8%. During this period, nominal interest rates were approximately 4% and the rate of *inflation* was roughly 1% per year.[92]

Interconnection was the second major strategic weapon used by the Bell and independent telephone systems. Both systems denied interconnections to each other and opposed legislative attempts to require interconnection of competing exchanges. Under its licensee contracts, AT&T could revoke the license and take over the property of any Bell exchange that connected with a non-Bell exchange. Nevertheless, even before 1907, AT&T made selective offers of interconnection to both competing and noncompeting independent exchanges.[93] MacMeal reports[94]

> One of the early instances of negotiations occurred this year (1898). In December a conference was held in Des Moines between the Iowa Telephone Company (Bell), the Nebraska Telephone Company (Bell) and two Independent companies, the Iowa and Nebraska Telephone Company, with five exchanges and 500 miles of toll lines, and the Western Electric Telephone Company, with 1200 miles of wire. The purpose of the meeting was to arrange for the joint use of wires and for uniform rates. This was said to be the first conference of the kind to be held. Following this, twenty-five representatives of Independent companies in Illinois and Indiana met in Paris, Illinois, at the invitation of the Central Union (Bell) to discuss interchangeable use of lines. After much discussion, the Independents rejected the proposal.

MacMeal claimed that independent opposition to interconnection with the Bell System was due to the fact that "Bell demanded an unduly large proportion of the message rate exchange in two-system towns."[95] In 1905, Hoge complained that Bell was "trying to disintegrate the interests of the various states and districts [organized into independent associations] by selecting one company here and another there that can be connected with its system."[96] Between 1899 and 1906, the percentage of independent telephones connected to the Bell System increased from 3% to 14%.[97]

The independents viewed connection with the Bell System as a sellout. An editorial in *The Telephone Magazine* warned[98]

> . . . it is more than ever the height of folly for any Independent telephone company to enter into a control with the Bell for long-distance service. In fact, there is nothing more suicidal than for an Independent company to thus voluntarily strip itself of individuality and place itself in the blighted power of the trust company. . . . [I]f you accept [Bell's] terms and lose your identity, Independent competition will spring up and you will have to compete against local influence and a much wider field. . . .

Like AT&T, the independents tried to prevent interconnection with the competing systems. For example, a meeting of the Ohio Independent Telephone Association resolved that the independent long-distance network should deny interconnection to any independent exchange that connected with the Bell System.[99] When an independent became a sublicensee of the Bell System, it was not uncommon for another independent exchange to take its place.

Beginning in 1907, AT&T became much more aggressive at seeking interconnections with strategically located independent exchanges. The independents complained that AT&T used a combination of threats of severe competition in the absence of merger or interconnection and offers to end competition in order to induce independents to merge or interconnect with the Bell System. According to Lindermuth,[100] " . . . as an 'inducement' to procure Independent telephone companies to enter into such combinations and agreements [Bell] threatened to introduce competition into cities and towns where no competition exists and offered to remove its competing exchanges in cities and towns where both systems are in operation." When an independent connected with the Bell System it usually had to forego connections with other independents. For example, one contract "provided that, after the consummation of the sale of the Central Union (Bell) Telephone Company's plant to the Marion company (thus eliminating competition in Marion), the Marion company, which has made connections with the toll lines of the Bell Company should thereafter pass every toll message originating in Marion for outside points over the lines of the Bell Company, which that company could carry to destination. This means that the Marion company has by the execution of this contract probably conspired not only against the United States Telephone Company, but also against such of its own subscribers as might prefer to use independent toll lines."[101] Similarly, when Bell purchased the Tristate Telephone Company the latter company cut off its connection with the Memphis Telephone Company and when Bell purchased the long-distance Telephone & Telegraph Company, the latter concern cut off connection with the Home Telephone Company.[102] During Vail's first year as President of AT&T—1907—the percentage of independent telephones connected with the Bell System increased from 14 to 27. By 1912, 65% of all independent telephones were connected with the Bell System.[103]

The independents also opposed legislative attempts to require interconnection between competing exchanges. In opposing a bill requiring interconnection introduced into the Ohio legislature, the president of the Ohio association of independents said, " . . . we are decidedly opposed to a law compelling an interchange of business, for this would be the first step in the elimination of competition, which has been the prime factor of the great telephone development of Ohio."[104] This same author had also argued that "Requiring or even permitting interconnection between telephone competitors is the final of a series of inevitable steps leading to monopoly."[105] Some independent telephone men believed that with interconnection the stronger system would always prevail.[106] Others believed that interconnection between competing and generally hostile systems was unworkable.[107]

After the wave of mergers and interconnections had decimated the ranks of the independent movement, the remaining independents endorsed government-controlled interconnections. In 1910, the president of the National Association of Independent Telephone Exchanges[108]

> declared for physical connections with the Bell System under proper legal supervision. He stated he believed the Interstate Commerce Commission would afford protection. Conditions, he said, had changed since the idea had first been proposed and had met with such bitter opposition. The public, he pointed out, demanded progress in the service. Universal service, in his opinion, was the desirable thing and he proposed that the Independents ask Congress and the several state legislatures for laws compelling interchange of service between all companies.

There are many appeals for supervised interconnection with the Bell System after 1910.

The Bell System and the independent telephone movement both used interconnection as a tool for establishing competitive national telephone systems. Each system tried to get as many exchanges as possible associated with its system and used access to these exchanges as an incentive to persuade subscribers in other exchanges to use its system. In this game, AT&T began with an enormous advantage. It had established service in numerous exchanges across the country and it was able to prevent the establishment of competitive exchanges in the major commercial centers of Boston, New York, and Chicago. The independents were able to capture half of the telephone market between 1894 and 1907 despite this handicap because there was little demand for long-distance service over the quality of lines then available. Bell's perfection of the Pupin coil, which made it possible to transmit intelligible speech across the continent, and the independents' continued lack of access to the major commercial centers made the development of a viable independent telephone system a much more dubious venture by the early 1910s.

The independents used their effective control over numerous rural exchanges as a selling point for obtaining franchises in areas where Bell had held monopoly franchises. This control was the major bargaining chip in, for example, the independents' efforts to acquire an independent franchise in Chicago. MacMeal noted, when the independents were refused a franchise by the Chicago City Council, "Chicago, under present conditions, is undoubtedly being deprived of a great volume of business which is flowing into other cities, having Independent systems and connections."[109] This bargaining chip was effective only through the independents' refusal to connect with the Bell System. The International Independent Telephone Association resolved that "we are unalterably opposed to any Independent connection with the Bell Telephone Company in the city of Chicago or elsewhere, as well as being opposed to any terminal for the Independent telephone companies to be used in connection with the Bell System."[110] The City Council of Chicago noted that " . . . it is our understanding that all of the demands [for toll connections by the independent telephone companies] that have been made have originated with a view to obtain such toll connections, not with a Bell company, but with an Independent company, to be established within the city."[111]

This strategy failed, in part because the independents did not have an effective cartel which could prevent defections to the Bell System. The major penalty for joining the Bell System was expulsion from the independent system and the consequent loss of access to independent telephone subscribers. The state independent associations could urge members not to connect with independents who had connected with the Bell System. But this penalty was only effective when the lost independent connections were more valuable than the gained Bell System connections. In the case of Chicago, discussed above, the value of extensive rural connections provided by the independents was presumably greater than the value of connection to Chicago provided by the Bell company in that city. AT&T could easily thwart this strategy by buying out independent telephone companies at crucial points in the independent system. It pursued this strategy to a lesser extent before 1907 and to a greater extent after 1907. In the absence of effective devices to enforce an independent cartel, the independents could have merged into a single corporate entity. This entity could have prevented AT&T from pursuing a strategy of attrition against the independents. Although there were numerous mergers to form larger regional independent telephone systems between 1894 and 1907, these mergers fell far short of establishing anything close to a national independent system. The fact that capitalists made no effort to merge the independent properties into a competing system but did, under the auspices of J.P. Morgan, seek to merge the independent properties into the Bell System may indicate that the telephone industry was a natural monopoly or that the Bell System's control over exchanges in the major commercial centers and its control over national, as opposed to regional, toll lines made a competing system unviable.[112] The next section considers these alternative hypotheses.

Natural Monopoly

In the early days of telephony, most observers acknowledged that the construction and operation of telephone exchanges was subject to diminishing returns to scale in the sense that the cost per subscriber increases with the number of subscribers. Anderson noted[113]

> But it's apparent on slight analysis, that the larger the number of subscribers in any given exchange, the more complicated, difficult and expensive is the process of adding on additional subscribers. The installation of the tenth hundred of the telephones costs much more than the installation of the first hundred. The installation of the tenth thousand costs much more per telephone than the installation of the tenth hundred. For instance, the cost *per subscriber,* for switchboard at the Central office, including the toll and trunking lanes, will be increased: From $5.00 in the exchange of 200 subscribers to $30.00 in an exchange of 15,000 subscribers.

He goes on to say that the installation and operating costs increase with the number of subscribers.[114] Stehman observed "an increased number of subscribers means increased fixed and operating expenses per subscriber."[115] These costs are due to increasingly complicated switchboards, increased lines, and increased number of lines. Bonbright said, "The telephone utilities, at least by their own contention and under the usual assumption of textbooks, are subject to increasing unit costs if the telephone subscribers or stations rather than the telephone call is taken as the unit of measurement."[116]

These cost characteristics are not necessarily inconsistent with the hypothesis that local telephone service is a natural monopoly. The value as well as the cost of telephone service increases with the number of subscribers. This fact has led many students of the telephone industry to argue that the appropriate test of natural monopoly is whether one system can serve a given number of subscribers more efficiently than two or more systems. Vail argued[117]

> The value of any exchange system is measured by the number of members of any community that are connected with it. If there are two systems, neither of them serving all, important users must be connected with both systems. . . . Given the same management, the public must pay double rates for service, to meet double charges, on double capital, double operating expenses and double maintenance.

Stehman claimed[118]

> Two telephone systems in a community were a great source of inconvenience and usually of expense to the subscribers. An individual who desires to talk to people on each of two systems is compelled either to install telephones of both companies or to go, from time to time, to some other place than his residence or place of business to use the telephone of the system to which he is not a subscriber.

More recently, Kahn argued that, with two systems serving a community, "each subscriber would have to have two instruments, two lines into his home, two bills if he wanted to be able to call everyone else."[119]

In rebutting Vail's argument, the independents argued that competition reduced the price of telephone service, increased the number of subscribers, and thereby enabled all subscribers to reach more subscribers at a lower cost with two telephones provided by two competing systems than with one telephone provided by a monopoly system. Their rebuttal has some merit. Even if telephone serivce were a natural monopoly, competition might have increased social welfare by reducing prices and stimulating output. Figure 2.1 illustrates this possibility. The average cost curve AC and the marginal cost curve MC are drawn to reflect the assumption that local telephone service is a natural monopoly. D denotes the demand curve and MR denotes the marginal revenue curve for telephone service. A profit-maximizing monopolist chooses a quantity Q_M so that marginal revenue equals marginal cost and charges a price P_M. Two competing firms both charge a price P_C equal to average cost AC_C which is greater than the monopolist's average cost AC_M but less than the monopolist's price P_M.[120] Output under competition is $2Q_C$ which exceeds output under monopoly Q_M. Society values the increased output by the shaded triangle. Because the average cost of producing Q_M units of output increases from AC_M to AC_C, competition imposes an additional cost on society equal to the crosshatched area. As drawn, the gains from com-

Figure 2.1 The gains and losses from competition vs. monopoly.

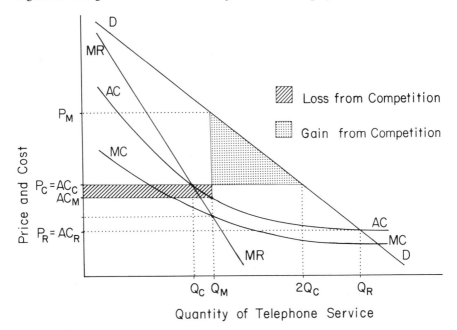

Quantity of Telephone Service

petition exceed the losses due to foregone scale economies from competition. Thus, at least in this example, society gains when two competitive firms rather than one profit-maximizing monopoly serves the market.

The problem with this analysis is that society could improve upon competition by conferring a monopoly upon a firm but forcing this firm to produce Q_R units of output and charge AC_R per unit. Given that there are scale economies and that the firm must break even, this level of output maximizes social welfare. The monopoly, regulated in this manner, produces more output at lower cost than competing firms. On the other hand, the costs of regulating the monopoly to behave in society's best interests may exceed the benefits from doing so.[121]

The argument advanced by Vail and other students of the telephone industry is, however, fundamentally flawed. Adding a subscriber to a system has two economic effects. First, it raises the cost of providing service to existing subscribers because there are diminishing returns to scale. Second, it raises the value of the service received by existing subscribers who now have the opportunity to talk to the new subscriber.[122] Existing subscribers favor adding the new subscriber only if the added benefit outweighs the added cost. There is no presumption that the added benefit always outweighs the added cost. As Bonbright wryly observed[123]

> Telephone company spokesmen insist that what they call "the value of the service" increases with an increase in the number of potential connections with other customers. This contention is more plausible for business users than for residential subscribers. I have heard several of the latter subscribers assert vigorously their wish that they could avoid being bothered with as many potential connections!

If local telephone service had been a natural monopoly, the existing Bell exchanges could have captured the entire local market by expanding their systems. This expansion would have raised their costs, on the one hand, and increased the value of their service, on the other hand. Their systems would have dominated the independent systems in terms of cost and value of service. Consequently, they would have been able to set prices that would have enticed subscribers to join their systems rather than the independent systems. The fact that, even after the Bell exchanges slashed their prices, by about half, to the levels offered by the independents, they were unable to establish a monopoly suggests that the addition of subscribers increased the cost of service more rapidly than the value of service. Groups of subscribers obviously found it to be in their self-interest not to join other groups of subscribers connected with the Bell System.[124]

The existence of two systems, rather than one, may have raised costs to some subscribers who would have liked access to the subscribers on both systems. Businesses were often duplicate subscribers and were the most outspoken advocates of interconnection between systems or establishment of local monopolies. But the fact that the existence of two systems inconvenienced some subscribers does not imply that one system would have benefited subscribers as a group.

Merger, even with rate-of-return regulation, may raise the cost of service more than the value of service for most subscribers and lower the cost of service more than the value of service for only a few subscribers. The fact that the number of duplications was small and that the competitive process did not establish prices that would entice subscribers into one system suggests that merger resulted in a net decrease in social welfare.

As Gabel observed, "All competition involves some duplication of plant facilities and work effort. The question is whether the pressures of competing market forces produces a better or cheaper product than a single supply service."[125] There is no fundamental economic difference between competition between two telephone systems and competition between two stores. Macy's, Bloomingdale's, and Brooks Brothers could economize on duplicate facilities by merging. You might even be able to purchase your Brooks Brothers' suits for less after the merger. But other consumers may have to pay more for their polyester leisure suits, video games, and fine china. Merger may thereby raise the aggregate cost of supplying the services offered by these stores. Two telephone systems could possibly economize on duplicate wires and duplicate telephones for subscribers who desire to reach subscribers on both systems. Duplicate subscribers gain from this merger. Nonduplicate subscribers who have little demand for reaching subscribers on the other system lose from this merger. In both cases, one would expect the competitive process to reveal the socially desirable configuration of businesses.

There are various methods available for interconnecting two exchanges, regardless of whether these exchanges are owned by the same or different systems. First, establish a trunk line between the exchanges which gives subscribers on each system access to each other. Telephone calls between exchanges over a trunk line are usually more expensive and lower in quality than telephone calls within exchanges.[126] Second, require subscribers who wish access to subscribers on both systems to have two telephones. Duplicate subscribers thereby obtain high-quality connections with both systems.[127] Third, replace two exchanges with a central exchange that can provide all subscribers the same quality interconnection with each other.

Which of these alternative methods of interconnection is more efficient from the standpoint of society depends on the costs of establishing and operating exchanges and trunk lines and the demand for interconnection between different subscribers. Where there is little demand for interconnection between different sets of subscribers and where the costs of establishing a trunk line connection are high, requiring those subscribers who wish access to subscribers on both systems to have two telephones and pay two telephone bills may well be desirable. Where there is modest demand for interconnection between different sets of subscribers, but the cost of establishing a common exchange is prohibitive, a trunk line may provide the least expensive method of interconnection.[128] Under these circumstances, one would expect two exchanges to establish a trunk line between each other even if they are competing for subscribers at the margin. Unfortunately, this was not the case.

Competing exchanges often refused to connect with each other. The Bell System and the independent telephone movement used their respective exchange monopolies as bargaining chips in obtaining franchises and in enticing subscribers onto their systems. The independents' control over rural exchanges in the Midwest helped them expand these systems through the Midwest. The Bell System's control over exchanges in the main commercial centers was an obvious selling point to business subscribers. The perfection of long-distance telephone service gave the Bell System an enormous advantage in this system-building competition. Prior to the development of the Pupin coil, only regional long-distance communications was possible. In 1906, long-distance service between cities as close as New York and Philadelphia was poor.[129] Bell's control over exchanges in New York, Boston, and Philadelphia made little difference to someone in Peoria. By 1915, several inventions enabled Bell to establish useable long-distance communications between New York and San Francisco.[130] Bell's ability to offer reliable connections to major cities across the country made the independents' system-building strategy futile. It is not surprising, then, that after it became apparent to the independents that interconnection refusal would no longer help them establish an independent system, they changed their strategy to one of supervised interconnections with the Bell System and its long-distance network.[131]

That the independent telephone companies and the Bell exchanges did not provide the optimal degree of interconnection in some circumstances does not, however, support the argument that competition was inefficient. The competing telephone companies, as the discussion above demonstrates, failed to interconnect because there was too little rather than too much competition. These companies tried to use local exchanges as strategic bottlenecks in developing telephone systems. Separating the exchanges from the companies (or associations) providing long-distance service might have fostered interconnection and prevented the Bell System from establishing a monopoly over the national telephone system. Lacking any system-building incentives, local exchanges would have had strong incentives to either interconnect with each other or interconnect with a common long-distance company. There is no reason to believe that local exchanges would have foregone these opportunities for mutually advantageous trades. This policy would have maintained a quasi-competitive local exchange market and, perhaps, a quasi-competitive long-distance market.[132] On the other hand, the incentive to collude between competitive local exchanges and between local exchanges and long-distance companies might have required vigilant antitrust oversight over such an industry.[133]

Competition between telephone companies for franchises and between franchises for subscribers clearly stimulated the extension of telephone service, the reduction of telephone rates, and technological improvements in telephony. Competition was far superior to unregulated monopoly. Whether unregulated competition—either between two separate national systems or between a multitude of competing local exchanges and long-distance companies—would have proved superior to regulated monopoly is more problematic. True telephone competition may have been impossible. Few markets could have supported more than two competing exchanges. These exchanges might have colluded, either directly or tacitly, on prices or territories.

There is some evidence, however, that local telephone markets were reasonably contestable. There are numerous instances where, after a Bell exchange acquired a competing independent exchange and raised prices, another competing independent exchange quickly appeared. Sunk costs of entry were not effective deterrents to entry. Long-term contracts between subscribers and independents, sometimes enforced by the franchise process, prevented Bell from lowering prices today in the hopes of raising prices tomorrow.

As new technologies emerge for providing local telephone service, there will be a fierce battle over the desirability of competition in local exchange service. Business users who presently pay discriminatorily high rates have strong incentives to find cheaper alternatives. Already some businesses use local cable distribution systems for communcation between local branches. Long-distance companies may have incentives to bypass the local exchanges in order to reach these often specialized groups of subscribers. The issues of natural monopoly, creamskimming, rate averaging, and cross subsidies which dominated the debate over the introduction of competition into intercity service will reappear in local debates over the desirability of entry into local exchange service. The historical evidence presented in this chapter suggests that the local exchange monopolies may never have been "natural" and that competition may improve service, reduce price, and encourage technological improvements.

NOTES

1. J. Warren Stehman, *The Financial History of the American Telephone and Telegraph Company* (Boston: Houghton Mifflin, 1925), p. 234.

2. Alfred E. Kahn, *The Economics of Regulation: Principles and Institutions,* Vol. II (New York: Wiley, 1971), p. 123.

3. *Ibid.,* p. 127.

4. For a comprehensive history of the telecommunications industry, see Gerald Brock, *The Telecommunications Industry* (Cambridge: Harvard University Press, 1981).

5. For an excellent discussion of the benefits of competition in the telephone industry, see Richard Gabel, "The Early Competitive Era in Telephone Communications, 1893–1920," *Journal of Law and Contemporary Problems,* Spring 1969, pp. 340–369.

6. This section is based on the following sources: AT&T, *Annual Report,* various years; Brock, *op. cit.;* John Brooks, *Telephone: The First Hundred Years* (New York: Harper & Row, 1975); Federal Communications Commission, *Telephone Investigation: Proposed Report* (Washington, DC: Government Printing Office, 1938), hereafter referred to as the Walker Report; and Stehman, *op. cit.*

7. Brooks, *op. cit.,* p. 47.

8. Brooks, *op. cit.,* p. 75.

9. Stehman, *op. cit.*, p. 26.

10. Stehman, *op. cit.*, p. 26.

11. Stehman, *op. cit.*, p. 43.

12. Stehman, *op. cit.*, p. 43.

13. Walker Report, *op. cit.* (in Note 6, above), p. 49.

14. Stehman, *op. cit.*, p. 54.

15. Walker Report, *op. cit.*, p. 96.

16. *Commerical and Financial Chronicle,* "American Bell Telephone Company," April 28, 1900, p. 817.

17. N.R. Danielian, *AT&T: The Story of Industrial Conquest* (New York: Vanguard, 1974), p. 49.

18. Walker Report, Table 32, p. 145.

19. Danielian, *op. cit.*, p. 58.

20. Walker Report, pp. 153–154.

21. Walker Report, Table 32, p. 145.

22. The Postmaster General, at the same time, advocated government ownership of the telephone system. See Harry B. MacMeal, *The Story of Independent Telephony* (Chicago: Independent Pioneer Telephone Association, 1934), p. 204.

23. Walker Report, Table 32, p. 145.

24. Walker Report, Table 32, p. 145.

25. This section is based on articles appearing in *The Telephone Magazine* and *Telephony* between 1898 and 1909; AT&T, *Annual Report,* 1900–1920; MacMeal., *op cit.*; Stehman, *op. cit.*; Brooks, *op. cit.*; Brock, *op. cit.*; Gabel, *op. cit.*; G.W. Anderson, "Telephone Competition in the Middle West and Its Lesson for New England," New England Telephone Company, September 1906; and miscellaneous historical sources.

26. Brock, *op. cit.*, p. 100.

27. Stehman, *op. cit.*, p. 52.

28. Walker Report, pp. 143–145.

29. Stehman, *op. cit.*, p. 56.

30. Brock, *op. cit.*, p. 102.

31. "American Bell Telephone and Proposed Competition," *Commercial and Financial Chronicle,* December 16, 1899, pp. 1222–1224.

32. AT&T, *Defendants' Third Statement of Contentions and Proof,* p. 135.

33. MacMeal, *op. cit.*, p. 221.

34. Bureau of the Census, *Telephones and Telegrams 1912* (Washington, DC: Government Printing Office, 1915), p. 38.

35. James B. Hoge, "National Inter-State Telephone Association," *The Telephone Magazine,* July 1905, p. 34. Hoge was a major advocate of cartelizing the independents.

36. *Telephony,* April 1908, p. 243.

37. W.F. Laubach, "Standard Toll Signs," *The Telephone Magazine,* July 1905, pp. 60–62.

38. James B. Hoge, "National-Interstate Telephone Association," Address made at the Annual Convention of Ohio Independent Telephone Association, *The Telephone Magazine,* April 1905, p. 292.

39. *Ibid.,* pp. 292–293.

40. MacMeal, *op. cit.*, p. 221.

41. There were some mergers to form regional systems. See Note 58.

42. James B. Hoge, "Necessity of State and National Organizations," *The Telephone Magazine,* June 1905, p. 374.

43. MacMeal reports that in 1916 there were 6000 Bell and 14,500 independent exchanges. Of the independents, 1200 competed with Bell and 600 competed with other independents. (MacMeal, *op. cit.*, p. 221.)

44. Frederick S. Dickson, "Independent Finances," *The Telephone Magazine,* January 1905, p. 29.

45. Gabel says, "Early Bell System telephone development took place at the business core of large urban communities. Territorial extension by competing independents was for the most part to contiguous rather than overlapping areas" (Gabel, *op. cit.*, p. 342). He refers to G. Johnston, "Some Comments in the 1907 Annual Report of AT&T," published by the International Independent Telephone Association, September 1908.

46. *Telephony,* March 1907, pp. 185–186; December 1907, p. 338; and February 1908, pp. 267–268.

47. See G.W. Anderson, "Telephone Competition in the Middle West and its Lesson for New England," prepared for New England Telephone Company, September 1906. In Pittsburgh there were 44,721 Bell subscribers and 120,000 independent subscribers including duplicate subscribers. In Detroit the Bell and Independent systems had 10,000 subscribers each with 2000 duplicate subscribers who were mainly businesses. In Toledo there were 10,000 independent subscribers, 6700 Bell subscribers, and 3400 duplicate subscribers. In Cleveland, there were 28,000 Bell subscribers, 22,000–24,000 independent subscribers and 9000 duplicate subscribers who were mainly businesses.

48. AT&T, *Annual Report,* 1907, *op. cit.*, pp. 17–18. Compare this statement with the statement by Kahn quoted at the beginning of this chapter. Vail included a section discrediting the competition in the next several issues of the *Annual Report*.

49. J. Ainsworth and G. Johnston, *A Discussion of Telephone Competition,* (Chicago: International Independent Telephone Association, 1908), p. 8, quoted by Gabel, *op. cit.*, p. 349.

50. Of course, this does not entirely meet Vail's criticism since he viewed the duplicate cost as the cost of providing everyone with connections to everyone else.

51. Ainsworth and Johnston, *op. cit.*, p. 9. Also see Frederick Dickson's discussion of competition in Cleveland, *op. cit.*, p. 29.

52. In modern economic terms, competition eliminates the deadweight loss to society of monopoly restrictions on output. This argument holds even if local exchange service were a natural monopoly. Competition and its attendant duplication of investment may yield greater social welfare than unfettered natural monopoly with its attendant restriction of output. We return to this point in the third part of this section.

53. Theodore Gary, *Telephony,* December 1907, p. 338.

54. Competition in some cities was not possible because the independents were unable to obtain a franchise. In 1907, the Chicago City Council renewed Bell's franchise and refused to allow an independent exchange to promote telephone service. See F.P. Tiffany, "The Chicago Situation," *Telephony,* October 1907, pp. 202–204. The independents accused Bell of a variety of tactics to prevent them from obtaining competitive franchises. They report that Bell officials had been indicted and convicted of bribery in San Francisco, and intimate that corruption may have been involved in Chicago as well. They also claimed that Bell would create a dummy company which would underbid the legitimate independent bid for a franchise and then go bankrupt. See *Telephony,* December 1907, p. 355. Of course, given the opportunities for corruption created by the franchise-bidding process it would not be surprising if both independents and Bell companies resorted to bribery.

55. *The Telephone Magazine,* January 1901, p. 64.

56. Z.G. Houck, "Long Distance Telephone Lines," *The Telephone Magazine,* March 1905, pp. 181–183.

57. Hoge, *op. cit.*, (July 1905), p. 32.

58. AT&T, *Defendants' Third Statement of Contentions and Proof,* in *US* v. *AT&T,* p. 115.

59. Bureau of the Census, *op. cit.*, p. 38. The increase between 1907 and 1912, during which time the share of telephones operated by the independents decreased, is probably due to the interconnections of many independents to the Bell System.

60. For example, the Lincoln Telephone Company acquired more than 100 exchanges in 22 Nebraska counties and formed a toll company for providing long-distance service between these exchanges. See MacMeal, *op. cit.*, p. 176.

61. Special Tariff Committee, "Interchanged Toll Business," *The Telephone Magazine*, July 1905, pp. 103–108.

62. See, for example, discussions in *Telephony*, May 1907, pp. 306–308 and September 1907, p. 173. AT&T claims that these traffic associations were largely unsuccessful. See AT&T, *Defendants' Third Statement of Contentions and Proof*, in *US* v. *AT&T*, p. 113. AT&T cites several articles in *Telephony* that were not available to us.

63. For example, during 1900 and 1901, the Long-Distance Telephone Company of Virginia sought "arrangements with the numerous local telephone companies, whereby the long-distance current can be strung in the poles already erected" and placement to give local telephone subscribers throughout the state "the benefit of the long-distance facilities by having the long-distance circuit 'tapped' in every county at the county exchange." *The Telephone Magzine*, March 1901, p. 132. In 1903, the Nebraska Toll Line Association was organized "for promoting the volume and quality of toll service between Independent companies in that state." *The Telephone Magazine*, December 1903, p. 241.

64. See, for example, *US Telephone Co.* v. *Central Union Telephone Co.*, 202 Fed 66, 6th Cir. (1913), p. 7055, for a presentation of the facts and the reasoning by which the court rejected their contracts.

65. Stehman, *op. cit.*, pp. 26–27, provides data on average rental rates.

66. The Walker Report, *op. cit.*, Chart 9, p. 149, provides data on revenue and operating expenses.

67. Walker Report, *op. cit.*, Table 33, p. 151.

68. AT&T, *Annual Report*, 1909, p. 29.

69. For a discussion of technological change during the competitive period see Gabel, *op. cit.*, pp. 346–347. There were apparently no major improvements during this period which would account for the rapid reduction in prices and none certainly to account for the abrupt reduction in rental rates made by AT&T.

70. An article in *Telephony*, January 1901, p. 113, stated that the Bell rental rate was $.50.

71. Economies of scale in manufacturing telephones may account for some of the reduction in rental rates.

72. Stehman, *op. cit.*, pp. 85–89, based on figures quoted by Anderson, *op. cit.*

73. Dickson, *op. cit.*, p. 29.

74. Ainsworth and Johnston, *op. cit.*, p. 9.

75. AT&T, *Annual Report*, 1909, p. 8.

76. *Telephony*, January 1901, p. 16.

77. *Telephony*, June 1901, p. 113.

78. AT&T *Annual Report*, 1909, chart, p. 26. Vail says rates were higher in noncompetitive cities because the exchanges in these cities serviced 50% more subscribers than the exchanges in competitive cities. See p. 25.

79. *Ibid.*, p. 25.

80. F. P. Tiffany, "The Chicago Situation," *Telephony*, October 1907, pp. 202–204; "The Chicago Bell Franchise," *Telephony*, December 1907, p. 359; and Committee on Gas, Oil and Electric Light, *Telephone Service and Rates*, Report to the City Council of Chicago, September 3, 1907.

81. The Bell exchange also agreed to pay the city 3% of operating income and to pay the city any excess over a 10% return on investment.

82. For a description of commitment strategies see Avinash Dixit, "Recent Developments in Oligopoly Theory," *American Economic Review*, May 1982, pp. 12–17 and the references cited therein.

83. For a derivation of this result see the first section of Chapter 4.

84. See the discussion of first-mover advantages in the third section of Chapter 8.

85. Walker Report, *op. cit.*, Table 33, p. 151. The expansion of telephones was determined partly by the rate AT&T charged the Bell franchises for Bell telephones. The rental rate was relatively constant between 1885 and 1894 and declined gradually between 1894 and 1905. See Stehman, *op. cit.*, pp. 26–27.

86. During the 1880s, Vail reportedly advocated aggressively expanding the Bell System but failed to persuade the Boston capitalists. According to John Brooks, "When the question arose as to whether to form a company or to open a new exchange in some doubtfully profitable farm area, the Bostonians tended to say: 'Why bother, if it probably won't make money?' Vail . . . would reply, in effect, 'Let's take a chance. That area needs telephones as much as any other and moreover, if we move in there now, it will be one more area where we will be established and operating when our patents run out.' It was this philosophical difference, reducible, perhaps, to immediate versus long-term self-interest that brought Vail and Forbes increasingly into conflict." Vail resigned in September 1887. See Brooks, *op. cit.*, pp. 84–85. Gerald Brock, *op. cit.*, p. 122, argues that Bell underestimated the elasticity of demand for telephone service during the period of patent monopoly. His only evidence for this conclusion, however, appears to be that demand increased when price decreased, a fact which happily confirms the law of demand but little else.

87. Committee on Gas, Oil and Electric Light, *op. cit.*, pp. 56–58.

88. The margin is overstated because an entrant who did not expect to remain in the market would invest in a less permanent and therefore "less sunk" exchange. Obviously, sunk costs and the technology used are endogenous to the entry decision. Also real interest rates were probably lower than 4%. In 1907, the yield on railroad bonds was 4.27% and the rate of inflation was approximately 1%. See Bureau of the Census, *Historical Statistics of the United States: Colonial Times to 1970* (Washington, DC: Government Printing Office, 1975), p. 1003.

89. Stehman, *op. cit.*, p. 256.

90. For example, Anderson, *op. cit.*, argues that independent telephone companies had an easy time raising rates. Furthermore, there is no reason to believe that franchise authorities would prevent reductions in rates. On the other hand, the fact that the Bell exchanges were controlled by AT&T and were assessed a common rate for their use of the Bell telephone may have limited the flexibility of individual Bell exchanges to respond to competition.

91. Many of the early independent telephone companies were in fact started by dissatisfied subscribers. See MacMeal, *op. cit.*, for example.

92. See Gabel, *op. cit.*, p. 352 for rate-of-return data and Bureau of the Census, *Historical Statistics, op. cit.*, p. 153 and p. 1003, for inflation and interest rate data. The inflation rates are based on average yearly changes in the consumer price index. The interest rates are based on the nominal yield on railroad bonds.

93. In some cases interchange of service was accomplished by having each company rent a telephone from the other. The operator would then relay messages from subscribers on the competing system.

94. MacMeal, *op. cit.*, p. 94.

95. MacMeal, *op. cit.*, p. 183.

96. Quoted in *The Telephone Magazine*, June 1905, p. 375.

97. Walker Report, Table 32, p. 143. Figures as of December 31.

98. *The Telephone Magazine*, March 1905, pp. 188–189.

99. *Telephony*, April 1908, p. 253.

100. A.C. Lindermuth, "Address Before the West Virginia and Indiana Telephone Conventions," *Telephony*, June 1908, p. 383.

101. Ainsworth and Johnson, pp. 10–11. Of course, it is difficult to feel much sympathy for the independents since (*a*) Central Union had required the exchanges on its system to sign contracts that forbade them from connecting with Bell—contracts that were subsequently struck down

and (*b*) the Ohio association of independents basically outlawed any independent who connected with Bell. It is possible the independents were merely reacting to similar Bell tactics.

102. MacMeal, *op. cit.,* p. 189.

103. Walker Report, Table 32, p. 143. Figures as of December 31.

104. Frank Beam, "Address," *Telephony,* April 1908, p. 253.

105. Frank Beam, "Preface" to Ainsworth and Johnston, *op. cit.,* p. 4.

106. Ainsworth and Johnston, *op. cit.,* p. 26, provide a most peculiar scenario in which the smaller exchange inevitably loses its subscribers to the larger exchange.

107. An article in *Telephony,* said, "It is unnatural to expect that any company would give its best service to its competitor and injure its good name." July 1907, p. 19.

108. MacMeal, *op. cit.,* p. 183.

109. Quoted by Tiffany, *op. cit.,* p. 202.

110. Tiffany, *op. cit.,* p. 203.

111. Committee on Gas, Oil and Electric Light, "Telephone Service and Rates," Report to the City Council of Chicago, September 3, 1907, p. 121.

112. For a discussion of J.P. Morgan's role in consolidating the telephone industry see John Brook's discussion, *op. cit.,* pp. 122–126.

113. Anderson, *op. cit.,* p. 7.

114. Also see James Thomas, "Address," *Telephony,* July 1900, p. 126, for a discussion of diseconomies of scale in constructing exchanges.

115. Stehman, *op. cit.,* p. 236.

116. James C. Bonbright, *Principles of Public Utility Rates* (New York: Columbia University Press, 1961), pp. 16–17. Also see City Council of Chicago, *op. cit.,* pp. 12–13.

117. Vail, AT&T, *Annual Report,* 1907, p. 18.

118. Stehman, *op. cit.,* p. 234.

119. Kahn, *op. cit.,* p. 123.

120. We ignore the question of how this competitive equilibrium might arise.

121. Government regulations may be less effective than competition at disciplining the natural monopoly, a view that prompted the FCC to allow entry into specialized communication markets. See FCC, *Specialized Common Carriers Decision,* 29FCC2d870 (1971).

122. In economic jargon, adding a subscriber generates a negative pecuniary externality (i.e., it raises the cost of serving inframarginal demanders) and a positive pecuniary externality (i.e., it makes the service more valuable).

123. Bonbright, *op. cit.,* p. 17, fn. 16.

124. It is possible, however, that we have an instance of a nonsustainable natural monopoly due to the fact that some group of subscribers feels it to be in its self-interest to secede from the natural monopoly coalition. See Chapter 9 and the reference cited therein for a discussion of sustainability concepts.

125. Gabel, *op. cit.,* p. 342.

126. The quality was lower because the waiting time for completing a call was longer.

127. Of course, nonduplicate subscribers could, with some minor inconvenience, use the telephones maintained by friends or neighborhood businesses with two telephones.

128. Indeed, even a monopoly telephone company has an incentive to construct the optimal size exchange and to establish the optimal level of interconnection between these exchanges. See the first section of Chapter 6 for a discussion of some related issues in the theory of natural monopoly.

129. Brooks, *op. cit.,* p. 121.

130. Brooks, *op. cit.,* pp. 138–139.

131. MacMeal, *op. cit.*, p. 183.

132. The possibility of a competitive long-distance market is more speculative. An innovation such as the Pupin coil might have enabled a single long-distance company to achieve a natural long-distance monopoly. Moreover, there is little historical evidence on whether multiple long-distance companies were viable. On the other hand, it is clear that the present market can maintain numerous long-distance companies.

133. Of course, it is not obvious why antitrust enforcement would be more expensive than direct regulation for preventing socially undesirable behavior.

Chapter 3
Predation:
A Critique of the Government's Case
in *US* v. *AT&T*

William A. Brock and David S. Evans

After Vail took charge in 1907, AT&T gradually embraced government regulation of the telephone industry. He noted in AT&T's Annual Report for 1907 that "It is contended that if there is to be no competition, there should be public control [over the telephone industry]. It is not believed that there is any serious objection to such control, provided it is independent, intelligent, moderate, thorough, and just. . . . "[1] In 1910 he advocated permanent commissions, with quasi-judicial powers, whose members should have the "time and opportunity to familiarize themselves with the questions coming before them" and which should not "dictate what the management or operation should be beyond the requirement of the greatest efficiency and economy."[2] In 1915 he observed that these commissions should think of themselves as juries charged with "protecting the individual member of the public against corporate aggression or distortion, and the corporate member of the community against public extortion and aggression."[3]

Although Vail was undoubtedly an idealistic man concerned with the public's welfare, he was ultimately responsible for pursuing the interests of AT&T's stockholders. It is not surprising, therefore, that embracing government regulation was probably the most sensible strategy AT&T could have pursued at the time. First, by 1915 telephone regulation was pervasive albeit seldom stringent.[4] Most states had commissions that regulated telephone rates and practices. Congress, through the Mann–Elkins Act of 1910, had empowered the Interstate Commerce Commission to regulate interstate telephone service. By embracing regulation, AT&T was probably able to promote a more favorable regulatory climate for itself. Second, regulation was better than nationalization. Most of the European countries had nationalized their telephone systems. The British government nationalized the British telephone system in 1912. The U.S. Postmaster General, in the spring of 1913, advocated the nationalization (or "postalization") of this country's telephone and telegraph system.[5] In this climate, AT&T might have encouraged nationalization had it opposed regulation at the same time it was eliminating effective competition. Third, regulation could encourage the elimination and prevent the renewal of competition. Vail believed the suppression of competition was one purpose of the public control he advocated.[6]

> A public utility giving good service at fair rates should not be subject to competition at unfair rates. It is not that all competition should be suppressed but that all competition should be regulated and controlled. That competition should be suppressed which arises out of the promotion of unnecessary duplication, which gives no additional facilities or service, which is in no sense either extension or improvement, which without initiative or enterprise tries to take advantage of the initiative and enterprise of others by sharing the profitable without assuming any of the burden of the unprofitable parts or which has only the selfishly speculative object of forcing a consolidation or purchase.

By 1920, AT&T was firmly entrenched as a regulated monopoly. All but three states regulated telephone rates and practices.[7] The ICC, although it apparently did not actively regulate interstate telephone rates,[8] had established a uniform system of accounts for telephone companies and regulated mergers and interconnections between Bell and independent telephone companies under the terms of the Kingsbury Commitment.[9] The Willis–Graham Act of 1921, by permitting mergers between competing telephone exchanges, enabled AT&T to eliminate the few remaining vestiges of competition and to establish its complete control over the operation of the telephone system. Most state regulatory commissions opposed local telephone competition.[10] Thus, in return for accepting regulation, AT&T gained tacit approval for its monopoly over the telephone industry.

Under the Communications Act of 1934, the Federal Communications Commission assumed the Interstate Commerce Commission's authority over the telephone industry. The FCC was empowered to require advance approval for new facilities and services (a power the ICC did not have), to compel interconnection, to suspend rates pending an investigation, and to allocate frequencies.[11] As a result, the FCC could control most aspects of telephone competition by controlling prices, interconnection, and entry.[12]

Beginning in the late 1940s, the FCC increasingly became the scene for skirmishes between businesses that desired to enter AT&T's markets and AT&T, which insisted on maintaining control over all aspects of the telephone business.[13] AT&T won a major skirmish in 1948. The FCC restricted competitive entry into the developing market for the transmission of television signals. In late 1956, the FCC began an inquiry into the allocation of microwave frequencies. Three years later, its *Above 890 Decision* ruled that there were sufficient microwave frequencies above 890 megacycles for the development of private point-to-point microwave communications systems. AT&T had lost a major skirmish. In 1969, the FCC approved an application filed by Microwave Communications, Inc. (MCI) in 1963 to establish a common carrier microwave system between Chicago and St. Louis. AT&T faced direct competition. Following on MCI's heels, numerous companies filed applications to establish microwave systems for providing intercity private line service. Responding to these applications the 1971 *Specialized Common Carriers Decision* announced that the FCC favored increased competition in private line service and would approve entry by qualified specialized common carriers. Several companies, including MCI and Southern Pacific Communications Corporation (SPCC), subsequently developed national private line systems.

Although the FCC had formally approved competition only in private line service, MCI began offering ordinary long-distance telephone service (i.e., message toll or switched long distance) in 1975. AT&T asked the FCC to require MCI to stop offering this service. The FCC complied on the grounds that it had not authorized entry into the intercity message toll service market. The Appeals Court subsequently reversed the FCC's decision. It argued that the FCC had to show that competition would be contrary to the public interest before restricting the services provided by MCI. MCI, SPCC, and several other companies began offering intercity message toll service to residential and business subscribers. AT&T faced its first competitive challenge in more than 50 years.

The manner in which AT&T responded to the competitive challenge became the subject of an antitrust suit filed by the U.S. Department of Justice in 1974. The Justice Department claimed[14]

> . . . during the past three decades, and particularly since the late 1960's, AT&T has engaged in a comprehensive course of conduct to willfully maintain its monopoly in the intercity telecommunications services market. As each new competitor has appeared, AT&T has carried the battle against competition outside the regulatory arena and into the marketplace where it has exploited its enormous monopoly power to maintain its monopoly position in the market. Broadly, the major features to AT&T's exclusionary conduct in the intercity services market have been the manipulation of the terms and conditions under which competitors are permitted to interconnect with AT&T's existing services and facilities, including those of the local exchange operators, and the repricing of AT&T's own intercity services in competition with the new entrants.

This chapter presents an economic analysis of the predatory behavior alleged by the Justice Department.[15] The first section examines the argument that AT&T, by imposing interconnection restrictions and abusing the regulatory process, erected entry barriers which impeded competition. The second section examines the argument that AT&T engaged in a form of predatory pricing called *pricing without regard to cost*. Both sections examine whether AT&T's alleged practices reflected plausible strategies for deterring socially desirable competition and eschew any discussion of the legality of the alleged practices.[16]

Barriers to Entry

The older industrial organization literature viewed a barrier to entry as any cost borne by an entrant but not by an incumbent.[17] By this definition, scale economies, brand loyalty, learning curve effects, the availability of cheap capital to larger firms, lack of bottleneck facilities, and "certificate of need" requirements imposed by regulators upon entrants were considered barriers to entry.[18] The newer industrial organization literature views a barrier to entry as any cost borne by an entrant but not by an incumbent which distorts the socially preferred allocation of resources.[19] By this definition, strategic advantages such as scale economies which do not deter socially desirable entry are not considered barriers to entry.

The older industrial organization literature argued that fixed costs for entry created especially onerous barriers to entry which impeded competition. An entrant who has to invest in expensive facilities, the costs of which are largely independent of sales volume, has several strategic disadvantages. If capital markets are imperfect, he or she may be unable to finance the initial investment for entry or be able to finance the investment for entry only at a prohibitive price. If he or she fails to capture enough of the incumbent's markets, it may mean bankruptcy. The newer industrial organization literature distinguishes between reversible investments and irreversible investments. Fixed costs which require merely reversible investments do not pose barriers to entry. If the entrant fails the fixed capital equipment can be resold, and initial investment less depreciation be recovered. The classic example is the airline market. Airlines have to make large investments in aircrafts in order to provide service. But if an airline fails to capture a profitable share of, say, the Boston–New York market, it can sell its aircraft to another airline or fly its aircrafts to another market.[20] Fixed costs that require irreversible investments may pose barriers to entry. If the entrant fails, he or she loses not only the operating losses but the irreversible investment as well.

By erecting barriers to entry, incumbents can deter competition which would reduce their monopoly profits. Consequently, they may have incentives to invest resources in raising entry barriers. Suppose the amount of output produced by entrants q_E is a function of the stock of barrier to entry capital k, $q_E = q_E(k)$, where q_E is smaller the larger is k. Also suppose that the discounted value of future profits to the incumbent V is a function of the output produced by entrants, $V = V(q_E(k))$, where V is larger the smaller is q_E. Then the incumbent will invest in barrier-to-entry capital until the marginal cost of investment in barrier to entry capital equals the marginal return, in the form of increased profit, of barrier-to-entry capital.[21] From society's standpoint, investment in barrier-to-entry capital that deters socially desirable entry is bad for two reasons: (a) it prevents society from realizing the benefits of increased competition which leads to greater output at lower prices; and (b) the resources spent on the production of barrier-to-entry capital, like those spent on war, are pure waste from society's standpoint.

As a matter of taxonomy, barriers to entry can raise the entrants' costs in three ways: (*a*) by raising the sunk (irreversible) costs of entry, (*b*) by raising the fixed (reversible) costs of production per period, and (*c*) by raising the operating costs of production per period. A simple example illustrates these costs. The incumbent has the following costs per period

$$F_I + c_I q_I \tag{3.1}$$

where F_I is the fixed cost per period (perhaps the rent for her plant), c_I is the operating cost per unit of output per period, and q_I is the output per period. It has incurred sunk costs S_I but these costs are irrelevant to its present business decisions. The entrant, likewise, has the following costs per period

$$F_E(k) + c_E(k)q_E \tag{3.2}$$

In addition, she must incur a sunk cost $S_E(k)$ in order to enter; these costs are relevant to her business decisions before she has incurred them through entry and are irrelevant to her business decisions after she has incurred them through entry. We have written S_E, F_E, and c_E as functions of the barrier-to-entry capital created by the incumbent. The incumbent can always undercut the entrant's price and still make an operating profit if either F_E or c_E exceeds F_I or c_I, respectively. Even if the entrant's operating costs are competitive with the incumbent's, the level of sunk costs may be so large as to outweigh any conceivable profits from entry. Thus, the incumbent may be able to deter entry by engaging in strategies that raise S_E, F_E, or c_E.

The Justice Department presented a persuasive case that AT&T tried to deter entry into (and encourage exit out of) the market for intercity telecommunications services by erecting barriers to entry.[22] First, AT&T used the regulatory process to impose sunk costs on the entrants. Second, AT&T made interconnection into the local exchanges artificially expensive thereby raising the entrants' fixed costs of operating. Third, AT&T charged the entrants discriminatorily high rates for each call which passed through the local exchange. The remainder of this section describes the manner in which AT&T was alleged to have imposed these costs.

Sunk Costs and the Abuse of the Regulatory Process

In order to construct facilities and offer services, potential entrants into intercity telecommunications had to obtain the consent of the FCC. This quasi-judicial agency had to hold hearings at which AT&T and its challengers could present their respective cases in opposition to and in favor of entry. The fact that many of the issues pertaining to the desirability of entry were extremely complex, the fact that the Commissioners did not seem to have strong predilections for or against entry, and the fact that the FCC had to regulate several large industries in addition to telecommunications with rather limited resources made these hearings expensive and lengthy for all parties involved. Nevertheless, AT&T had several advantages in this regulatory game. First, AT&T had a staff of lawyers, economists, and other experts who were already well versed in regulatory matters (due to the constant and pervasive regulation of the telephone industry) and who could be used to present AT&T's case before the FCC at probably little added cost to AT&T. The entrants had to assemble their regulatory experts from scratch. If they failed to get into the industry, the investment in developing this expertise would have had little value. Second, AT&T prevented entry during these regulatory hearings. It had an incentive to delay entry so long as the costs of pleading its case before the FCC exceeded the profits which would be lost from entry. Third, to the extent the entrants proposed new services which AT&T had not thought to offer, the regulatory delay gave AT&T time to develop competitive offerings and plan its competitive strategy.

AT&T consistently argued before the FCC that entry was neither necessary nor desirable. In seeking to deny MCI's application for installing a private line microwave system between Chicago and St. Louis, AT&T argued that "MCI had made no showing of need for the proposed services; that the proposed services were of 'questionable adequacy and reliability,' . . . and that MCI had 'not fully stated in its applications the investment that will be required,' listing several necessary features of its system for which MCI had not indicated the costs," and that MCI was a creamskimmer which "would impose an economic burden on Bell System customers."[23] MCI did not obtain final approval from the FCC until early 1971. In order to obtain permission to construct a microwave system which cost $2 million and took 7 months to complete, MCI spent $10 million in regulatory and legal costs and waited 7 years.[24] MCI's investment paid off handsomely for the specialized common carriers who surged into the industry in response to MCI's success. Nevertheless, AT&T continued to oppose enty and required extensive FCC hearings to plead its case. AT&T opposed MCI's entry into message toll service and succeeded in suspending MCI's service for a period of 2 years. AT&T also opposed Datran's application to construct a digital communication network, arguing that Datran's application would result in "uneconomic duplication of common carrier facilities," that there may not be sufficient demand for Datran's services, that AT&T could meet foreseeable demand at lower cost, and that Datran's proposal was neither economically nor technically feasible.[25] AT&T's opposition may have increased the time the FCC needed to rule on Datran's application.

In approving entry into intercity telecommunications, the FCC required AT&T to provide the entrants with access to the local exchange facilities[26]

> . . . established carriers with exchange facilities should, upon request, permit interconnection or leased channel arrangement on reasonable terms and con-

ditions to be negotiated with the new carriers, and also afford their customers the option of obtaining local distribution service under reasonable terms set forth in the tariff schedules of the local carrier. Moreover, . . . where a carrier has monopoly control over essential facilities we will not condone any policy or practice whereby such carriers would discriminate in favor of an affiliated carrier or show favoritism among competitors.

According to Gerald Brock, "AT&T denied the FCC's juridiction over interconnection and filed tariffs for local distribution of specialized common carrier communications only with the various state regulatory commissions."[27] Had AT&T succeeded in this strategy, it would have forced the specialized common carriers to seek regulatory hearings in each state in order to oppose potentially discriminatory AT&T tariffs. The FCC, however, prevailed. Nevertheless, AT&T allegedly resisted interconnection, thereby forcing the specialized common carriers to incur additional legal and regulatory costs in order to obtain redress from the FCC.[28]

From an economic standpoint, it is apparent that AT&T could use the regulatory process to impose sunk costs on entrants and thereby discourage entry and encourage exit. But AT&T was simply using the forums available to it. After all, this industry had been closely regulated for decades before the specialized common carriers came along. Moreover, AT&T had every right to participate vigorously and express its opinions freely in the regulatory process. The *Noerr–Pennington* doctrine, it may be argued, shields businesses from liability for their participation in governmental proceedings.[29] According to Justice Black, in his opinion on *Noerr,* making businesses liable under the antitrust laws for their attempts to influence government proceedings would "impute to the Sherman Act a purpose to regulate, not business activity, but political activity, a purpose that would have no basis whatever in the legislative history of that Act" and, since the right of petition is protected by the Bill of Rights, would "raise important constitutional questions."[30] Thus, AT&T's legitimate use of the regulatory process to deter possibly socially desirable entry was protected.

The *Noerr–Pennington* doctrine does not protect businesses that make "sham" use of the regulatory process for anticompetitive purposes. As one court noted, "no actions which impair the fair and impartial functioning of an administrative agency would be able to hide behind the cloak of an antitrust exemption."[31] The Justice Department made use of this "sham exception" to *Noerr–Pennington* to argue that, "AT&T has abused and manipulated the regulatory process in various ways, including making knowingly false statements, for the purpose of maintaining its monopoly position. Such conduct cuts across all substantive areas in the case since it arises in connection with entry proceedings, tariff investigations and interconnection disputes."[32] Judge Greene, in his opinion on AT&T's motion to dismiss the case, ruled that the Government, in order to prove its sham exception theory applied, had to show that AT&T "subverted the integrity of

the government process through misrepresentations or similarly unprotected conduct, or that they effectively barred their competitors' access to that process."[33]

This said, Judge Greene ruled that, with one exception, AT&T's participation
in the regulatory process was protected by the *Noerr–Pennington* doctrine. He
argued that the Government's evidence concerning MCI's 1963 application, for
example, showed at best that AT&T failed to reveal all relevant facts but did
not engage in overtly corrupt behavior and that its actions did not rise to the
level of a sham.[34] His finding demonstrates the difficulty of using the antitrust
laws to prevent regulatory abuses, as recommended by Bork.[35]

Two lessons may be learned from AT&T's ability on the one hand to impose
enormous sunk costs on its challengers and the Justice Department's inability
on the other hand to make AT&T liable for deterring possible socially beneficial
competition. First, the regulatory process is antithetical to the competitive process. True competition cannot work when entrants have to prove that there is a
need for their service and that they will not injure themselves financially, when
incumbents can impose lengthy delays and enormous costs on entrants, and when
regulators rather than self-interested businessmen have to grope their way towards
competitive equilibrium. Second, given the difficulty of proving a sham exception
to *Noerr–Pennington* and the enormous waste of resources involved in erecting
regulatory barriers to entry, there should be extremely stiff penalties for sham
use of the regulatory process. But the most direct method for preventing these
sorts of abuses is to rescind the power of regulatory agencies over pricing and
entry decisions which are best determined by the competitive market.

Judge Greene concluded that, on the basis of the Government's evidence,
AT&T had made sham arguments before the FCC concerning Datran's application to construct a nationwide system for transmitting digital information.[36]

> It appears from the evidence that AT&T opposed an FCC application by Datran
> for authority to construct a nationwide network, claiming that Datran had not
> demonstrated a need for the proposed service and that the economic and technical
> feasibility of the proposal was highly questionable. Internal Bell documents
> introduced into evidence revealed, however, that defendants recognized at that
> very time that the Datran proposal was "carefully planned" and "well financed";
> that they viewed it as a threat to AT&T's monopoly in network transmission;
> and that, in their opinion, a strategy of delay was necessary, to be implemented
> by a petition to the FCC requesting a general inquiry into the public interest
> aspects of the subjects raised by the Datran application. The Court can reason-
> ably infer from this evidence that AT&T's sole purpose in opposing the Datran
> application was to preserve its monopoly and that it well knew that the positions
> it took before the FCC were baseless.

Ironically, the Datran case is the one instance that an economic argument can
be made that entry was not socially desirable even by an entrant who was as
efficient as the incumbent. A digital network requires a large investment in fixed
capacity. Under plausible assumptions, it can be shown that there is a strong

"first mover" advantage in building a digital network capable of meeting all market demand. Competition between two firms causes a race to build first.[37] There is a tendency to build too soon and capture the market. Arguably, AT&T recognized the advantages of a digital network but tried to delay construction until the socially optimal time. Its delaying tactics at the FCC may, therefore, have been in society's best interests.[38]

Interconnection Restrictions

AT&T did not accede to the FCC's authority over interconnection until May 1974 when AT&T filed interconnection tariffs for various cities. These tariffs restricted interconnection to customer premises, thereby preventing the specialized common carriers from using the local distribution system to provide service to customers outside the range of the carriers' facilities. In order to serve these customers, the specialized common carriers would have had to develop their own local distribution systems. Local regulatory authorities would probably have prevented them from doing so. Moreover, to the extent that AT&T realized scale economies in the local distribution systems, the specialized common carriers would have been at a cost disadvantage in providing private line service to customers removed from their microwave facilities. From society's standpoint, arguably, the preferable arrangement would be to require AT&T to provide interconnection to the specialized common carriers at a price equal to the long-run incremental cost of interconnection.[39] The FCC ultimately struck down AT&T's interconnection restrictions and required AT&T to provide interconnection to the specialized common carriers on the same bases as it provided interconnection to AT&T Long Lines.

According to the Government, AT&T also "imposed a number of cumbersome and unnecessary technical and operational practices on its competitors which increases their costs and lowered the quality of their services."[40] For example, the specialized common carriers can only serve customers with push-button telephones; their customers have to dial 20 digits versus 11 for Bell customers to complete a call; they receive poorer transmission quality than Bell; and they cannot receive answer supervision.[41] Moreover, despite the FCC requirement of nondiscriminatory interconnection, AT&T failed to negotiate in good faith with the specialized common carriers.[42] These actions served to raise the cost to the specialized common carriers of offering intercity service.

AT&T's behavior, as reprehensible as it may have been, was certainly not a surprising business response to competitors. The FCC required AT&T to make its facilities available on a nondiscriminatory basis to businesses who openly planned to attack several of AT&T's markets. Competing businesses negotiate successfully when there is a mutual gain to be made. AT&T had nothing to gain from negotiating with the specialized common carriers. It seems clear that, had AT&T retained ownership of the local exchange facilities, AT&T and the specialized common carriers would have faced perpetual conflict and that the FCC would have spent its resources trying to resolve these conflicts. Moreover, once the specialized common carriers were entrenched in the market and had developed their own staffs of regulatory experts they would have been able to use the regulatory process to impose costs on AT&T.

Predatory Pricing

To an economist, a pricing strategy is predatory if it decreases social welfare by deterring socially desirable entry or promoting socially undesirable exit. The most commonly cited predatory pricing strategy involves pricing below cost. According to Posner, "Predatory price cutting is the practice whereby a firm having a monopoly position in a number of local markets sells below cost in those markets in which it has competitors. After the competitors are bankrupted by being forced to sell below cost for a prolonged period, the monopolist raises his price in the market to the monopoly level."[43] This price cutting reduces social welfare by (a) enabling the monopoly to restrict output below the socially desirable level after eliminating competition and (b) possibly preventing more efficient competition from meeting market demand.

The literature on predatory pricing strategies has burgeoned in the last 10 years. This literature addresses (a) the circumstances under which predatory pricing reduces social welfare and should therefore be dealt with severely by the antitrust laws and (b) whether predatory pricing strategies are plausible and therefore a form of business behavior that our antitrust laws should deal with in the first place. With regard to (a), economists generally argue that only prices below marginal cost could deter socially desirable entry.[44] Suppose the incumbent's marginal cost is c_I and the entrant's marginal cost is c_E. If the incumbent's price p_I exceeds the incumbent's marginal cost c_I and if the entrant's cost c_E is no higher than the incumbent's cost c_I, the entrant can meet the incumbent's price and still turn a profit. A entrant who cannot make a profit when p_I exceeds c_I must, therefore, be less efficient than the incumbent. If the incumbent's price p_I is less than the incumbent's marginal cost c_I then the incumbent loses money. The only rational explanation for this behavior is that the incumbent expects to reap monopoly profits after driving his competitor to bankruptcy.

With regard to (b), some economists argue that pricing below marginal cost is never a rational strategy and therefore probably one that businesses do not use.[45] If both the entrant and the incumbent can borrow money at the same interest rate, they can both hold out until the day monopoly profits role in. If the incumbent can borrow at a lower interest rate than the entrant, the entrant's customers have an incentive to sign a long-term contract with the entrant in order to avoid having to pay the incumbent the monopoly price after the entrant fails. The incumbent's losses from price cutting may be enormous. His gains are distant and uncertain. The argument that pricing below marginal cost is unlikely has been buttressed with the argument that the courts are particularly unsuited to detecting this behavior. It is extremely difficult to estimate marginal costs reliably. Predatory intent is too hard to gauge. The social losses due to inhibiting nonpredatory price cutting in response to competitive entry may exceed the social gains due to discouraging predatory pricing which may not exist anyway. To anyone familiar with cases in which the courts tried to evaluate economic evidence of predatory price cutting, these arguments are extremely persuasive.

Pricing Without Regard to Cost

The Government argued that AT&T pursued a pricing strategy called *pricing without regard to cost*. It did not claim that any of AT&T's prices were necessarily below cost.[46] Rather, it claimed that AT&T had set prices "with the sole intent of excluding competition" and without regard to whether these prices covered cost.[47] It concluded that AT&T's pricing behavior reflected predatory intent.

In order to gain a clearer idea of the meaning of "pricing without regard to cost" it is useful to review the testimony of Bruce Owen, the economist who presented the economic theory of this concept to the court.[48]

> GOVERNMENT ATTORNEY: I am going to ask you to assume now, Dr. Owen, that in fact AT&T has priced services in which it faces competition only with regard to the effects of that pricing on its market share, and not taking into account the costs of those services. If that has been the case, what would you conclude with respect to AT&T's intent with respect to those prices?

> BRUCE OWEN: It seems to me that that is evidence bearing on their intent to drive out competitors and to maintain their monopoly. Pricing without regard to cost isn't explicable on any other basis than an intent to drive out competitors. It is irrational behavior that can't be explained on any other basis.

> JUDGE GREENE: Even if it is not below cost?

> BRUCE OWEN: Even if it is not below cost. They don't know whether it is below or not. Even a monopolist would take account of cost in making pricing decisions, unless its only objective was to drive out competitors. . . . The question is not whether prices were above or below some cost standard, that's not the issue. The issue is whether pricing behavior which fails to take account of cost is itself evidence of illegal intent. There is no other inference in this situation that I can see than can be drawn from it, other than illegal intent.

Moreover, pricing without regard to cost demonstrates predatory intent even if subsequent evidence shows that prices were not below marginal cost.[49]

GOVERNMENT ATTORNEY: Dr. Owen, assume that in fact AT&T has priced in the way that we say that they have, and that after the fact someone comes along, for example AT&T, and can demonstrate that the prices were above cost, however measure [sic]; would that show, in your opinion, that there was no anticompetitive intent at the outset?

BRUCE OWEN: No, that's just an accident. What matters is whether they set the prices with relationship to a knowledge of costs in the beginning, and not what turned out to be the case *ex post*.

As presented by Owen, it is unclear whether pricing without regard to cost is a meaningful concept. A simple example illustrates the theory. The regulated firm produces two services and has costs

$$C = F + c_1q_1 + c_2q_2 \tag{3.3}$$

where F is common cost, c_1 is the cost per unit of service one, c_2 is the cost per unit of service two, q_1 is the amount of service one, and q_2 is the amount of service two. Before competition, the firm earned R by charging prices p_1 and p_2 such that

$$R = p_1q_1 + p_2q_2 - c_1q_1 - c_2q_2 - F \tag{3.4}$$

and thereby earned its allowed rate of return on the rate base. It may have determined p_1, p_2 such that equation (3.4) holds by performing cost studies to determine F, c_1, and c_2 or by a process of trial and error. Service one is opened to competition. The firm decides to respond vigorously by slashing the price for service one. If it follows a predatory pricing without regard to cost strategy, it identifies a price p_1^*, which will destroy the competition, without determining whether p_1^* exceeds c_1 and therefore covers the marginal cost of production. If it were an unregulated firm charging profit-maximizing prices this strategy would result in at least temporary losses. But because p_2 is less than the profit-maximizing price and because it is guaranteed a return R, it can raise the price p_2 and thereby offset the losses from its predatory price reductions.

As a result of the price decrease on service one from p_1 to p_1^*, the demand for service one increases from q_1 to q_1^*. Revenue changes by

$$\Delta R_1 = [p_1^*q_1^* - c_1q_1^*] - [p_1q_1 - c_1q_1] \tag{3.5}$$

Assume that ΔR_1 is negative so that the price reduction causes a revenue loss. The regulated firm can offset this loss by increasing price on service two from p_2 to p_2^*. For simplicity, suppose that the higher price on service two leads to no change in the demand for service two. Then revenues increase by

$$\Delta R_2 = [p_2^*q_2 - c_2q_2] - [p_2q_2 - c_2q_2] \tag{3.6}$$

The firm must find a p_2^* such that $\Delta R_1 + \Delta R_2 = 0$ (i.e., so that the new prices p_1^*, p_2^* ensure that it will earn its allowed rate of return). Simple algebra shows that, in order to calculate p_2^*, given p_1^*, the regulated firm has to know the costs of both service one and service two

$$p_2^* = \frac{R + c_2 q_2 + c_1 q_1^* - p_1 q_1^*}{q_2} \tag{3.7}$$

One could argue that the firm will find a price p_2^* by a process of trial and error. But prices are more costly to change in a regulated market than in an unregulated market.[50] The firm has to file tariffs and cost justifications. It is unclear whether the firm could change its tariffs quickly enough and often enough to discover a new price for service two that will raise enough revenues to offset the losses on service one. During this period of groping, the regulated firm may fail to earn its allowed rate of return. If a regulated firm is to pursue the strategy described by Owen, it clearly must have reliable estimates of the costs of both of its services. Therefore, the assertion that the firm does not know whether its prices are above or below cost is nonsensical.[51]

The Government presented four types of evidence that AT&T priced intercity services without regard to cost. First, AT&T failed to present cost studies which justified its tariff reductions on competitive services. Summarizing the evidence, Judge Greene said "sometimes no cost studies were filed; sometimes costs were inaccurately estimated; sometimes cost studies were deliberately terminated; and at other times no account was taken of predictable revenue effects of user shifts from one service to another."[52] Second, the FCC often found that AT&T's cost studies were defective and therefore inadequate to support competitive price reductions. Third, various internal documents demonstrate that AT&T believed that its prices were lower than its costs on some of the services opened to competition. Fourth, "there was some evidence that Bell failed to consider alternative prices and to select among the alternatives the price that maximized the contribution from competitive services to the recovery of common costs, and thus, according to the government, evidences an intent to maximize market share by excluding competition."[53]

The fourth type of evidence does not demonstrate that Bell's prices deterred socially desirable entry or encouraged socially undesirable exit. If Bell's marginal costs were less than the entrant's marginal costs, where both marginal costs are calculated on the basis of prospective long-run incremental costs, then society gets more from its scarce resources when Bell rather than the entrants provide the service. There is no economic basis for requiring Bell to charge a price that "maximizes the contribution from competitive services to the recovery of common cost." If entry was sufficiently easy, then competition would have forced Bell to charge a profit-maximizing price no higher than marginal cost. The third type of evidence supports the standard predatory-pricing-below-cost argument.

Presumably, by itself, this evidence was not sufficient to support the Government's predatory-pricing case. Otherwise, the Government would not have presented the more novel, and hence more risky, pricing without regard to cost case.

The first and second types of evidence provide the major support for the pricing without regard to cost case. To an economist, this type of evidence is not persuasive. First, in explaining how competitive markets operate, economists do not assume that businessmen actually calculate marginal costs by taking the first derivative of their econometrically estimated cost function with respect to output or by taking some other highly technical approach.[54] We assume businessmen know how to maximize profits—we do not say how—and that, by doing so, they end up producing at the point where price equals marginal cost. The relationship between the economist and the businessman is like the relationship between the physicist and the golfer. A physicist could calculate the direction and intensity of the force that would propel a golf ball to a hole. But a professional golfer can make a hole-in-one without being a Ph.D. in classical mechanics. An economist can derive the mathematical conditions for profit maximization. Yet a businessman can make money without having a Ph.D. in economics. Physicists and economists, who are interested in obtaining a scientific understanding of how the world operates, do not necessarily make good golfers or profitable businessmen, professions which require instinct rather than calculus. Thus, the fact that AT&T's tariffs were not adequately supported by cost studies may demonstrate a failure to comply with regulatory requirements but certainly does not demonstrate that AT&T ignored costs in setting prices or intentionally set prices below cost.

Second, the cost standards adopted by the FCC for regulating the telecommunications industry differed, over the period in question, from the cost standards usually espoused by economists.[55] The FCC generally advocated some form of fully distributed cost standard which required that the price for each service cover a share of common cost as well as marginal cost. The FCC may have found that it could better regulate the telecommunications industry by requiring carriers to price according to some fully distributed cost rules than by the long-run incremental cost rules recommended by economists, but AT&T's failure to comply with these standards does not imply that AT&T priced so as to deter socially desirable entry or to encourage socially undesirable exit. Although AT&T's failure to justify its costs to the FCC may reflect antiregulatory intent, it does not necessarily reflect anticompetitive intent.[56]

Predatory Pricing and the Regulated Firm

According to the Government, regulated firms have greater incentives to engage in predatory pricing than unregulated firms.[57]

AT&T is subject to rate base rate of return regulation under which an overall rate of return is applied to the sum total of AT&T's investment. Unlike an unregulated firm, which must forego current profits to engage in below cost pricing, AT&T is subject to an overall earnings ceiling. It therefore has the ability, virtually indefinitely, to subsidize the prices of its competitive services with earnings from services in which it faces little or no competition—without sacrificing current profits.

Judge Greene apparently accepted this argument. He noted that "the opportunity which a multiproduct firm subject to rate of return regulation has to cross-subsidize low prices for one product across other products (rather than across time) renders it far more likely to engage in anticompetitive pricing than the firm that must hope to recoup its losses."[58]

In fact, the proposition that a regulated firm has a greater incentive to indulge in predatory cross subsidization than an unregulated firm is false under plausible assumptions concerning the regulatory process. Let us review the government's argument. Consider a profit-maximizing monopolist that provides two services. It maximizes profits Π by charging profit-maximizing prices p_1^* and p_2^*. An entrant appears in service one. The monopolist responds by lowering price to $\hat{p}_1$. Since p_2^* was the profit-maximizing price before entry it will be the profit-maximizing price after entry (assuming that the cross elasticity of demand between service one and service two is zero and that there are no cost complementarities in production between services one and two). Since p_1^* was the profit-maximizing price before entry, $\hat{p}_1$ must reduce profits by, say, $\Delta\Pi$. Therefore, the monopolist must forego profits of $\Delta\Pi$ in order to engage in predatory pricing. It may recover these foregone profits by raising prices after the entrant is destroyed. As mentioned earlier, this strategy is risky since profits are distant and uncertain. Now consider a rate-of-return regulated monopolist that provides two services. It achieves its revenue target R and thereby earns its allowed rate of return by charging prices p_1 and p_2 which must be less than the profit-maximizing prices p_1^* and p_2^*. In response to entry, the regulated monopolist lowers price to $\bar{p}_1$ on service one thereby reducing revenues by ΔR. Since p_2 is less than the profit-maximizing price p_2^*, the regulated monopolist may be able to offset lost revenues ΔR by raising the price on service two from p_2 to $\bar{p}_2$. If so, the regulated monopolist can engage in predatory pricing without foregoing current profits. Therefore, predatory pricing is costly to an unregulated firm but costless to a regulated firm.

The flaw in this argument is that an unregulated firm may recoup losses it incurs this period by reaping monopoly profits next period, whereas a regulated firm can earn no more than its allowed rate of return in subsequent periods.[59] The regulated firm neither gains nor loses from predatory pricing. Moreover, to the extent that regulatory lag makes it difficult for the regulated firm to raise prices in order to cover losses and easy for the regulators to reduce prices in response to excess earnings, the regulated firm may, on net, lose by pursuing a predatory pricing strategy.[60]

The proposition that a regulated firm has a greater incentive to engage in predatory cross subsidization than an unregulated firm can be salvaged by arguing that the regulated firm engages in predation for nonmonetary reasons. Owen, for example, suggested at one point that AT&T may have had other incentives besides maximizing profits.

> GOVERNMENT ATTORNEY: Have you assumed in your analysis of AT&T's incentives that AT&T itself is a profit-maximizing firm?

> BRUCE OWEN: Well, that assumption is sufficient to derive the results I have described, that is, it is sufficient to explain their incentives to monopolize markets and discriminate against competitors. But, because AT&T doesn't have this kind of competitive pressure that a competitive firm does, it is free to pursue other objectives to serve the interests and tastes of its managers, for example, and those other objectives are also consistent at any time and might also be consistent with the behavior I have described.

It may be that AT&T desired to crush the competition for the sake of preserving its monopoly power rather than preserving its profits *per se*. This argument, however, is highly speculative and not directly articulated in the trial record. The most reasonable conclusion we can draw from economic theory and from trial record is that both regulated and unregulated firms have strong profit incentives *not* to engage in predatory pricing.[61]

NOTES

1. AT&T, *Annual Report*, 1907, p. 18.
2. AT&T, *Annual Report*, 1910, p. 32.
3. John Brooks, *Telephone* (New York: Harper & Row, 1975), p. 144, quoting a speech entitled "Some Observations on Western Tendencies" given by Vail at the National Association of Railway Commissioners meeting in San Francisco.
4. *Ibid.*, p. 145.
5. *Ibid.*, p. 148.
6. AT&T, *Annual Report*, 1910, p. 33.
7. AT&T, *Defendants' Third Statement of Contentions and Proof*, in *US* v. *AT&T*, p. 137.
8. Brooks, *op. cit.*, p. 146.
9. See Chapter 2 for a discussion of the Kingsbury Committment.
10. J. Warren Stehman, *The Financial History of the American Telephone and Telegraph Company* (Boston: Houghton Mifflin, 1925), p. 234.
11. Gerald Brock, *The Telecommunications Industry* (Cambridge: Harvard University Press, 1981), p. 232.
12. The FCC's control over entry results from its control over the construction of service facilities and the provision of new services.

13. This chapter focuses on competition in intercity telecommunications services. The FCC was also the scene of major skirmishes concerning entry into the telecommunications equipment market. See Brock, *op. cit.*, Chapter 9, for a discussion of the introduction of competition into terminal equipment.

14. U.S. Department of Justice, *Plaintiff's Memorandum in Opposition to Defendants' Motion for Involuntary Dismissal Under Rule 41(b)*, in *US* v. *AT&T*, August 16, 1981; hereafter referred to as *Plaintiff's Memorandum*.

15. This chapter also provides a nontechnical introduction to Brock's analysis, in Chapter 8 of this volume, of pricing and predation in regulated industries.

16. Some legal scholars would, of course, argue that any predatory practice that does not deter socially desirable competition (i.e., competition that enables society to realize a greater value of goods and services from its scarce resources) should not be considered illegal under the antitrust law. For example, see Robert Bork, *The Antitrust Paradox* (New York: Basic Books, 1978).

17. See, for example, George Stigler, *The Organization of Industry* (Homewood, IL: Irwin, 1968), p. 67.

18. Joe S. Bain, *Barriers to New Competition* (Cambridge: Harvard, 1956).

19. See C. C. von Weizsäcker, *Barriers to Entry* (Berlin: Springer-Verlag, 1980). For a related discussion see William Baumol, John Panzar, and Robert Willig, *Contestable Markets and the Theory of Industry Structure* (San Diego: Harcourt, Brace, Jovanovich, 1982).

20. E. E. Bailey and J. C. Panzar, "The Contestability of Airline Markets During the Transition to Deregulation," *Journal of Law and Contemporary Problems*, Winter 1981, pp. 125–145.

21. For a treatment of investment in barriers to entry see Gerald Brock, *op. cit.*, Chapter 2 and Chapter 8 of this volume.

22. For a discussion of the evidence see *Plaintiff's Memorandum, op. cit.* and Judge Greene's *Opinion on Defendants' Motion for Involuntary Dismissal Under Rule 41(b)*. For AT&T's rebuttal of this evidence see AT&T, *Defendants' Third Statement of Contentions and Proof;* AT&T, *Defendants' Memorandum in Support of Defendants' Motion for Involuntary Dismissal Under Rule 41(b);* and the trial record of AT&T's defense. This chapter presents an economic analysis of the Justice Department's arguments but does not pretend to assess whether the evidence presented in support of the Justice Department's case is factually correct or whether there are explanations other than the anticompetitive behavior theory advanced by the Justice Department for the actions taken by AT&T.

23. AT&T, *Defendants' Third Statement of Contentions and Proof,* p. 583 quoting "Petition to Deny Applications," filed by AT&T before the FCC, February 12, 1964, pp. 3–6.

24. Brock, *op. cit.*, p. 351.

25. AT&T, *Defendants' Third Statement of Contentions and Proof,* p. 593.

26. Brock, *op. cit.*, p. 356 quoting from the FCC's *Specialized Common Carriers Decision,* 29FCC2d870 (1971), p. 940.

27. Brock, *op. cit.*, p. 356.

28. Once in, however, AT&T's challengers could impose costs on AT&T by demanding detailed justifications for new, lower AT&T tariffs and changes in AT&T's service offerings.

29. *Eastern Railroads Presidents' Conference* v. *Noerr Motor Freight, Inc.* 365US127(1961) and *United Mine Workers* v. *Pennington* 381US657(1965). See Judge Greene's *Opinion, op. cit.*, for a review of the relevant decisions.

30. Quoted by Judge Greene, *op. cit.*, p. 37.

31. Quoted by Judge Greene, *op. cit.*, p. 38.

32. *Plaintiff's Memorandum , op. cit.*, p. 288.

33. Judge Greene, *op. cit.*, p. 40.

34. Judge Greene, *op. cit.*, p. 40.

35. Bork, *op. cit.*, pp. 347–349.

36. Judge Greene, *op. cit.*, p. 42.

37. For a discussion of the first-mover advantage see Chapter 8 of this volume.

38. From a legal or ethical standpoint, this argument certainly does not justify making sham use of the regulatory process. Moreover, competition for the first mover advantage may encourage innovative behavior. The gains from competition may offset the losses from building too soon.

39. Of course, there was much debate at the FCC over the appropriate interconnection price.

40. *Plaintiff's Memorandum*, p. 79.

41. Judge Greene, *op. cit.*, p. 26.

42. Judge Greene, *op. cit.*, p. 27.

43. Richard Posner, *Antitrust: Cases, Economic Notes, and Other Materials* (St. Paul: West Publishing Co., 1974), p. 361.

44. See P. Areeda and D. Turner, "Predatory Pricing and Related Practices Under Section 2 of the Sherman Act," *Harvard Law Review,* February 1975, pp. 697–733.

45. See Frank H. Easterbrook, "Predatory Strategies and Counterstrategies," *University of Chicago Law Review,* Spring 1981, pp. 263–311 and John McGee, "Predatory Price Cutting: The Standard Oil (N.J.) Case," *Journal of Law and Economics,* October 1958, pp. 137–168, and Bork, *op. cit.*

46. It did suggest that the Telpak tariffs were probably set below cost.

47. *Plaintiff's Memorandum*, p. 190.

48. Bruce Owen, testifying in *US* v. *AT&T,* Tr. 10962–10963.

49. *Ibid.,* Tr. 10968–10969.

50. This is not to say that AT&T was unable to change its price in response to entry. As we show in Chapter 4, there is considerable evidence that AT&T lowered its prices on services opened to competition. But, AT&T would have attracted considerable attention and ire from the FCC had it filed frequent tariff changes and forced frequent FCC hearings. AT&T did not, in fact, change prices frequently on noncompetitive services after lowering prices on competitive services.

51. *Ibid.,* Tr. 10963, lines 7–11. In Chapter 8 of this volume, we describe a version of pricing without regard to cost which may deter socially desirable entry. In this version, a dominant firm knows its costs but randomizes prices in order to impose artificial risks on entrants.

52. Judge Greene, *op. cit.*, p. 46.

53. Judge Greene, *op. cit.*, p. 48.

54. A point stated eloquently by Kenneth Arrow in testimony submitted on behalf of AT&T.

55. See Walter G. Bolter, "The FCC's Selection of a 'Proper' Costing Standard after Fifteen Years— What Can We Learn from Docket 18128?" in H. Trebing, ed., *Assessing New Pricing Concepts in Public Utilities* (East Lansing, MI: Institute of Public Utilities, Michigan State University, 1978), pp. 333–372 for a discussion of costing standards at the FCC. The FCC shunned marginal cost pricing because it believed that AT&T had too much leeway in manipulating the cost estimates.

56. The FCC was not the first regulatory agency to express dismay over being unable to determine the relationship between telephone rates and telephone costs. A report prepared for the City Council of Chicago in 1907 complained,

> We have endeavored to obtain data bearing on the cost of specific classes of telephone service from other telephone companies, including the Bell Telephone Companies of New York and elsewhere, and also from certain of the so-called Independent telephone companies, but in no instance have we been able to obtain records kept in such detail or in such manner as to afford appropriate data for fixing rates, and in every instance the rates seem to have been dictated by estimates based on experience or the requirements of business expediency, instead of being founded on a knowledge of the costs of the different classes of service. The telephone companies seem to go on the belief that their business is satisfactory if the total results of each year's business show a profit, and that it is unsatisfactory if the year's business does not show a profit. . . .

See Committee on Gas, Oil and Electric Light, *Telephone Service and Rates,* Report to the City Council of Chicago, September 3, 1907, p. 49.

57. *Plaintiff's Memorandum, op. cit.,* p. 206.

58. Judge Greene, *op. cit.,* p. 51.

59. For further discussion of this point, see Kenneth Arrow's testimony in for AT&T.

60. See Chapter 8 for further discussion of predatory cross subsidization when the tightness of the regulatory constraint varies over time or over service lines. Neither of these situations reverses the conclusion that regulated firms are no more likely to engage in predation than unregulated firms under any plausible scenario.

61. The argument that a regulated firm has a greater incentive to engage in predatory cross subsidization than an unregulated firm could be defended on the following grounds. Averch and Johnson have argued that regulators permit public utilities to earn a rate of return on their capital (their so-called rate base) in excess of their cost of capital. Therefore, public utilities have an incentive to use capital excessively. To the extent competition reduces the amount of capital that the public utility can install for providing competitive services and to the extent regulation guarantees that the public utility will get its allowed rate of return on all capital in use, the public utility has an incentive to squash competition in order to preserve and enlarge its capital stock. The cost of such predation is zero since, by assumption, the public utility is guaranteed its allowed rate of return on its capital stock. The gain from such predation is positive since the public utility will be able to replace the competitors' capital with its own capital and obtain a rate of return on this capital in excess of the cost of this capital. See Harvey Averch and Leland Johnson, "Behavior of the Firm Under Regulatory Constraint," *American Economic Review,* December 1962, pp. 1052–1069. There are two problems with the Averch–Johnson model. First, there is considerable anecdotal evidence that the regulatory process does not work in the manner assumed in the model. Second, several empirical studies have found little evidence of the major effect predicted by the Averch–Johnson model, namely that the public utility will use too much capital and too little labor. See Paul Joskow and Roger Noll, "Theory and Practice in Public Regulation: A Critical Overview," in G. Fromm, ed., *Studies in Public Regulation* (Cambridge: MIT Press, 1981) and the references cited therein.

Chapter 4
Creamskimming

William A. Brock and David S. Evans

In arguing against competitive entry into their markets, regulated public utilities often claim that their competitors are uninnovative and inefficient businesses merely seizing profit opportunities created by discriminatory ratemaking. An example illustrates this argument. A public utility provides two services. It charges $10 per unit for the first service and $5 per unit for the second service. It incurs a cost of $7 per unit for both services. It thereby earns a profit of $3 per unit on the first service and takes a loss of $2 per unit on the second service. Its profits on the first service offset its losses on the second service. A competing firm which incurs a cost of $8 per unit for the first service, and which is therefore less efficient than the public utility, but which has no obligation to provide the second service, could offer the first service for $9.50 and thereby capture the market for the first service. The competing firm in this example is said to creamskim the public utility's markets. The competing firm skims the cream— the profitable first service—and leaves the skimmed milk—the unprofitable second service—for the public utility.

Most public utilities have raised the specter of creamskimming. The Motor Carrier Act of 1935 tried to prevent creamskimming by restricting entry into trucking.[1]

> The contract carrier may differ from the common carrier only in the fact that he undertakes to skim the cream off the traffic and leave the portion which lacks butterfats to his common-carrier competitor. Obviously such operations can have very unfortunate and undesirable results.

Airlines claimed creamskimming entry would force them to abandon service to remote and sparsely populated areas. Taplon and Gerwitz argued[2]

It is surprising to some that trunk carriers . . . provide service to the smaller cities in the United States. The trunklines are able to do this because the losses sustained at these marginal cities, which also deserve the advantages of air transportation, are compensated for by the revenues developed at the greater traffic-producing areas. . . . With free entry every airline would tear up its timetables, disregard its certificates, forget that it has franchise responsibilities, and do what business it pleased in the interest of greater profits and not public convenience. . . . The industry, in such a chaotic struggle for survival, would then have to abandon service to roughly some 500 of the cities to which it is now certificated, and operate only between the 50 most profitable points.

Spokesmen for the electric utility industry have argued that small cogenerating plants creamskim the electric rate structure.[3] These plants allegedly reduce the availability of low-cost power by diverting demand from companies with scale economies.

In its defense against Government charges that it tried to thwart competition in intercity telecommunications, AT&T argued[4]

The Bell System's fundamental defense is that defendants' behavior reflects not a continuing course of conduct to monopolize any market but instead a good faith effort to . . . compete fairly and thereby mitigate the creamskimming effects of new entry in segments of the market that have been opened to competition and to refuse to become a party to the expansion of creamskimming into segments of the market where responsible regulatory agencies have not yet certified entry to be in the public interest.

It used creamskimming to refer to entry into markets in which its "rates . . . have been established substantially above costs" and in which its "real economic costs . . . are less than the new entrants' costs of providing comparable services." Its rates exceeded costs because of "regulatory ratemaking policies adopted to reduce the price of basic service and to allow the extension of relatively lower cost service to high cost areas."[5] Creamskimming, it claimed, would encourage inefficient investment in communications and thereby impose additional costs on society and undermine regulatory policies favoring low, presumably subsidized, prices for certain telephone services.[6]

It is important to distinguish creamskimming, which may damage the public interest, from competition, which advances the public interest. According to a theorem that dates back to Adam Smith, competition between selfish, profit-hungry businesses provides society with the greatest wealth of goods and services for the lowest possible cost.[7] A business that prices its product above cost and thereby makes a profit will be beset by other businesses willing to sell at a lower price albeit at a lower profit. So long as entry is unimpeded, businesses will continue to lower price until they earn no more than a competitive rate of return. This will happen when each business charges a price equal to the smallest cost per unit of output. Businesses that raise their prices above this competitive level will rapidly lose sales. Businesses that use inefficient production techniques or do not operate at the least-cost scale of operation will incur unit costs in excess of unit price. They will lose money and eventually close down. The invisible hand of competition thereby forces businesses to serve the greater good. This hand moves most swiftly when businesses are many, the least-cost scale of operation is small relative to the size of the market, and there are no barriers to new businesses. Market forces themselves ensure that these conditions hold true at least in the long run. Profits encourage new business formation and increase the number of businesses competing with each other. Profits also encourage technological innovations that can remove the stranglehold a large, efficient firm may have on the market.

Competition in public utility markets may, arguably, harm social welfare. According to most advocates of public utility regulation, public utilities have two characteristics that differentiate them from other industries. First, they are natural monopolies so that a single firm can provide public utility service more cheaply than several firms. Competition would lead to wasteful duplication. Second, they provide essential services which everyone should have access to at reasonable prices. Competition would lead to price gouging. Regulation can improve on competition by simultaneously preventing inefficient entry, requiring the provision of essential services at reasonable prices, and preventing the utility from earning more than a competitive rate of return.

Both characteristics of public utilities lead to discriminatory ratemaking. The prices charged by a natural monopoly must exceed marginal costs for at least some services provided or some customers served.[8] Regulators may require the public utility to provide essential services at rates less than marginal costs. In order to offset the losses from these essential services, regulators must permit the public utility to provide other services at rates greater than marginal costs. Because of both natural monopoly and essential service considerations, the rates charged by public utilities bear little relationship to costs.

Under these circumstances, creamskimming is competition turned on its head. In competition, prices inform potential entrants about the profits and costs of incumbents. High prices signal excessive profits or excessive waste on the part of incumbents. In creamskimming, prices misinform. High prices mean high cross subsidies for low-priced services. In competition, entrants promote social welfare by expanding output, lowering costs, and eliminating excessive profits or excessive waste. In creamskimming, entrants promote social waste, because they are less efficient and less innovative than the incumbent, and disrupt socially desirable price discrimination. In competition, the invisible hand looks out for society's best interest. In creamskimming, the public utility and its regulators do so.

This chapter has four sections. The first section examines the economic theory of creamskimming entry. It shows that creamskimming is more likely the less flexible the public utility's prices and the smaller the level of sunk costs necessary to enter the market. The second section presents a case study of MCI's entry into AT&T's intercity telecommunications markets. It suggests that MCI was not a creamskimmer. The third section describes the optimal regulation of entry into public utility markets. It argues that entry prohibition is a needlessly restrictive method for preventing the adverse effects of creamskimming. It describes the revenue-surcharge method for discouraging inefficient entry and preserving socially desirable subsidies. The fourth section examines AT&T's arguments against competitive entry into intercity telephone services. It finds that AT&T's arguments had little merit.

Economic Theory of Creamskimming

Let us continue the creamskimming example presented in the first paragraph of this chapter. The public utility has lost its lucrative first market but has retained its revenue-draining second market. In order to avert bankruptcy, it must seek a rate hike from its regulators. It will probably request a rate increase for the second service. But it will probably also request a rate reduction for the first service. It could lower the rate on the first service to $7.99 per unit, earn $.99 profit per unit, and drive its inefficient competitor out of the market.

Faced with eventual failure, why would a competitor challenge a more efficient public utility to begin with? There are several explanations. First, the public utility may take some time to reduce its price in response to competitive entry. During this time, its competitors might earn some profits. Second, regulators might not permit the public utility to lower its prices. Third, there are circumstances under which a public utility may not be able to find prices that deter socially inefficient entry and thereby sustain its natural monopoly.

As usually told, the creamskimming story says little about the strategic responses available to and the strategic responses anticipated by the creamskimmer and the public utility. This section views creamskimming as a strategic dynamic game between an entrant and an incumbent.[9] The object of each player in the game is to maximize profits. The rules of the game place constraints on the strategic responses available to each player at each point in time. Both the entrant and the incumbent would like rules that favor their own sides. The entrant would like regulators to freeze the incumbent's prices. The incumbent would like regulators to block entry altogether. Society prefers rules that ensure that the game maximizes social welfare. The game has an equilibrium if there is a sequence of moves from which no player has an incentive to deviate.[10]

Real world games are exceedingly complex. The game may follow a multitude of paths. Some of the players may be unable to determine the choice made by some other player at a particular stage. Finally, the rules of the game may not be precisely defined. An entrant may not know if the incumbent's tariff, filed at an earlier stage of the game, is above or below the incumbent's marginal cost for that service. The regulators may be unable to enforce a rule such as "price must exceed long-run incremental cost" because of auditing problems. An incumbent may not know whether regulators will permit tariff reductions at a future stage of the game.

This section is divided into two parts. The first part reviews the strategic aspects of creamskimming entry. Although it does not derive the equilibrium path for the game, it shows how the strategic responses available to and the strategic responses anticipated by the players affect this equilibrium. The second part analyzes the special case of an incumbent that has a natural monopoly. It shows how alternative assumptions concerning the strategic responses available to the incumbent affect the outcome of the game.

Strategic Aspects of Creamskimming Entry

Entrepreneurs are in the business of making money. They make cold, hard calculations of the profitability and riskiness of alternative ventures. They seek profits and shun risks. They insist on potentially large profits to offset possibly large risks. They are not myopic. Their calculations incorporate rational forecasts of future events. One study reports that "the only characteristics that successful entrepreneurs, venture capitalists, and behavioral scientists say are important for success include the ability to set clear goals and objectives that are challenging, yet realistic and attainable and a preference for moderate, calculated risks, where the chances of winning are not so small as to make the effort a gamble, nor so large as to make it a sure thing, but which provide a reasonable and challenging chance of success."[11]

Before such an entrepreneur—the entrant—would challenge an established firm—the incumbent—she would ask herself some basic business questions. Are my costs competitive with the incumbent's? With other potential entrants? If my costs are not competitive, will my product command a price premium from buyers because it is superior to my competitors' products or because it appeals to some submarket of buyers? Will the incumbent meet or undercut my price? Would I be able to recover any of my investment should I subsequently withdraw from this business? Her hopes of "making a killing" and attracting investors to the venture in the first place hinge upon the answers to these questions.[12]

A simple example shows why these questions are relevant. Anyone can buy a widget maker for F dollars. The incumbent can make widgets for c_I dollars per widget with its widget maker.[13] The entrepreneur can make widgets for c_E dollars per widget with her widget maker. The incumbent charges p_I dollars per widget, sells q_I widgets per year, and earns a substantial profit. Should the entrepreneur buy a widget maker? She would capture the entire widget market if she charged less than p_I dollars per widget *and* if the incumbent were unwilling or unable to lower its price. If she charged p_E dollars per widget—where p_E is less than p_I—she would sell q_E widgets per year—where q_E exceeds q_I because quantity demanded increases as price decreases. She would earn revenues of $p_E \times q_E$ dollars per year, incur costs of $c_E \times q_E$ dollars per year, and earn operating profits of

$$\pi_E = p_E \times q_E - c_E \times q_E$$

dollars per year. If she earned these profits in perpetuity, standard investment theory shows that the present discounted value of her operating profits would be

$$V_E = \frac{\pi_E}{r}$$

where r is the discount rate for future earnings.[14]

These calculations show that the entrepreneur could receive a stream of operating profits whose present discounted value is V_E dollars, in return for investing F dollars in a widget machine. Standard investment theory tells us she should buy the widget machine if and only if V_E exceeds F. For then she would make more money by investing her funds in the widget machine than by investing her funds in a financial instrument paying an interest rate of r.

When does V_E exceed F? Simple arithmetic shows that V_E exceeds F when

$$p_E \text{ exceeds } \frac{rF}{q_E} + c_E$$

The price that equates V_E and F is the lowest price at which the entrepreneur would be able to recover her initial investment in the widget maker. Call this price p_E^*. If p_E^* exceeds p_I, the entrepreneur would have no chance whatsoever of profitably capturing even one widget of the incumbent's market. No one would pay more than p_I for a widget and the entrepreneur would never recover her investment if she charged less than p_E^*. If p_E^* is less than p_I, the entrepreneur could charge some price between p_I and p_E, attract all of the incumbent's widget customers, and make a handsome profit.

But the incumbent is hardly likely to stand by and watch this upstart entrepreneur steal its widget business. What will it do? If its costs per widget are lower than the entrepreneur's costs per widget (c_I less than c_E), it could reduce its price to slightly less than p_E^* but above its minimum profitable price p_I^*, and continue to make a profitable albeit smaller return on its investment in the widget maker. Moreover, the incumbent could deter an entrepreneur from invading its market simply by revealing its costs to the entrepreneur and her investors and announcing it would lower its price to below p_E^* if the entrepreneur persisted in entering. Students of business strategies call this announcement a *credible threat*. The threat is credible because the incumbent has an incentive to lower its price below p_E^* in response to competitive entry. These considerations show the importance of price flexibility.

> *An incumbent with flexible prices can make and carry out a credible threat to crush an inefficient and uninnovative entrant. An entrepreneur would be extremely foolish to challenge an incumbent if the incumbent has lower costs than the entrepreneur and is willing and able to lower its price.*

What would happen to an entrepreneur with pie-in-the-sky expectations who challenged a more efficient incumbent? The answer to this question depends on the extent to which the entrepreneur and the incumbent could recover their investment in the widget maker if they withdrew from the widget business. The unrecoverable portion of an investment is called a *sunk cost*. The recoverable portion of an investment is called its *resale value*. Consider two extreme cases.

First, sunk costs are zero so that a widget maker can be sold for its original purchase price. Upon belatedly realizing her foolishness, the entrepreneur could sell her widget maker and thereby recover her initial investment in full. She foregoes the interest she would have earned by originally investing her funds in an alternative venture and she may incur some operating losses between opening and closing her business. These costs would not be substantial if she acts quickly enough.

Second, because widget makers are worthless to anyone but their original owners and then only for the purpose of making widgets, all of the investment in the widget maker is sunk cost. Upon discovering her investment in the widget maker will never pay off, the entrepreneur will try to minimize her losses. Because sunk costs are sunk, she should only consider the prospective costs of remaining in the business. Her prospective costs are c_E dollars per widget produced. She would make an operating profit if she succeeded in selling widgets for any price higher than c_E. But the incumbent, whose widget maker also has no resale value, can play this game too. Suppose the entrant lowered her price to slightly above c_E and attracted all of the incumbent's customers. Rather than close down its business, the incumbent would lower its price to below c_E but above c_I, regain its customers, and drive the entrant out of business once and for all. The entrant whose operating costs exceed the incumbent's is inevitably forced to withdraw. She sustains not only operating losses but also forfeits her sunk investment.

This discussion illustrates the importance of sunk costs.

An entrepreneur would face certain and devastating bankruptcy if she incurred substantial sunk costs in order to challenge a more efficient incumbent whose prices are flexible.

These same considerations concerning price flexibility, sunk costs, and relative costs apply to the entrant vis-á-vis other entrants. If the entrant is more efficient than the incumbent but less efficient than other entrants, she has limited hopes for success. If the entrant and other entrants produce comparable products for comparable costs, these entrants may have to share the market with each other and thereby diminish the profits available to any single entrant.[15]

The real world of business is much more complicated than this simple example reveals. The incumbent could be a large, lethargic corporation unable to respond quickly to entry threats by small upstart entrepreneurs or its regulators might require lengthy hearings on requested tariff reductions. Consequently, the incumbent might maintain its price in the face of price-cutting incursions by an entrant. The entrant might earn substantial profits while bureaucrats in the incumbent's organization trade memos on how to respond to competitive entry. The incumbent, which sees its market share eroding, will eventually respond by lowering its price. But the entrant might be able to more than recoup her investment before the incumbent drives her out of the market. This hit-and-run strategy is more likely to be profitable the longer it takes the incumbent to lower its price, the closer the entrant's operating costs are to the incumbent's, the longer it takes for other hit-and-run entrants to appear, and the greater the resale value of widget makers.

The entrant faces a great deal of uncertainty. Having never operated in the widget market, she has no direct knowledge of demand, she is unsure whether she will be able to produce widgets for c_E dollars per unit, she has no definitive information about the incumbent's costs and therefore about the incumbent's ability to survive a price war, and she is unsure whether there are other entrepreneurs eager to enter the widget market. Consequently, an entrepreneur or investor who wagers her money on a new venture must expect to reap large profits to offset the enormous risks she bears. Stanley Rubel, an expert on financing new ventures, says "Most venture capitalists . . . want to see a profit of three to five times the money or even more (before capital gains taxes) in a three to five year period. Some shoot for tenfold appreciation. . . . "[16] Other things being equal, the entrepreneur and her investors must convince themselves that their costs are so much smaller or that their product is so much better than the incumbent's and other entrants' that they will be able to capture a large, profitable share of the market. Rubel goes on to say[17]

> . . . the venture capitalist will estimate the chances of the business's achieving its stated goals and also the chances of its failing. The venture firm may attempt to estimate the financial results of success and failure to arrive at a risk–reward ratio. In the final analysis, the venture capitalist must make a horseback guess as to these probabilities and the caliber of management and then act accordingly. There is really no science in making a successful venture capital investment. It is simply based on seat-of-the-pants business judgment and hard work.

Our example illustrates several important points which apply just as well to sophisticated market rivalries that occur in the real business world. An entrant would be foolish to challenge a more efficient incumbent, and consequently creamskimming is unlikely, when

> *the incumbent can quickly respond to price-cutting incursions by entrants;*
>
> *the entrant must incur substantial sunk costs in order to challenge the incumbent;*
>
> *many entrants share in the fleeting profit opportunities created by the temporary rigidity of the incumbent's prices.*

These considerations suggest a useful classification of entrants into public utility markets. (1) *Hit-and-Run Entrants:* They take their profits and run before the incumbent can reduce prices. Their profits are higher the lower their sunk costs and the longer the incumbent takes to reduce prices. (2) *Protected Entrants:* They remain profitable despite the fact that they are inefficient because regulators impede the incumbent from lowering prices on competitive services. (3) *Efficient Entrants:* They make money because they are more efficient or more innovative than the incumbent. Hit-and-Run Entrants and Protected Entrants reduce social welfare because they use more of society's scarce resources than the incumbent does to provide the same service. They are true creamskimmers. Efficient Entrants increase social welfare because they require less of society's scarce resources than the incumbent does to provide this same service. They may, however, divert revenues that the incumbent uses to subsidize essential services. Regulators can alleviate this problem by requiring entrants to either provide the essential service themselves or pay the incumbent to do so. In the next section we analyze the strategic responses available to MCI and AT&T in order to infer whether MCI was a creamskimmer or an efficient entrant. Before doing so, we examine the conditions under which a natural monopoly is sustainable against competitive entry by inefficient and uninnovative firms.

Entry into Natural Monopoly Markets[18]

We begin with a parable. American Toll and Travel (AT&T) operates a canal system which has been the only means of travel within Panzania for more than a century. The Federal Canal Commission (FCC) regulates the tolls charged, the quality provided, and the rate of return earned by AT&T. Motorways Carrier, Inc. (MCI) recently petitioned the FCC for permission to operate a highway that would compete with AT&T's canal between Chicago and St. Louis. The FCC held hearings at which AT&T and MCI made the following arguments:

AT&T: Numerous econometric and engineering studies show that we have a natural monopoly over travel services. That is, we can provide travel services at the least cost to society. The FCC should therefore deny MCI's entry petition.

MCI: AT&T would be able to undercut our price and thereby drive us out of business, if, indeed, they were the least costly provider of travel services. The FCC should allow the competitive process to reveal whether AT&T truly is a natural monopoly.

AT&T: But it is possible that we will not be able to establish a price which simultaneously prevents MCI from making a profit, guarantees that we will at least break even, and enables us to satisfy demand at the least cost to society. Figure 4.1 demonstrates this possibility. The curve shows the average cost of providing travel services. We are satisfying market demand Q_S at the least possible cost to society P_S. MCI proposes to "piece out" the quantity Q_E by only serving the Chicago to St. Louis market. That leaves us with Q_M. The average cost of our producing Q_M and MCI producing Q_E exceeds the average cost of our producing $Q_S = Q_M + Q_E$. If MCI were permitted to enter, (1) society would pay more, on average, for travel services and (2) we would go bankrupt—our average cost P_M would exceed the price we are allowed to charge P_S—unless the FCC permitted us to raise our price.

MCI: AT&T has not shown that its costs correspond to those shown in Figure 4.1. The FCC should disregard AT&T's speculation that it is an unsustainable natural monopoly.

AT&T: It is true that we do not have hard evidence that our costs correspond to those shown. But, if they do, MCI's entry will result in drastically higher travel prices and may force the best travel system in the world out of business. Does the FCC wish to risk these dire consequences in return for highly speculative and at best miniscule cost savings from allowing MCI to build a highway that merely duplicates the services we already provide? We think not.

Figure 4.1 Socially inefficient creamskimming entry.

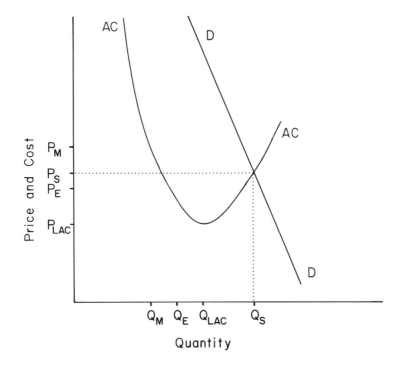

In this parable, AT&T tried to prevent MCI's entry by using an argument based on the sustainability theory developed by Baumol, Panzar, and Willig, but which has not been used, to our knowledge, in any regulatory hearings.[19] Under the assumptions made by these authors the conditions for AT&T's argument to be valid are stringent. It is unlikely that any public utility could bear the burden of demonstrating that these conditions hold in its particular circumstances.[20] Nevertheless, if valid, this argument raises the risks of deregulating public utilities and permitting free entry into heretofore monopolized markets. Panzar has argued that the provision of air services was a nonsustainable natural monopoly when the airline market was smaller. He concluded that deregulation of airlines would have been a major mistake earlier in this century.[21] Panzar and Willig have suggested that, even though nonsustainability is a highly abstract concept, "public policy analysis of entry into a regulated natural monopoly industry should not be heedless of considerations of sustainability."[22]

Most economic theories often rest on unverifiable assumptions about market behavior. For this reason, it is important to examine whether the conclusions reached by a theory hold under alternative assumptions. The sustainability theory developed by Baumol, Panzar, and Willig assumes that the incumbent's prices are rigid. Price maintenance is the nicest response a natural monopoly could make to an entrant. The sustainability theory developed in this section assumes that the incumbent's output levels are rigid. Quantity maintenance is one of the nastier responses a natural monopoly could make to an entrant. Truth, no doubt, lies somewhere between these polar assumptions.

In the traditional theory of competitive markets, the equilibrium price equals two different measures of cost. First, price equals the cost of supplying an additional unit, the so-called marginal or incremental cost of production. Second, price equals the average cost of providing the quantity demanded at this price. This twofold equality guarantees that society uses its scarce resources as efficiently as possible and that businesses earn no more than a competitive rate of return. Part of the reason for this desirable property is that average cost is smallest when it equals marginal cost. When there are a large number of identical firms, the industry and each firm in it operates at the so-called efficient scale of operation where the average cost of production is smallest.

The equality of price, average cost, and marginal cost follows from three basic assumptions. First, there are no entry restrictions which would prevent firms from entering the market, competing price down to least average cost, and thereby eliminating excess profits. Second, the average cost of production eventually increases with increases in production so that scale economies are exhausted before market demand is satisfied. Third, the output supplied by all firms, each producing at a common efficient scale of operation, equals the output demanded when price equals the least average cost. Economists recognize that firms may have different average costs and different efficient scales of operation. But we have found it reasonable to abstract from these differences when least average costs are roughly the same for all firms and when the firm with the largest efficient scale of operation is still small relative to market demand when price equals least average cost.

The traditional theory of competitive markets breaks down when firms have a large efficient scale of operation relative to market demand. Figure 4.2 illustrates the problems that arise. For firm A, least average cost is P_{LAC} and the efficient scale of operation is Q_{LAC}. At a price equal to P_{LAC} market demand is 1.5 times Q_{LAC}. Firm B, which has the same costs as firm A, could not profitably supply the additional units of Q for price P_{LAC}. The least expensive way to satisfy market demand in these circumstances is for firm A to expand production past the efficient scale Q_{LAC}. Economists have shown that society gets the most from its scarce resources by giving firm A the sole rights to this market but regulating firm A so that it is unable to exploit its inherent monopoly power. Socially desirable output is Q_S and price P_S.[23] Notably, marginal cost exceeds average cost at the preferred level of output.

Panzar and Willig asked whether regulators have to bar entry into firm A's market in order to ensure that output is maintained at Q_S and sold for P_S. By carefully examining Figure 4.2, we see that a competitive entrant could disrupt the socially preferred arrangement. Firm A is not producing at least average cost but at a cost in excess of P_{LAC}. Firm B could produce Q_{LAC} units—which would

Figure 4.2 Price nonsustainable natural monopoly.

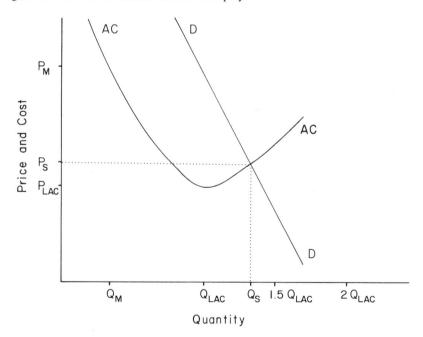

not be sufficient to satisfy market demand at price P_S—charge slightly less than P_S, and earn a handsome profit. Its profits would equal $(P_S - P_{LAC})Q_{LAC}$ less its small price discount. What would happen to firm A? If it maintained its price P_S it would lose its market for Q_{LAC} units and retain a market for only $Q_M = Q_S - Q_{LAC}$ units. But it spends P_M per unit to produce Q_M units; P_M exceeds P_S so it loses money. It is not clear where Panzar and Willig's story goes from here. Firm A must raise its price or go out of business. Once firm A changes its price firm B may change its price as well. In any case, this example supports AT&T's argument in the parable that entry by MCI might destroy the most socially efficient provision of travel services. Prices rise and the natural monopoly may be forced out of business. In the language of Panzar and Willig, firm A's natural monopoly is not sustainable against competitive entry. For reasons which will become clear, we say that firm A's natural monopoly is not price sustainable.

Let us now try to complete the Panzar and Willig story. The monopolist loses money if he continues to sell Q_M for P_S. He could retract his scale of operation to Q_{LAC} rather than to Q_M and try to undercut the entrant's price. If the entrant continued to produce Q_{LAC}, the monopolist and the entrant together would produce $2Q_{LAC}$. But, as we see from Figure 4.2, demanders will not buy this amount of output at any positive price. The monopolist and the entrant would both lose money if they continued to produce Q_{LAC} each. One of both firms would have to retract its output. The reader who continues this process of conjecturing how the drama will unfold will quickly conclude that the story may never end. The market may never reach a point where neither the entrant nor the monopolist has an incentive to alter his price or output. The game may not have an equilibrium.

Why did the entrant challenge the monopolist in the first place? Supposedly because he saw an opportunity to make a quick profit by piecing out an especially profitable segment of the monopolist's market. But, in the story related above, it is unclear whether he actually reaps any rewards from entry. His rewards depend on how quickly he and the monopolist respond to each other's sales plans. Therefore, it is unclear whether a true profit opportunity existed which would have induced an entrant to attempt to annihilate the natural monopoly in the first place. The chief weakness with the sustainability literature is that it says little about the strategies the monopolist and entrant adopt and how these strategies work out over time. It does not model the "game" played by the entrant and incumbent in a manner which makes explicit the strategies used and the dynamics involved. We now consider the strategies available to these players but ignore dynamic interactions between them.

The competitive model developed by economists assumes that there is a large number of firms none of which can individually affect the market price. Free entry in response to profit opportunities forces all firms to charge a price equal to the least average cost of production. The competitive model breaks down when there are only a few firms. These firms may not choose prices and quantities independently of each other, as our discussion has demonstrated. Economists have not yet reached a consensus on how to model interactions among a small group of firms. But they have reached some tacit agreement on how to approach this problem. Since Cournot's work, a popular modeling strategy has been to concentrate on games in which firms choose how much to produce—hence the name quantity game—and take price, which is determined by demand and the sum of the quantities chosen by all firms, as given.[24] They have done so because the predictions obtained from quantity games are more plausible than predictions obtained from price games in which firms announce their prices and supply whatever quantity is demanded at these prices. The quantity game predicts a positive relationship between profits and market shares; dominant firms are more profitable largely because they can exert more market control. The price game predicts that prices and quantities constantly fluctuate (there is no market equilibrium) or that, for example, two firms that produce the same service and together supply the whole market do not earn any monopoly profits. Common sense and some empirical evidence supports the predictions of the quantity game and not the price game.

The sustainability story we related in the parable is based on a price game. The monopolist announces his price $\bar{P}$. The entrant then announces a lower price under the assumption that the monopolist will maintain his price at $\bar{P}$. If the entrant can make a profit at any price less than $\bar{P}$, then the natural monopoly is not sustainable according to Panzar and Willig. But our earlier discussion showed that the story cannot possibly end here. The monopolist must either go broke or revise his prices. As we would expect with a price game, the monopolist and the entrant may never settle on prices and outputs that are mutually satisfactory in the sense that neither firm has an incentive to alter its price or output. The game, therefore, may never reach equilibrium.

It is easy to tell a sustainability story where the monopolist and entrant play a quantity game. Refer to Figure 4.3. The monopolist produces the socially preferred output Q_S and charges P_S. The entrant assumes that the monopolist will continue to produce Q_S regardless of what the entrant produces. The entrant produces Q_{LAC}. Total output is $Q_S + Q_{LAC}$. The added output depresses market price from P_S to P_E. The entrant cannot make a profit at P_E. The monopoly output of Q_S is thus sustainable against an entrant who produces Q_{LAC}. Therefore, the monopolist who was not "price sustainable" against an entrant producing Q_{LAC} is "quantity sustainable" against this same entrant.

Can the entrant find any output level that would enable him to profitably supply some portion of the market given that the monopolist continues to produce Q? Because the average cost curve has a U shape, he would have to receive more than P_S in order to profitably supply less than Q_L or more than Q_S. If he supplies any additional output, price would fall below P_S. Therefore, if he can profitably supply any quantity of output, this quantity must be between Q_L and Q_S. He will be able to supply some output q profitably if the demand price for $Q_S + q$ exceeds the average cost of q. By inspecting Figure 4.3, the reader may

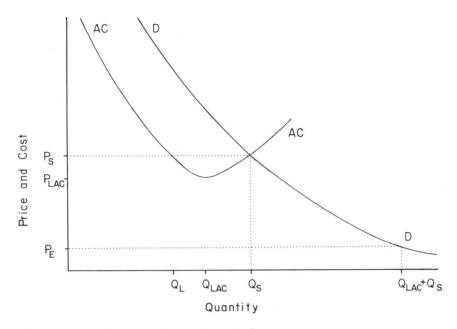

Figure 4.3 Quantity sustainable natural monopoly.

verify that there is no sales plan between Q_L and Q_S that would enable the entrant to make a profit given that the monopolist maintains its sales plan of Q_S. Therefore, the monopolist is quantity sustainable against all possible entrants when he produces the socially preferred quantity Q_S.[25]

What is the economic difference between quantity sustainability and price sustainability? Producing Q_{LAC} is attractive to a potential entrant only when he is certain that price P_S will remain in effect for long enough for him to make a profit. But, P_S must change upon his entry into the market. Whether the postentry price will enable him to make a profit depends on the strategic behavior of both the entrant and the monopolist. The monopolist could continue to produce Q_S and let price go where it will. This would be an especially plausible strategy for a monopolist who has large sunk costs and small variable costs. If so, the entrant would depress price and would lose money under the circumstances shown in Figure 4.3. The entrant could expect to make a profit only if the monopolist reduces his output to $Q_S - Q_{LAC}$. This would be a charitable monopolist indeed.

A monopolist who plays a *quantity game* with potential entrants is more likely to be sustainable than a monopolist who plays a *price game* with potential entrants. When firms have access to a common technology that requires sunk costs, economic models that assume firms engage in a quantity game provide more realistic predictions of market behavior than economic models that assume firms engage in a price game. Consequently, we believe that quantity sustainability is more relevant than price sustainability for evaluating entry into markets where sunk costs loom large.

Neither the price game nor the quantity game discussed above, however, are plausible representations of how markets involving small numbers of firms operate. Both games assume that players are myopic and irrational. For example, in the price game upon which the sustainability theory developed by Baumol, Panzar, and Willig is based, the monopolist does not change the price that induced the entrant into the market and thereby loses money. The monopolist is myopic because he charged a price that induced an entrant to virtually annihilate his monopoly position. He is irrational because he does nothing to counter the entrant who undercuts his price. If the monopolist does alter his price in response to competitive entry, then the entrant may be irrational in chasing ephemeral profits. Whether an entrant can dislodge a natural monopolist who engages in sophisticated and realistic strategic behavior remains to be seen. We have shown, however, that if the natural monopolist engages the entrant in a quantity game he is more likely to be able to sustain his monopoly.

Creamskimming Entry into Intercity Telecommunications

This section uses the theory developed in the previous section to infer from the strategic behavior of AT&T and MCI whether MCI was a creamskimmer or an efficient entrant. This may seem a circuitous approach.[26] The most direct method to determine whether AT&T's competitors were creamskimmers is to compare their costs with AT&T's costs of providing comparable service. Several problems arise. (1) The Justice Department has pointed to "AT&T's control over cost information and its demonstrated ability to frustrate any attempt to penetrate the relationship between its costs, by whatever definition, and its prices." They also point to "the FCC's longstanding inability to obtain reliable accurate information regarding cost relationships."[27] (2) Competitive entrants are still extremely young. They are probably engaged in the learning and experimentation every new business goes through. But the relevant cost comparison is between AT&T's current costs, which reflect its accumulated experience, and the competitors' costs after they have had a chance to work the bugs out of their operations. Obviously, it is not possible to make this comparison yet. (3) Some of the services provided by competitive entrants are not strictly comparable to AT&T's services thereby making straight cost comparisons difficult.

MCI filed an application with the FCC on December 31, 1963 to construct and operate a microwave system between Chicago and St. Louis. The FCC approved the application, over AT&T's objection that MCI was trying to creamskim Bell's high-density routes, on August 13, 1969. The FCC was subsequently innundated with applications to provide private line services and established the *Specialized Common Carriers* inquiry in order to evaluate the common policy issues raised by these applications. As of March 15, 1971, 33 applicants, including 17 MCI affiliates, sought to provide specialized common carrier services. MCI and its affiliates wanted to construct a microwave system connecting a number of major cities. Datran proposed a nationwide digital communications network for data transmission. SPCC initially wanted to establish microwave service between San Diego and Seattle and between St. Louis and Los Angeles. WTCI wanted to expand its video business system, which primarily served CATV systems, to provide other services and serve other users. Pushing AT&T's cream-skimming argument aside, the FCC determined "there is sufficient ground for a reasonable expectation that new entry here will have some beneficial effects."[28] This decision was based on a conclusion that revenue diversion would be unlikely to have a great effect given the markets being entered, and that any such adverse effect would be exceeded by the potential benefits of entry. In January 1975, MCI began offering its Execunet service which allowed subscribers access to its microwave system from local telephones. The FCC ordered MCI to cease providing Execunet because this offering was a public message telephone service (MTS), which MCI was not authorized to provide, and not a private line service, which MCI was allowed to provide. The FCC claimed that its *Specialized Common Carriers* decision granted AT&T a monopoly over MTS and WATS service with which Execunet allegedly competed. The Circuit Court overturned this ruling noting, "The Commission's analysis of creamskimming in the *Specialized Common Carriers* decision gives no evidence that the Commission made an affirmative finding that revenue diversion would be a problem if specialized carriers were allowed to compete on the fringes of its message telephone service market. . . . "[29] The Court's finding that "there is simply nothing in *Specialized Common Carriers* that would suggest a conclusion that revenue diversion required restrictions on MCI's facility authorizations" was based in part on the FCC staff's skeptical view of AT&T's creamskimming argument.[30]

Its creamskimming pleas spurned or ignored, AT&T moved quickly to meet the competition head on. It considered lowering its private line tariffs between St. Louis and Chicago in early 1971.[31] The proposed tariffs approximately equaled MCI's rates and were to be filed as soon as possible after MCI filed its rates.[32] Its operating company presidents " . . . generally agreed that a decisive response to the entry of specialized common carriers is required to provide a clear signal—inside and outside the business—of the Bell System's competitive spirit. What remains at issue are questions of timing and strategy."[33] It filed these rates on supposedly higher density routes on November 15, 1973. It timed these rates to become effective when the private line services were beginning to attract substantial numbers of customers.[34] Prior to filing these rates, it publicized its intent to respond vigorously to competition. It used interviews in important trade journals and press conferences for this purpose. John deButts, its chairman, told the *Chicago Tribune* that it was going to fight and defeat the specialized common carriers, by cutting prices if necessary.[35] DeButts reiterated these intentions in an article in the October 30, 1972 issue of *Telephone:*[36]

> Our . . . determination is to remove any doubts among our customers, our
> competitors, and ourselves as to whether we really intend to compete. We must.
> We don't intend to abdicate any sector of our business—not the terminal area,
> not the private line area—where we are convinced, and by our performance
> can prove, that we can do a better job than anybody else. If there is to be
> competition, let it be *real* competition, and not division of the market, arbitrarily
> established and artificially maintained.

It postponed the effective date of the HiLo Tariff until June, 1974, allegedly
because it expected to earn greater total revenues without this tariff until roughly
that time.[37] Its tariff amounted, in practice, to a selective rate cut on routes
subjected to competition by the specialized common carriers. After the FCC
rejected these rates, in June of 1976, AT&T quickly filed even lower rates.
Thus, AT&T possessed considerable price flexibility.

MCI and its investors were aware that AT&T intended to compete vigorously
and that the FCC would allow AT&T to file competitive tariffs. MCI's *Prospectus*
of June 22, 1972, noted[38]

> Many of the private line communications services to be offered by the MCI
> Carriers have been offered commercially by AT&T, Western Union and in-
> dependent telephone companies for a considerable period of time. These es-
> tablished carriers can be expected to compete vigorously with the MCI Carriers
> and to offer services comparable to those proposed to be offered by the MCI
> Carriers. . . . The FCC had indicated that AT&T and others already serving
> the private line market would be allowed to compete fairly and fully and could
> be expected to gain a very substantial portion of the potential market. . . . The
> FCC's current policy appears to be that competition will determine the prices
> that the specialized common carriers will be able to charge subscribers. . . .

Its *Prospectus* of February 19, 1975 says it "found it necessary to file new, lower
tariffs in order to meet . . . price competition" resulting from AT&T's HiLo
Tariffs and similar tariffs filed by Western Union.[39] In its *Prospectus* of December
5, 1978, MCI commented on its continued price competition with AT&T.[40]

> AT&T has in the past lowered its rates over the routes, among others, where
> MCI competes. When these rates were declared illegal by the FCC in January
> 1976, AT&T filed a tariff which became effective in August 1976 providing
> for even lower rates forcing MCI, as a matter of competitive necessity to lower
> its rates.

MCI realized early in its formation that AT&T enjoyed considerable price flexibility which AT&T intended to exercise in order to "fight and defeat" the
competition. MCI subsequently matched AT&T's rate reductions.

MCI anticipated AT&T's rate reductions; believed the FCC would permit fair
price competition; and was informed in no uncertain terms that AT&T would

compete vigorously. Yet it spent more than $8 million by July 31, 1973, more than $116 million by March 31, 1974, and more than $134 million by September 30, 1974. Would it have done so if it suspected that its economic costs exceeded AT&T's costs for comparable service? Most probably not, unless the entrepreneurs and investors behind MCI were characterized by a foolishness uncommon to hard-nosed successful businessmen.[41]

Two considerations argue against this conclusion. (1) MCI might have honestly relied on erroneous estimates of its costs relative to AT&T's costs. But given the amount of money MCI's investors were gambling on this venture, the enormous uncertainty surrounding AT&T's willingness to provide reasonable interconnection with the local operating companies, the delays AT&T could impose on MCI by insisting on lengthy regulatory hearings, and the amount of information available on two well-known technologies—microwave and cable—the latter of which had been in operation for more than a century, it is hard to believe that entrepreneurs and investors could misjudge relative costs as badly as AT&T claims. Indeed, MCI's investors would probably have convinced themselves that MCI's costs were so much lower or its service so much better than AT&T's that MCI would be sufficiently profitable, if it successfully met the hurdles AT&T would place before it, to offset the extraordinary risks entailed in challenging the largest corporation in the world.[42] (2) MCI might have counted on the FCC and the courts finding AT&T's rate reductions predatory. But surely MCI could not have anticipated in the early 1970s that AT&T would file largely unsupported cost studies and that the FCC and the courts would find these reductions predatory. A business venture based on the vagaries of regulatory commissioners and several levels of judges ruling on abstruse technical questions would be a gamble indeed.

The conclusion that MCI was probably more efficient and innovative than AT&T in providing private line service becomes even more compelling when the facts that (*a*) a substantial portion of the funds invested in MCI were sunk costs and (*b*) numerous other entrants followed MCI's footsteps and thereby may dissipate the profits from breaking AT&T's monopoly over private line services are taken into consideration. AT&T claims that MCI's system was inefficient and unnecessarily duplicative of its own network. If AT&T had succeeded in driving MCI out of business through fair price competition, no business would have been interested in buying the MCI system *in toto*. The resale value of the MCI system would be the resale or scrap value of the capital equipment used in constructing the system. As of September 30, 1974 MCI had spent $105,900,854. Of this amount, $47,771,147 was spent on "system development" including general and administrative costs, communication systems and engineering, marketing and customer services, professional services, other costs, and interest. All of these costs are sunk and unrecoverable. An additional $58,129,707 was spent on system construction of which $23,080,408 was spent on site selection and acquisition, construction and engineering services, engineering construction salaries and overhead, and other costs. Most of these costs are probably sunk unless there are buyers for completed microwave sites. The remaining $30,397,643 was used to purchase communications system equipment. Some of these costs are probably recoverable through reselling the equipment piecemeal or as part of a completed microwave site. Therefore, substantially less than $58 million worth of the $106 million spent through September 30, 1974 was recoverable. More than half, and probably closer to two-thirds, of MCI's investment would have been unrecoverable if AT&T had succeeded in driving MCI and the other specialized common carriers out of business.[43] MCI had about $59 million in equity and about $48 million in long-term debt as of March 31, 1974. Therefore, MCI's equity holders would have lost all of their money if MCI had been driven into bankruptcy.[44]

Sixteen enterprises had filed applications to offer private line services as of March 15, 1971. In 1975, MCI commented that it[45]

> . . . competes with others which have filed, or may at some future time file applications with the FCC. Applicants proposing to provide services similar to MCI's include video common carrier companies and others which have extensive experience in operating microwave facilities. One of the applicants plans and has built a portion of an all-digit network designed for the transmission of data only. Another applicant has built portions and has acquired other portions of a planned national network similar to MCI's system. One very large international conglomerate recently received authority to construct an extensive regional system between New York City, Atlanta and Houston, but has stated that it does not expect its system to be in operation prior to 1976. The FCC's policy of permitting competition in the business and data communications market will allow others that have not yet filed applications with the FCC to enter the market if they meet the FCC's technical, legal, and financial requirements.

MCI therefore anticipated stiff competition from many other entrants as well as AT&T.

Let us summarize. MCI anticipated that AT&T would enjoy considerable price flexibility. MCI also anticipated that other firms would follow its lead into the intercity telecommunications market. Yet, MCI continued to make unrecoverable and irreversible investments in order to enter this market. AT&T quickly responded to MCI's competitive incursion by lowering prices on intercity private line services. Despite complaints by the FCC, MCI, and other specialized common carriers that the reduced rates were predatory, AT&T successfully maintained lower prices. Many firms followed on MCI's heels into both the private line market and the message toll market. These facts are not consistent with the theory that MCI was a creamskimmer. MCI was not a hit-and-run entrant because it incurred substantial sunk costs and because it correctly anticipated that AT&T would be able to lower prices dramatically and rapidly. MCI was not a protected entrant because the FCC, whatever its intentions, could not effectively prevent AT&T from reducing rates on an intercity service. These facts are consistent with the hypothesis that MCI was either more efficient than AT&T or provided an innovative service that AT&T could not provide.

Optimal Regulation of Creamskimming Entry

This section analyzes the optimal regulation of entry into markets where a public utility has common carrier obligations to provide unremunerative essential services. We assume that the public utility charges rates that exceed costs on some services in order to lower rates on essential services. We also assume that the public utility is subject to hit-and-run entry, perhaps because entry requires no sunk costs and the public utility's prices are temporarily rigid. Under these circumstances, creamskimming entry poses a serious problem. The public utility may face repeated competitive incursions into its profitable markets by inefficient hit-and-run entrants who dissipate the profits it uses to subsidize essential services. An obvious solution to the problem is to prohibit entry. The drawback to this solution is that it excludes efficient entrants as well as creamskimmers and makes the public utility immune to the discipline of the competitive market. Regulators could impose common carrier obligations on entrants. The disadvantage of this method is that entrants may be efficient at serving certain profitable public utility markets but not at serving unprofitable public utility markets. This method prevents entrants from realizing economies of specialization.

The optimal regulation of creamskimming entry should (*a*) ensure that output is produced by the most efficient firm and therefore at the least cost to society and (*b*) maintain services deemed essential by society. Regulators could simultaneously encourage socially beneficial entry, discourage socially wasteful creamskimming entry, and maintain subsidies to socially desirable but unprofitable services by requiring entrants to replenish the cream they skim from profitable public utility markets. Let us give an example. Suppose the incumbent provided the profitable service for $5 per unit, incurred a cost of $2 per unit, and earned a profit of $3 per unit which he used to subsidize service to the needy. The regulator could require the entrant to contribute $3 per unit sold to the incumbent for the purpose of subsidizing service to the needy. This surcharge would make it unprofitable for entrants with costs above $2 per unit to compete with the incumbent, make it profitable for entrants with costs below $2 per unit to compete with the incumbent, and maintain subsidies to the needy at their precompetition level. Hunt has proposed surcharges for encouraging competitive entry into private line service.[46] This section analyzes certain aspects of the surcharge method for discouraging inefficient entry.[47]

Suppose for simplicity that demands are independent, costs are linear, and the regulator wants to maximize the following social welfare function

$$S = \sum_{i=1}^{n} (\lambda_i B_i(q_i) - c_i q_i) \qquad (4.1)$$

where B_i is the benefit from q_i units of service i and λ_i indicates the degree of cross subsidy with $\lambda_i < 1$ indicating that service i provides cross subsidies and $\lambda_j > 1$ indicating that service j receives cross subsidies.[48] The benefit $B_i(q_i)$ equals the value received by consumers from q_i units: $B_i(q_i) = \int_0^{q_i} D(q_i) dq_i$ where P_i

$= D(q_i)$ is the inverse demand function for q_i. The λs weight consumer surplus according to the value regulators place on particular services. When firms break even, using marginal cost pricing, competition maximizes

$$\sum_{i=1}^{n} B_i(q_i) - c_i q_i \tag{4.2}$$

Necessary conditions are

$$c_i = \frac{\partial B_i}{\partial q_i} = D_i(q_i) = P_i, \; i = 1, 2, \ldots, n \tag{4.3}$$

giving the familiar result that price equals marginal cost. When the market is contestable, competition ensures that this social optimum is attained even if there are only a few competitors with extensive scale economies.[49]

We specialize equation (4.1) to the case of two services and leave the general case to later. We define a feasible surcharge–subsidy scheme as a pair of numbers (τ_1, τ_2) such that

$$D_1(q_1)(1 + \tau_1) = c_1 \tag{4.4}$$
$$D_2(q_2)(1 - \tau_2) = c_2$$

$$\tau_1 p_1 q_1 = R_1(\tau_1) = R_2(\tau_2) = \tau_2 p_2 q_2 \tag{4.5}$$

where

$$R_1(\tau_1) = \tau_1 D_1^{-1} \frac{c_1}{(1 + \tau_1)} \frac{c_1}{1 + \tau_1} \tag{4.6}$$

$$R_2(\tau_2) = \tau_2 D_2^{-1} \left(\frac{c_2}{1 - \tau_2} \right) \frac{c_2}{1 - \tau_2} \tag{4.7}$$

express revenue as a function of the tax rate and where D_i are inverse demand functions and D_i^{-1} are demand functions.

Figures 4.4 and 4.5 display this scheme and its properties. Figure 4.5 shows that, for any target level or revenues R_1 desired by the regulators for subsidies to the first service, there are two possible surcharges on the second service. Obviously, regulators should use the lower surcharge. Figure 4.4 displays the welfare loss triangle, revenues, quantities, and effective prices generated by the surcharge–subsidy scheme. There are two revenue areas of equal size displayed in Figure 4.4. Notice that the output distortion $q_2 - \hat{q}_2$ for the poorly cho-sen surcharge $\underline{\tau}_2$ is much larger than the output distortion $q_2 - \hat{q}_2$ for the chosen surcharge $\overline{\tau}_2$.

Figure 4.5 suggests an iterative scheme for finding the correct surcharge. Choose a purposely small surcharge $\underline{\tau}_2$. Adjust it according to the scheme

$$\frac{d\tau_2}{dt} = -\mu(R_2(\tau_2) - R_1). \tag{4.8}$$

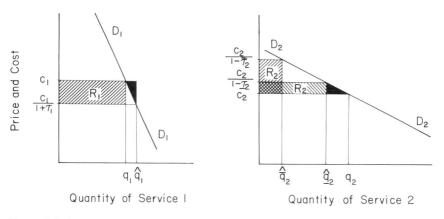

Figure 4.4 A tax-surcharge scheme for two services.

Here $\mu > 0$ is called the speed of adjustment. Figure 4.5 shows that $\underline{\tau}_2$ is stable and $\bar{\tau}_2$ is unstable for the adjustment scheme given by equation (4.8).

Turn now to the general case. Suppose that a regulator wishes to place a surcharge τ_i (a subsidy will be treated as a negative surcharge) on service i. Then revenue from that service is given by

$$R_i = p_i(1 - \tau_i)q_i \qquad i = 1,2, \ldots ,n. \qquad (4.9)$$

Suppose that $C(q)$, $q = (q_1, \ldots q_n)$ is subadditive so that we have natural monopoly. How will the prices be determined? In the case of independent demands, we can argue, as do Baumol, Bailey, and Willig, that the weak invisible

Figure 4.5 Tax revenue function for Service 2.

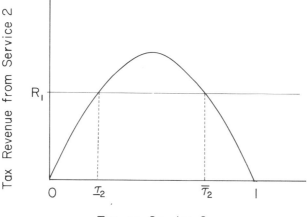

hand in a perfectly contestable market will force effective prices (i.e., surcharge-inclusive prices)[50]

$$p_i^* = p_i(1 - \tau_i) \qquad (4.10)$$

to satisfy the modified Ramsey relation for all i, j

$$r_i^* = \frac{p_i^* - C_i}{p_i^*} e_i = \frac{p_j^* - C_j}{p_j^*} e_j = r_j^* \qquad (4.11)$$

$$\sum_{i=1}^{n} p_i q_i(1 - \tau_i) = C(q) \qquad (4.12)$$

where r_i^* denotes the modified Ramsey number, C_i denotes marginal cost of i, and e_i denotes absolute value of elasticity of demand for i. Equation (4.11) is just the break-even condition. In the case of interdependent demands, we can follow Baumol, Bailey, and Willig and write out the conditions for tangency of the effective (i.e., surcharge-inclusive) pseudorevenue hyperplane

$$R^*(y) = \sum_{i=1}^{n} h_i(1 - \tau_i) y_i q_i \qquad (4.13)$$

to the effective hyperbagel of points where effective revenue

$$\tilde{R}(q) = \sum_{i=1}^{n} p_i q_i(1 - \tau_i)$$

equals cost $C(q)$. In this way, Ramsey-like conditions emerge for the appropriate concept of elasticity. Alternatively, put $\lambda_i = 1 - \tau_i$ in equation (4.1) and maximize equation (4.1) subject to the break-even constraint (4.12). Call the solution to the regulator's maximization problem the modified Ramsey point.

The conditions derived by Baumol, Bailey, and Willig (e.g., transray convexity and decreasing ray average cost) are sufficient to show that the modified Ramsey point is price sustainable. Brock and Scheinkman have shown that this point is also quantity sustainable.[51] Thus, the modified Ramsey point given by equations (4.11) and (4.12) has nice properties. These prices, and the associated surcharges, deter inefficient entry even when hit-and-run entry is a serious problem (which is the most hostile environment that the public utility could face) provided that the sufficient conditions for the weak invisible hand theorem hold. The public utility is likely to be even more immune to entry when the market is not perfectly contestable and the public utility can erect entry barriers.

Figure 4.6 displays a situation where a public utility marks up prices on service two in order to generate revenues for subsidizing service one. For service two, the public utility has unit costs c_2, charges p_2, produces $\bar{q}_2$, and earns revenues R_1 for subsidizing service one. There is a potential entrant who has unit costs $\underline{c}_2$ which are considerably lower than the incumbent's unit costs c_2. If allowed to enter, the entrant would drive the incumbent from the market, expand output

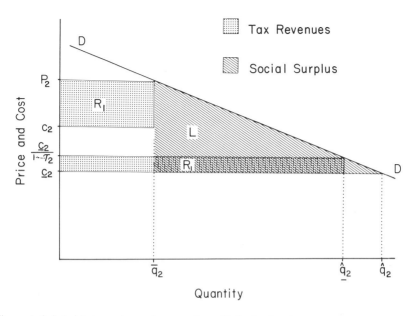

Figure 4.6 Subsidy-tax scheme for revealing efficient entrants.

from $\bar{q}_2$ to $\hat{q}_2$, and thereby increase social surplus by an amount equal to the area L. The social cost of prohibiting entry in the market for service two in order to maintain the public utility's source of revenues for subsidizing service one is obviously enormous. By permitting entry and imposing a revenue tax of τ_2 on all providers of service two, regulators could simultaneously raise tax revenues R_1 for maintaining the subsidy to service one and increase social surplus by an amount almost equal to the area L. A subsidy-tax scheme such as the one displayed in Figure 4.6 would enable the competitive process to reveal efficient entrants who can advance social welfare.

A Critique of AT&T's Creamskimming Argument

In his classic treatise, Frank Knight said, "Uncertainty is one of the fundamental facts of life. It is as ineradicable from business decisions as from those in any other field."[52] Businesses are seldom sure of their prospective production costs because input prices fluctuate over time and because it takes years to develop a smoothly working productive organization. They are seldom sure of the market for their products because demand fluctuates over time as peoples' tastes change and new products appear on the market. And they seldom are sure of the competitors' costs and strategic opportunities. Businesses may reduce uncertainty, however, by acquiring information and using foresight. They have the incentive to do so because they stand to make profits from accurate predictions of future events. As Knight said, "Viewing society, then, as a want-satisfying machine and applying the single test of efficiency, free enterprise must be justified if at all on the ground that men make decisions, exercise control, more effectively if they are made responsible for the results of the correctness, or the opposite, of those decisions."[53]

The power of the invisible hand to promote the greatest social welfare resides in the profit incentives businesses have to discipline each other and the freedom of businesses to pursue these incentives. Profit opportunities are created whenever some businesses charge excessive prices, incur excessive costs through inefficiencies, fail to satisfy the spectrum of social wants, or fail to produce innovations that could lower costs or satisfy latent social wants. Other businesses seize these profit opportunities and, in so doing, eliminate these market deficiencies and move society closer to achieving the greatest wealth available from society's scarce resources. Restrictions that prevent businesses from challenging other businesses, whatever ancillary benefits such restrictions may have, impede the power of the invisible hand to achieve the greatest good for all.

The FCC recognized the salutary effects of competitive entry in several decisions. In approving MCI's application to provide microwave service between Chicago and St. Louis, the FCC noted approvingly that the service would meet unmet demand "by potential users who have no need for and cannot afford the [services provided by AT&T]."[54] In approving entry by the specialized common carriers, the Commission noted, "The applicants are seeking primarily to develop new services and markets, as well as to tap latent but undeveloped markets for existing services, so that the effect of new entry may well be to expand the size of the total communications market."[55] For example, Datran proposed a switched digital data system—a wholly new service—which encouraged AT&T to speed up its development of its own digital data system.

AT&T tried to forestall entry by the specialized common carriers. It petitioned the FCC to deny MCI's pending application in February 1964.[56] It argued MCI "had made no showing of a need for the proposed services; had proposed services which were of 'questionable adequacy and reliability' and had 'not fully stated in its applications the investment that will be required.' "[57] It petitioned the FCC to deny Datran's application in April 1970.[58] It suggested Datran's application would result in "uneconomic duplication of common carrier facilities"; there might not be sufficient demand for Datran's proposed services; AT&T could meet foreseeable demand at lower cost; and Datran's proposal was neither economically nor technically feasible.[59]

According to AT&T's argument, the specialized common carriers faced enormous market uncertainties. Demand may not materialize as they expected. Their plans to construct and operate communications networks may have numerous unforeseen bugs. They may have miscalculated AT&T's ability to lower price and usurp their market. But every new venture faces uncertainties. The invisible hand operates in the face of these uncertainties by forcing businesses who take risks to bear the consequences of these risks and thereby encouraging businesses to use as much information and foresight as possible.

The risks of miscalculation in the presence of uncertainty fell, not on AT&T, but on the specialized common carriers. If demand were insufficient to support the specialized common carriers, investors in the specialized common carriers rather than in AT&T would have suffered the consequences. If AT&T were able to provide comparable service at lower costs than the specialized common carriers and thereby undercut them, the specialized common carriers' investors rather than AT&T's investors would have suffered the consequences. If the specialized common carriers were seriously engaging in uneconomic duplication of AT&T's facilities, the specialized common carriers' investors would have suffered the consequences. The consequences the specialized common carriers faced were severe. By the end of 1975, MCI had spent $106 million, of which more than $48 million was sunk and unrecoverable, to design and construct an intercity microwave system. By the end of 1974, Datran had spent $74 million, of which more than $30 million was sunk and unrecoverable, to design and construct a switched digital data communications system. Because of Datran's bankruptcy, in 1976 Wyly Corporation took a tax write-off of $53 million as a result of its unrecoverable investments in Datran. Fifty million potentially unrecoverable dollars are strong incentives for hard-nosed realism.

All of the information and foresight in the world could not eliminate market uncertainty. Many well thought out business ventures fail as a result of erroneous forecasts. Given inherent uncertainty, however, it is virtually impossible to predict which business venture will succeed and which will fail. Market economies work so well because the investors who bear the risks of failure have the incentives to form the best estimates of the likelihood of success or failure.

AT&T could not have known with certainty that the specialized common carriers were not tapping submarkets, or developing new markets it may have overlooked, or were not more efficient than itself.[60] Its incentives to make reliable estimates of the demand and costs for the services proposed by the specialized common carriers were surely weaker than those of the specialized carriers themselves. The fact that these carriers proposed to spend hundreds of millions of dollars to develop intercity telecommunications systems, despite the sunk costs involved and the likelihood of stiff price competition from AT&T, suggests that enormous profits were available because of untapped demand or inefficiencies on the part of AT&T. Because of these considerations, there is a strong presumption that entry into intercity telephone service will improve social welfare.

The specialized common carriers may nevertheless jeopardize AT&T's ability to provide essential services and thereby fulfill its common carrier obligations. Three points are worth noting in this regard. First, it is hard to reconcile the subsidies Bell wanted to protect with any sensible notion of equity. Supposedly, local callers benefit at the expense of long-distance callers; residential customers benefit at the expense of business customers; and rural users benefit at the expense of urban users. Local and long-distance callers generally are one in the same. People who make relatively few long-distance calls therefore benefit at the expense of people who make relatively many long-distance calls. Businesses pass telephone costs on to consumers. Consumers who spend relatively more on goods or services produced by telephone-intensive businesses lose compared with consumers who spend relatively less on goods and services produced by telephone-intensive businesses. Rural users may make relatively more long-distance calls and therefore pay for one subsidy with the proceeds of another.[61] Who ultimately benefits from these alleged subsidies it is virtually impossible to tell. Perhaps the subsidies were designed to foster particular classes of users. Bell mentions the regulatory goal of universal service which would be attained if everyone subscribed to basic service. But since the demand for basic exchange service is extremely inelastic with respect to price (i.e., people subscribe almost regardless of price) it is difficult to understand why substantial subsidies to lower prices are needed to attain this goal.[62]

Second, any group of demanders who have to pay a multiservice supplier more than it would collectively cost these demanders to obtain comparable service from a specialized supplier will spend resources to circumvent the multiservice suppliers. These resources are pure waste from a social standpoint. The endless FCC proceedings on competitive entry into voice, data, and video transmission were fostered by businesses who felt they could obtain less expensive service outside the Bell System. AT&T claims it is more efficient than specialized suppliers because it has economies of scope, which arise from the use of shared inputs to produce several outputs, as well as economies of scale, which result from aspects of network planning. If so, it could have provided all services more cheaply than the specialized suppliers and, by so doing, could have thereby prevented unsatisfied demanders from trying to break away from the system and deterred alleged creamskimming incursions. Moreover, a rate structure that distributes the benefits of economies of scale and scope so that some demanders pay more than it would cost them to collectively "go it alone" may violate some notions of equity.[63]

There is some circumstantial evidence that the customers who MCI attracted from Bell had been paying AT&T more than their go-it-alone cost of providing service. AT&T says its costs were lower than MCI's. Its marketing department found that it would retain 40% of its private line service customers regardless of MCI's price; 50% of its customers would consider switching to MCI if MCI's prices were 20% below AT&T's price; and more of its customers would consider switching to MCI if MCI's prices were 10% below AT&T's price. Before the HiLo Tariffs were established, AT&T's tariffs were substantially higher than MCI's tariffs on competitive routes. Therefore, about 50% of AT&T's customers paid more than their go-it-alone (or go-it-with-MCI) costs prior to MCI's entry.

Third, it is unclear that entry by the specialized common carriers jeopardized subsidies to essential services. For example, in *Specialized Common Carriers*, the FCC's conclusion that "there is sufficient ground for a reasonable expectation that new entry here will have some beneficial effects" was based on a finding that (a) revenue diversion would be unlikely to have a great effect given the markets being entered and (b) any such adverse effect caused by revenue diversion would be exceeded by the potential benefits of entry.[64] It pointed out that "applicants are seeking primarily to develop new services and markets as well as to tap latent but undeveloped submarkets for existing services, so that the effect of new entry may well be to expand the size of the total communications market."[65] It noted, "We do not see how there could be any diversion of revenues of a magnitude to have the impact claimed by AT&T, in view of the very small percentage of AT&T's existing total market that is vulnerable to competition of the kind proposed here, the growth rate of Bell's basic services, and the likelihood that AT&T would obtain a very substantial share of the potential market for specialized services."[66] When the FCC subsequently tried to limit MCI's offerings, the D.C. Circuit Court pointed out that "there is simply nothing in *Specialized Carriers* that would suggest a conclusion that revenue diversion required restrictions on MCI's facilities authorization."[67]

The FCC made the creamskimming argument moot by finding that revenue diversion, upon which this argument relied, was unlikely to result from competitive entry. Subsequent developments have proved the FCC right. Since 1970, AT&T's local service revenues have grown 8.7% per year, its toll service revenues have grown 10.9% per year, and its private line telephone service revenues have grown 9.7% per year.[68] Since 1975, its local service revenues have grown 7.3% per year, its toll service revenues have grown 10.3% per year, and its private line telephone service revenues have grown 10.9% per year. Its toll and private line service revenues have grown almost 1.5 times more rapidly than its local service revenues since 1975 and almost 1.2 times more rapidly since 1970. Its revenues and presumably its profits have therefore risen much more rapidly than necessary, even after the *Specialized Common Carriers* and *Execunet* decisions, to maintain subsidies for local service at a constant percentage of local revenues.[69]

Whether revenue diversion is real or imagined, whether regulators encouraged subsidies or not, whether the interests of subsidized users should give way to the interests of the general public or not, blanket entry prohibition is, as we discussed in the previous section, a needlessly restrictive method to maintain subsidies to essential services. A simple numerical example which follows AT&T's creamskimming story reveals less restrictive methods to prevent creamskimming. Suppose AT&T charged $2.00 per unit for message toll service.[70] Its long-run incremental cost for providing this service was $1.00 per unit. It used $1.00 per unit ($2 less $1) to lower rates on basic exchange service. MCI sought permission from the FCC to provide message toll service on certain routes served by AT&T. AT&T opposed MCI's application because it believed (1) MCI's costs were $1.50 per unit compared with its own costs of $1.00 per unit, (2) MCI would charge $1.75 compared with its price of $2.00, (3) all of its customers on the competitive routes would turn to MCI for service, and (4) it would lose revenue needed to subsidize basic business service.

AT&T could not have known MCI's costs with certainty and therefore could not have known that MCI was a creamskimmer rather than an efficient innovative entrant, with certainty, as discussed above. It presumably had better estimates of its own costs than it had of MCI's costs. It could have prevented creamskimming by the following less restrictive policy. It would lose $1.00 of revenues used for subsidies for each unit that it would have provided but that MCI provided instead, on competitive routes. It could have asked the FCC to require MCI to provide AT&T with $1.00 for each unit provided by MCI on routes also served by AT&T. This requirement would, in effect add $1.00 per unit to MCI's costs. If MCI's long-run incremental costs were $1.50 per unit in the absence of this requirement, MCI's effective long-run incremental costs would be $2.50 inclusive of the $1.00 surcharge per unit. Obviously, MCI would obtain few customers if it charged $2.50 or more while AT&T charged $2.00. If MCI's long-run incremental costs were $.90 in the absence of the surcharge, MCI's long-run incremental costs would be $1.90 inclusive of the dollar surcharge. MCI could charge $1.95 per unit, attract all of AT&T's customers on competitive routes, replenish AT&T's subsidies through the surcharge, and save MCI's customers $.05 per unit over what AT&T would have charged them.

Given reliable estimates of the subsidies required from above-cost services, it is straightforward to devise a surcharge that could simultaneously prevent creamskimming, ensure the availability of revenues that could be used to maintain desirable subsidies, and permit efficient innovative competitors into the market. The surcharge method for preventing creamskimming is obviously more desirable than the blanket entry-prohibition method for preventing creamskimming. But is the surcharge feasible? AT&T could have asked the FCC to explicitly guarantee the subsidies by requiring the specialized common carriers to pay access charges that would implicitly include the surcharge. Instead, AT&T initially asked for entry prohibition.[71] If AT&T believed the specialized common carriers were trying to creamskim and that the FCC wanted AT&T to subsidize unprofitable services, AT&T would have served the public interest better had it initially advocated the surcharge method rather than blanket entry prohibition.

NOTES

1. U.S. House of Representatives, Committee on Interstate and Foreign Commerce, 74th Congress, 1st Session, *Report of the Federal Coordinator of Transportation, 1934,* House Document No. 89, Washington, January 30, 1935, p. 17. Quoted in Alfred Kahn, *The Economics of Regulation: Principles and Institutions,* Vol. II (New York: John Wiley and Sons, 1971), p. 8.

2. Stuart G. Taplon and Stanley Gerwitz, "The Effect of Regulated Competition on the Air Transport Industry," *Journal of Air Law and Commerce,* Vol. XXII, No. 2, 1955. Quoted in Alfred Kahn, *op. cit.,* Vol. II, p. 9.

3. See Joseph Asbury and Steven Webb, "Decentralized Electric Power Generation: Some Probable Effects," *Public Utilities Fortnightly,* September 25, 1980, p. 21.

4. AT&T, *Defendants' Third Statement of Contentions and Proof,* in *US* v. *AT&T,* p. 2084.

5. *Ibid.,* p. 457.

6. *Ibid.,* pp. 229, 372, and 393.

7. Adam Smith, *Wealth of Nations* (New York: The Modern Library, 1937), p. 423.

8. See, for example, Kahn, *op. cit.,* Vol. I, pp. 123–158.

9. The incumbent and the entrant compete for the same market and do not collude. Consequently, the game is said to be noncooperative. The game is dynamic because it is played over time.

10. An equilibrium has a natural stability property in that it is in the self interest of each player to follow the equilibrium path if the other players follow this path.

11. Alexander L. M. Dingee, Jr., Leonard E. Snollen, and Brian Haslett, "Characteristics of a Successful Entrepreneur," in Stanley M. Rubel, *A Guide to Venture Capital Sources* (Chicago: Capital Publishing Corporation, 1977), p. 10.

12. One guide for preparing a business plan recommends the author " . . . identify and discuss the major problems and risks that you think you will have to deal with to develop the venture. This should include a description of the risks relating to your industry . . . including price cutting by competitors, any potentially unfavorable industrywide trends, and design or operating costs in excess of estimates." See Brian Haslett and Leonard E. Snollen, "Preparing a Business Plan" in Stanley M. Rubel, *ibid.,* p. 28.

13. More generally, unit costs could rise or fall as output increases.

14. A discount rate of 10% per annum, for example, means that $1 received a year from now is worth roughly 10% less than $1 received today. When the discount rate is 10%, $r = .10$ so that π_E dollars per year in perpetuity is worth $1/.10 = 10$ times π_E dollars received today.

15. This last statement applies fairly generally but not, as it happens, to the example in the text. With fixed entry costs and constant returns to scale as assumed above, only the most efficient entrant who enters first will survive; moreover, the sunk costs incurred by the incumbent could act as a barrier to entry which would make it unprofitable for a more efficient business to enter.

16. Stanley M. Rubel, "Guidelines for Dealing with Venture Capital Firms," in Stanley M. Rubel, *ibid.,* p. 36.

17. *Ibid.,* p. 37.

18. For extensions and proofs of the material in this section see Chapter 9.

19. See William Baumol, John Panzar, and Robert Willig, *Contestable Markets and the Theory of Industry Structure* (San Diego: Harcourt, Brace, Jovanovich, 1982).

20. See John Panzar and Robert Willig, "Free Entry and the Sustainability of Natural Monopoly," *Bell Journal of Economics,* Spring 1977, pp. 1–22, for the sufficient conditions for nonsustainability.

21. John Panzar, "Regulation, Deregulation, and Economic Efficiency: The Case of the CAB," *American Economic Review,* May 1980, pp. 311–315.

22. Panzar and Willig, *op. cit.,* p. 21.

23. In the single-product case, this point is the Ramsey-optimum where social surplus is maximized subject to the constraint that the firm just break even.

24. A. Cournot, *Researches into the Mathematical Principles of the Theory of Wealth,* Nathaniel Bacon, trans. (New York: Kelley, 1960).

25. This result was first shown by William A. Brock and José A. Scheinkman, "Free Entry and the Sustainability of National Monopoly: Bertrand Revisited by Cournot," SSRI Working Paper No. 8126. (Madison, WI: Social Systems Research Institute, University of Wisconsin at Madison, March 1981). A revised version of this paper appears as Chapter 9 of this volume.

26. This strategic behavior has been the subject of two major antitrust suits: *MCI* v. *AT&T* and *US* v. *AT&T.* The plaintiffs in both cases charged that AT&T tried to thwart competition in intercity

communication by refusing interconnection with the local exchanges and by engaging in predatory pricing. MCI was awarded approximately $1 billion in treble damages, a verdict which is under appeal. *US* v. *AT&T* was settled out of court. AT&T agreed to divest its local exchange facilities in return for permission to enter unregulated markets. Prior to this settlement, the trial judge rejected AT&T's motion to dismiss and noted that the Government had presented a strong case against AT&T. We have not digested the evidentiary basis of these exceedingly complex cases. Nor do we feel qualified to sort out the truth from the conflicting stories told by the Government and MCI on the one hand and AT&T on the other hand. We rely in this section on the evidentiary record established by the Government and MCI. Our justification for doing so is that this record led to a decisive guilty verdict against AT&T in the MCI case and to a major retreat by AT&T in the Government's case. We must emphasize, however, that the circumstances surrounding MCI's entry into intercity telecommunications and the FCC's willingness to permit true competition in this market are far from clear.

27. US Department of Justice, *Plaintiff's Memorandum in Opposition to Defendants: Motion for Involuntary Dismissal Under Rule 41(b)* in *US* v. *AT&T*, p. 207.

28. FCC, *Specialized Common Carriers*, 29FCC2d870 (1971), p. 918.

29. *MCI* v. *FCC*, 561F2d, p. 378.

30. *Ibid.*

31. U.S. Department of Justice, *Plaintiff's Third Statement of Contentions and Proof*, in *US* v. *AT&T*, p. 676.

32. *Ibid.*, p. 677.

33. *Ibid.*, p. 679.

34. *Ibid.*, p. 654.

35. *Ibid.*, p. 687.

36. *Ibid.*, p. 680.

37. *Ibid.*, p. 691.

38. MCI Communications Corporation, *Prospectus*, SEC Registered Statement No. 2-43717, June 22, 1972, pp. 19–20.

39. *Ibid.*, No. 2-48974, February 19, 1975, p. 9.

40. *Ibid.*, No. 2-62748, December 5, 1978, p. 10.

41. Of course, the possibility remains that AT&T is a nonsustainable natural monopoly subject to competitive incursions by uninnovative entrants. There is some empirical evidence, however, that AT&T does not have a natural monopoly over local and intercity service. See Chapter 6 for details.

42. There were numerous investors in MCI. In 1972, MCI's principal stockholders owned less than 25% of its stock. This fact suggests its stock was widely held by the investment community during its formative years. In 1975, almost 60 investment firms agreed to purchase MCI stock warrants. See MCI Communications Corporation *Prospectus*, SEC Registered Statement No. 2-48974, November 18, 1975.

43. *Ibid.*, No. 2-48974, February 19, 1975, pp. 40–44.

44. *Ibid.*, p. 39.

45. *Ibid.*, p. 20.

46. Carl Ellis Hunt, Jr., *Competition in Telecommunications: A Surcharge as a Method to Promote Competition in Private Lines Services*, Ph.D. Thesis, University of Colorado at Boulder, 1980.

47. The nontechnical reader may wish to proceed to the next section.

48. The demand function is therefore $q_i = D^{-1}(p_i)$. When $\lambda = 0$, equation (4.1) equals the traditional measure of social surplus which is the difference between the aggregate value received by consumers from q_i units and the aggregate cost to society of producing these q_i units. For a fairly nontechnical discussion of these kinds of welfare arguments, see E. Zajac, *Fairness or Efficiency* (Cambridge: Ballinger, 1978).

49. A market is contestable if prices adjust instantaneously and there are no sunk costs of entry. See Baumol, Panzar, and Willig, *op. cit.* The Baumol, Bailey, and Willig weak invisible-hand theorem implies that even a single firm with a natural monopoly may be forced by hit-and-run entry to price at the social optimum.

50. Baumol, Bailey, and Willig, *op. cit.*

51. See Chapter 9.

52. Frank H. Knight, *Risk, Uncertainty, and Profit* (Chicago: University of Chicago Press, 1971), p. 347.

53. *Ibid.*, p. 358.

54. US Department of Justice, *Plaintiff's Third Statement of Contentions and Proof,* in *US* v. *AT&T,* p. 393.

55. FCC, *Specialized Common Carriers,* 29FCC2d870 (1971), p. 916.

56. AT&T, *Defendant's Third Statement of Contentions and Proof,* in *US* v. *AT&T,* p. 582.

57. *Ibid.*, p. 583.

58. *Ibid.*, p. 593.

59. *Ibid.*, p. 593.

60. The invisible hand rewards entrepreneurs for discovering and satisfying latent demand. But, suppose the entrepreneur is less efficient than an incumbent business that had failed to discover the latent demand. Should society spurn the entrepreneur and set loose the incumbent who is now armed with the information provided by the entrepreneur? Fairness aside, society would thereby discourage other entrepreneurs from investing resources in discovering profit opportunities or revealing profit opportunities they might discover. Society would sow the seeds of long-term economic stagnation and inefficiency in return for fleeting cost savings. This conclusion may apply to Datran which apparently discovered a latent demand for an all-digital data system. Even if Datran would have preempted a market which AT&T could have served more efficiently, the long-run public interest might have been best served by Datran rather than AT&T. But, see the discussion of Datran in Chapter 3.

61. John Meyer has concluded, " . . . it is clear that AT&T is not heavily involved today in directly servicing truly low-density, rural locations and, for that matter, never was." Casual inspection shows that the 1500-odd independents and rural cooperatives generally serve more remote and sparsely populated areas than do the Bell operating companies. Meyer also finds, " . . . there is little empirical evidence to establish the validity of the underlying assumption that a great deal of rate uniformity does exist in several parts of the telephone tariff structure." See John R. Meyer *et al.*, *The Economics of Competition in the Telecommunications Industry* (Cambridge: Oelgeeschlager, Gunn and Hain, Publishers, Inc., 1980), p. 173.

62. Taylor estimates that the elasticity of demand for access to the telephone system is between .01 and .19 with respect to monthly charges for basic service and between .02 and .04 with respect to the initial hook-up charges. Lester D. Taylor, *Telecommunications Demand: A Survey and Critique* (Cambridge: Ballinger, 1980), p. 170.

63. See, for example, Kahn, *op. cit.*, Vol. II, p. 225.

64. FCC, *Specialized Common Carriers,* 29FCC2d870, p. 905.

65. *Ibid.*, p. 906.

66. *Ibid.*, p. 910.

67. *MCI* v. *FCC* 561F2.d365, p. 378.

68. Logarithmic growth rates based on 1970–1979 data reported in the *Bell System Statistical Manual,* May 1980.

69. One argument against entry that we have not considered is the carrier-of-last-resort argument discussed by Kahn, *op. cit.*, pp. 236–239. The specialized common carriers may be viable because their customers can rely on the Bell System for back-up service wherever the specialized common carrier service proves unsatisfactory. The solution to this problem requires the Bell

System to charge the specialized common carriers or their customers the long-run incremental cost of back-up service. It may be difficult to devise such charges.

70. These numbers are purely illustrative.

71. In *Specialized Common Carriers*, AT&T says it took the position "that competition in the provision of common carrier communications service should be encouraged where it would benefit the public through lower costs or through the provision of service that would not otherwise be available." See AT&T, *op. cit.*, p. 596. But one searches the record in vain to find an instance where AT&T failed to oppose competitive entry.

Chapter 5

Integration

David S. Evans and Sanford J. Grossman

This chapter addresses the assertion, made by several economists testifying for AT&T, that the planning, construction, and management of complex systems like telecommunications networks can be accomplished most efficiently by an integrated firm which owns the major "piece-parts" of the facilities network and maintains research, development, manufacturing, and systems engineering capabilities. It shows that market systems, in which the ownership over the means of production is dispersed among numerous enterprises, can coordinate the provision of complex and interactive goods and services. It describes market mechanisms that could alleviate the problems identified by AT&T's economists without common ownership over the telecommunications system. It shows that common ownership and central planning themselves create coordination problems. It concludes that the problems identified by AT&T's economists probably will not impede the telephone system created by the settlement of *US* v. *AT&T* in providing telephone service as efficiently as the Bell System has done in the past.

The production and distribution of any commodity entails many stages and much coordination. Perhaps your desk is illuminated by a lamp. Your lamp is designed to fit on your desk by a special attachment; has a plug which fits into an electrical socket on the wall of your office; has a frame colored to match your rug; was sold to you in a lamp store which enabled you to choose a lamp to match your tastes and which you were able to reach by car. The lamp manufacturer produced neither your wall socket nor your lamp's electrical parts. He simply thought of the lamp's design and supervised its assembly. Many things could have gone wrong but did not. The electrical plug is compatible with the same electrical socket you plug your radio and calculator into. The bulb socket is compatible with bulbs sold by several manufacturers at numerous convenient locations. A retail store, located in a shopping mall where you purchase many other goods, stocked the lamp you wanted. Yet the lamp manufacturer owns neither the shopping mall, the retail store, the companies that manufacture the wires and sockets, the electrical cables that deliver the electricity to light the lamp, the electricity generating plants, nor the roads you drove upon.

Market systems in which property ownership is dispersed among numerous self-interested businesses and individuals have demonstrated a remarkable ability to coordinate the provision of goods and services far more complicated than a lamp. But AT&T contended that, without common ownership of each and every component of the telephone system, the market system would fail to coordinate the provision of local and intercity telephone service. It argued that interdependence among the components of the telephone network necessitates a degree of coordination and standardization which only a central authority like the AT&T General Departments can attain.[1] It also argued that integration of the Bell operating companies and Western Electric eliminates the need for time-consuming and potentially expensive negotiations concerning specifications, quantities, prices, and other aspects of contracting with outside companies; that if separate unrelated entities were involved, proprietary concerns would limit information flow and make contractural arrangements difficult and costly, if not totally impossible; and that the Bell System is especially vulnerable to these kinds of transaction problems.[2,3]

It is useful to place these claims in some perspective. In terms of assets, AT&T was the largest corporation in this country in 1980.[4] It was as large as the next three largest corporations combined. Nineteen of its operating companies and Western Electric were large enough, individually, to make the Fortune 500, and four of its operating companies, as well as Western Electric, were large enough, individually, to make the Fortune 100.[5] Its yearly sales were larger than the gross national products of Finland, Greece, Norway, and Columbia.[6] It had offices in virtually every major city and town in this country.

The settlement will create 22 telephone companies which will own the local exchanges presently held by the 22 Bell operating companies. Even in the absence of mergers to form regional operating companies, 16 of the 22 telephone companies will be large enough to make the Fortune 500.[7] The settlement will transfer ownership of all intercity facilities presently held by the Bell operating companies to AT&T which will also retain ownership of Western Electric and Bell Labs. AT&T, divested of its local exchanges, will remain one of the four largest corporations in this country.[8]

The fundamental assertion made by AT&T is that common ownership over local exchanges, intercity facilities, and manufacturing facilities is necessary for the efficient provision of telephone services and that, by implication, several corporations with separate ownership over local exchanges, intercity facilities, and manufacturing facilities could not provide telephone service as efficiently as AT&T. Yet market systems in which ownership and authority are highly decentralized coordinate activities far more complex, interactive, and interdependent than the provision of telephone services. Common ownership and central planning create coordination problems themselves, which increase with the size of the enterprise. These problems are demonstrated in the extreme by the failure of socialist economies to prevent shortages and surpluses of goods and services, as well as by the failure to provide the quality and variety of consumer goods produced by decentralized economies of equal wealth.[9]

The first section of this chapter describes how markets coordinate complex activities without joint ownership of the resources used in these activities. It describes market mechanisms that solve actual and theoretical coordination problems like those which, AT&T claimed, arise in providing telephone service. AT&T's witnesses pointed to numerous problems that afflict the provision of telecommunications services. Schwartz said "there are certain characteristics of telecommunications (such as the interdependent nature of the network, interrelationship of development projects, and the technological and economic uncertainties of development projects) which create the need for cooperative action, either to provide a sufficient level of funding, to share operating costs and capital investment, to ensure the availability of necessary technology, and/or to maintain compatible plant and operations."[10] Teece and Phillips argued that an integrated firm can accomplish these "cooperative actions" more efficiently than several nonintegrated firms can, dealing with each other on an arms-length basis, because the integrated firm lessens transactions costs associated with contracting between firms and ensures a freer flow of "proprietary information" between the parties to the cooperative action. The second section of this chapter shows that integration does not make these cooperative activities easier. Market interaction between several self-interested telephone service providers can alleviate the problems raised by Teece, Phillips, and Schwartz at least as well as integration.

This chapter does not deny that the problems raised by AT&T's economists are serious business problems. It suggests that market mechanisms can alleviate these problems at least as well as integration. Thus the presence of these problems may not justify integration. This chapter does not deny that integration may yield efficiencies. It argues that the particular gains from integration identified by AT&T's economists are nebulous. Finally, although we provide real-world illustrations of the principles we discuss, we have not performed rigorous empirical research on the causes of integration or the gains from integration. We hope that this essay will stimulate empirical tests of alternative theories of integration.

Market Coordination

The most complex, interactive, and interdependent system imaginable is this country's economy. Each day 2.7 million-odd businesses and 220 million-odd individuals engage in millions of transactions with each other. Through these transactions, scarce resources such as labor and raw materials are combined to produce goods and services which become embodied in more complex goods and services which consumers ultimately combine in various proportions to satisfy their basic wants and desires. Many goods and services are "compatible" and "interconnective" with each other. As a result, consumers can combine goods and services to provide themselves with basic commodities, such as light, travel, food, music, lodging, and entertainment.

You purchase light for your desk at home by purchasing a lamp from a manufacturer who, in turn, probably purchased the parts for the lamp from several other manufacturers; by purchasing a house, with electrical sockets and wires made by a number of companies; by buying electricity from your local utility; and by buying a lightbulb from one of several manufacturers. You do not have to install a different type of wall socket for every appliance you use. You do not have to find separate power sources or buy expensive adaptors to operate your appliances. If you move, your new home will have wall sockets that accommodate your old appliances and receive the same voltage of electricity as your old home. You do not have to travel far from home to purchase a lightbulb that fits your lamp. You are able to light your desk because numerous businesses found it in their self-interest to produce commodities that are mutually compatible and interconnective.

You can purchase travel from Chicago to Washington by purchasing an airplane ticket from a travel agency; using your credit card; hiring a taxi to transport you to O'Hare Airport; and leasing a car to drive from National Airport to the center of town. The travel agency, the taxi company, the taxi manufacturer, the airplane manufacturer, the credit card company, and rent-a-car company, and the car manufacturer probably have no common ownership between them. Your flight departs at a reasonably convenient time. Your taxi gets you to the airport on time. A bus takes you from the terminal to the rent-a-car company. Coordination is achieved, not by common ownership, but by each party realizing that it can make a profit by becoming more compatible with other parties.

Consumers who closely examine the goods and services they use will discover two things. First, many of these goods and services are mutually compatible with each other. Second, collections of these goods and services constitute systems and networks which enable consumers to obtain basic services like light, travel, and lodging. Light over your desk is provided by a network. The piece-parts of this network include your wall socket, the wires that connect your wall socket with the electricity-generating plant, the electricity-generating plant, the lamp, and the bulb. These piece-parts are mutually compatible and interconnectible. Travel is provided by a network. The piece-parts of this network include the roads, waterways, airspace, cars, trains, buses, boats, airplanes, and other tangible and intangible assets that afford the consumer enormous possibilities for travel. It is difficult to imagine any good or service that consumers purchase that is not used interdependently with other purchases. Similarly, every manufacturer produces items that are used interdependently with the items sold by other manufacturers. The physical and economic coordination problems that the market solves in the presence of these interdependencies are not intrinsically different from the physical and economic coordination problems that AT&T claimed the market could not solve if the Bell System were broken up.

Before discussing how the market solves simple coordination problems such as those identified by AT&T, it is useful to show how prices and self-interest allow the economic system to exhibit a large degree of coordination without central planning.

The Role of Prices in Producing Coordination

The distribution of finished and intermediate goods to their best uses and the transformation of raw materials and intermediate goods into the final goods and services that consumers want is the most essential role of an economic system. That people realize and respond to a single good's scarcity is easily conceivable. That, because they do so, the millions of commodities bought and sold each year are distributed in a rational way is almost inconceivable and probably inconsistent with any single organizing entity. Adam Smith coined the term the *invisible hand* for the implicit organizing force behind the allocation of commodities in a free market economy. He used this term to suggest that no real central planning was occurring but, instead, individuals, all acting in their own self-interest, trying to maximize profit, will achieve allocations that appear as if these individuals were coordinated.

The price system is the mechanism by which individual self-interest yields coordinated allocations. An increase in the demand for an item leads its price to rise which signals producers that it has become scarce. Producers do not have to process this signal. The price increase makes it profitable for them to produce more. The desire for profit drives the invisible hand which allocates resources.

It is worthwhile to give a detailed description of how this works for the example of the desk lamp mentioned in the introduction. From her market research, an entrepreneur learned of the need for a desk lamp. Her research may tell her the prices of existing lamps are excessive or that consumers need a whole new lamp. After examining the cost of producing the lamp from existing components, she may find production profitable. The important point is that the price of lamps and the price of the components used to produce lamps provide her with information about which components to use and which lamps to produce. To obtain this information, she has to own neither the plants that produce the components nor the stores that retail the lamps.

F. A. Hayek provided the classic explanation of how the price system, by efficiently conveying information between market participants, coordinates the enormous number of transactions necessary in a large economy.[11] He argued that the production and distribution of wealth is difficult because information is dispersed throughout the economy. Each individual obtains little pieces of information in the course of day-to-day activity. Each individual *could* relay this information to a "central board which, after integrating all knowledge, issues its orders."[12] The central board, however, would have to have millions of people to receive, synthesize, and act upon these many pieces of information. But, "[e]ven the single controlling mind, in possession of all the data for some small, self-contained economic system, would not—every time some small adjustment in the allocation of resources had to be made—go explicitly through all the relations between ends and means which he might possibly affect."[13] The price system, through the incentives it provides producers and consumers, is a decentralized "mechanism for communicating information." He noted,[14]

> The most significant fact about this system is the economy of knowledge with which it operates, or how little the individual participants need to know in order to be able to take the right action. In abbreviated form, by a kind of symbol, only the most essential information is passed on, and passed on only to those concerned. It is more than a metaphor to describe the price system as a kind of machinery for registering change, or a system of telecommunications which enables individual producers to watch merely the movement of a few pointers, as an engineer might watch the hands of a few dials, in order to adjust their activities to changes in which they may never know more than is reflected in the price movement.

The general principle described by Hayek is that a scarcity or a surplus anywhere in the economy puts pressure on prices. A producer observing a rise in price need not know that there is a scarcity or the degree of scarcity. Assuming that his costs are unchanged, he will find it profitable to produce more. The essential point to realize is that prices aggregate the information of consumers and producers. There is no need for people to send their information to a central authority, which could then decide on the distribution of goods, because markets substitute for the central authority. The fact that producers and consumers trade on markets which are coordinated by entrepreneurs makes prices transmit information across people. The law of supply and demand causes prices to convey to people exactly the information that an omniscient central authority would have conveyed to them.[15]

Hayek's article was written in response to the popular argument of the 1930s and 1940s that centralized planning can allocate scarce resources as efficiently as the private sector. Further, if reliance is removed from the private sector then the distribution of income across individuals will no longer be at the whim of the price system but, instead, could be adjusted in a way that the legislature considered equitable.

Hayek pointed out that a production or distribution problem is difficult because information is decentralized throughout the system. Each individual, going about his or her activity, picks up a little piece of information. Information does not originate at the Home Office. It originates out in the field in an enormously dispersed manner. The "need" for an improvement in automobiles might arise because (a) a rise in the price of oil changed the type of transportation people desire (say they want small rather than large cars), (b) people are moving from areas in the East to less congested areas in the West, (c) divorce rates have gone up so couples date more and have a desire for more automobile trips (or less?) (d) weather changes people's desire to travel, etc. The list is infinitely long. Virtually everything affects the desire of people to use automobiles.

A central authority could, perhaps, employ millions of people to analyze the impact of everything on automobile transportation. As Hayek pointed out, however, there is another way in which peoples' information gets transmitted and coordinated. When people desire small cars as opposed to large cars, the price of small cars will rise relative to the price of large cars. This increase will (*a*) signal to manufacturers that people want small cars rather than large cars and (*b*) create an incentive for the manufacturers to produce small rather than large cars. It is important to note that, when consumers' preferences shift towards smaller cars, smaller car prices do not rise relative to large car prices because this information gets communicated by individuals to a central authority who then decides it would be a good idea to raise prices. The law of supply and demand operates instead, thereby negating the need for a central authority. Consumers simply start buying more small cars and fewer large cars. This shift in demand forces dealers to cut the price of large cars relative to small cars in an attempt to eliminate their inventories. Dealers do not have to do a market survey. The demands of consumers, in terms of the price that they are willing to pay, transmit all the information that the dealer needs to know.

Take a more complex example. A substantial number of people have migrated from the East to the Southwest. This shift affects the demand for automobiles in many complicated ways. People who live in Eastern cities use public transportation relatively intensively. To the extent cars are used for short commutes, people may prefer smaller cars to larger ones. Migration makes these cities less congested and may make commuting easier. Public unions have increased the price for transportation by demanding large wage increases and featherbedding. Because numerous factors are important for analyzing the impact of migration on automobile demand, the list could go on *ad infinitum*. The price system makes it unnecessary to analyze all these factors. The quantities of car models purchased by consumers will cause changes in dealers' inventories. Dealers will adjust their prices to deplete inventories of less desirable cars and to eliminate shortages of more desirable cars. The models that consumers desire most will rise in price relative to the less desirable models. Seeing the changing prices for different models of cars, automobile manufacturers will find the model whose price has risen most relatively more profitable to produce. Thus, they will expand the production of the cars consumers find relatively more desirable.

This discussion has examined the impact of an isolated shift such as migration. However, at any given point in time an enormous number of events are taking place. A centralized authority would not only have to analyze a particular event's impact on optimal car production but, even more importantly, have to decide which events are worth analyzing in the first place. In a decentralized economy, producers only have to look at prices. Society learns of relative scarcity when an event makes a particular price relatively high. The producer need not care about the source of the scarcity. The fact that his price is higher, given costs, creates an incentive for him to increase production. This transmittal of information in the form of an increased incentive to produce occurs immediately, without the intervention of a Home Office which decides what events are worth transmitting to production units. For this reason, many centralized economies have made increasing use of the price system and other decentralized mechanisms.[16]

The profit motive is the main engine that drives the price system. An entrepreneur who receives information that consumers desire a new product, that a resource he or she uses to produce a commodity has become scarcer, or that consumers would find this commodity more valuable if it were compatible with other commodities, reaps profits by acting upon this information. She or he could demand a higher price or incur a lower cost and thereby make a higher profit by doing so. As Adam Smith noted,[17]

[The entrepreneur] generally . . . neither intends to promote the public interest, nor knows how much he is promoting it. . . . [B]y directing industry in such a manner as its produce may be of the greatest value, he intends only his own gain, and he is in this, as in many other cases, led by an invisible hand to promote an end which was not part of his intention. Nor is it always the worse for the society that it was no part of it. By pursuing his own interest he frequently promotes that of society more effectually than when he really intends to promote it.

Physical Coordination

In order to illustrate its robustness, we show how the market system coordinates the provision of complex goods and services despite three problems, each of which would appear to make the price system unworkable: physical coordination, standardization, and public goods. This part addresses physical coordination. Workers may have to coordinate their physical movements in order to screw together the parts to a lamp. Consequently, the lamp manufacturer employs supervisors to monitor and coordinate her workers. This coordination requires the close proximity of workers and supervisors and thereby determines the optimal size plant and the tasks performed in each plant.[18] Travel services have to coordinate the physical and temporal locations of their services in order for passengers to make timely connections. Consequently, airlines, bus companies, and taxi services have traffic managers to plan scheduling. Real estate developers have to coordinate the installation of plumbing, wiring, painting, and plastering. Often, developers have a prime contractor to coordinate several subcontractors.

Physical coordination usually requires physical integration. Men and machines work side by side in many manufacturing plants. Airline terminals, bus routes, and taxi stops are close to one another. Subcontractors work side by side at construction sites. But, physical coordination does not necessarily require common ownership over productive resources. Airlines do not own taxi cabs or buses. Real estate developers do not have to own plumbing, electrical, and masonry firms. Nevertheless, some firms own equipment. Office building owners own their buildings' elevators. Manufacturing companies often own machinery that is permanently fixed in place. But long-term leases could replace ownership in these instances with no obvious detriment to economic efficiency. Many businesses lease computers and other office equipment, for example. Many office building owners have maintenance contracts for elevators. Thus, the relationship between physical coordination and common ownership is tenuous.[19]

The market system provides strong incentives for physical coordination without common ownership. A simple example illustrates these incentives. Company A sells service a. Company B sells service b. A earns a profit of $100 and B earns a profit of $150. If A and B physically coordinate their activities their joint profit rises from $250 to $350. If each party can implement coordination separately and if each party gains from coordination, they will obviously both coordinate their services. For example, coordination may require businesses to take the interdependence between their services into account when they separately plan and schedule their services. Taxi drivers probably take airline schedules and symphony performance times as given when they plan their schedules to meet customer needs.

If each party can implement coordination separately but if each party does not bear the full cost or does not reap the full benefits of coordination, the parties may have to agree on a division of benefits and costs between them. For example, A may increase its profits by $50 and B's profits by $50 if it hires a scheduling supervisor, costing $80, to coordinate its schedule with B's schedule. Separately, this coordination is unprofitable to A. Jointly the profits to A and B increase by $20 ($50 + $50 − $80 = $20). It is reasonable to assume that A and B will eventually agree on how to divide the gain between them. Physical coordination in these circumstances involves a mutually beneficial trading opportunity. Suppose a seller has an item that he is willing to sell for as little as $2. He meets a buyer who is willing to pay as much as $3. Because it is in their self-interest, we would expect them to haggle for a while and then trade at some price between $2 and $3. Likewise, we would expect A and B to allocate costs and benefits between themselves so that each party gains some portion of the $20 available in increased profits.

The experience of the railroads in the 19th century and the experience of the electric utility industry in the 20th century demonstrate how the market system encourages physical coordination. Many separate companies built segments of our railroad system during the mid-19th century. By interconnecting with each other and enabling produce and passengers to transfer easily between railroads, these companies were able to increase revenues and profits. In their classic study of railroad coordination, George Taylor and Irene Neu found that by 1861[20]

New England [railroads] were . . . well unified internally [and] were also relatively free of gaps or obstructions at points of intersectional connection. Traffic
between New York City and Boston via New Haven, Hartford, and Springfield
moved without physical obstacle over tracks owned by four different companies.

According to Alfred Chandler,[21]

> By the 1880's a rail shipment could move from one part of the country to
> another without a single transshipment. By then the traffic departments of the
> major roads had become responsible for moving a large share of the long-
> distance traffic within the United States. This internalization of the activities
> and transactions previously carried out by many small units, well under way
> in the 1850's, was completed by the 1880's. . . . As the nation's rail network
> expanded, as interconnected lines became completed, and as the roads became
> physically and organizationally integrated, through traffic grew rapidly.[22]

According to T. J. Nagel, "electric power companies saw the benefits from
interconnecting separately managed power systems" by the early part of this
century.[23] Improved technology, such as "high voltage transmission that permitted higher capacity interconnections over greater distances and power control
technology," encouraged interconnections. For example, American Electric Power
Company interconnected with Chicago's Commonwealth Edison Company after
World War II by building a tie-line between the two systems. Their interconnection agreement "included mutual generating capacity support, economy power
interchange, short-term power sales and purchases, the interchange of maintenance energy, and diversity power exchange." By making this agreement, they
secured "economies of scale in investment and operation with improved reliability
of service." Long-term agreements for transferring capacity between systems
"permit staggered construction of generating facilities, allowing installation of
units larger than otherwise would be the case, thereby providing economies of
scale in construction and operation." According to Nagel[24]

> Over a long period, this nation has developed the necessary control equipment
> and operating practices and tools to make interconnection practical. The eastern
> two-thirds of the United States has an extensive network, and development is
> progressing in the western states. These networks are extremely complex, and
> their size and reliable operation requires careful technical analysis and the full
> cooperation of all utilities. The industry has developed regional councils and
> an umbrella organization, the National Electric Reliability Council: the regional
> organizations review the plans of each system to ensure coordination.

The parties may have to take some joint action in order to physically coordinate
their activities. The following example illustrates how market transactions can
accomplish joint physical coordination without joint ownership. Company A
provides service a separately. Company B provides service b separately. Company A earns \$50 profit. Company B earns \$200 profit. If a common factor of
production x linked services a and b, A could earn profits of \$100 and B could
earn profits of \$400. In order to link a and b to x and thereby gain the profits
from a linked service, A and B could engage in several alternative transactions.

A and *B* could form a joint venture which could construct and own facility *x* and link *a* and *b* to *x*. *A* could build *x* and lease *B* a connection to *x*. A third party *C* could build *x* and lease both *A* and *B* connections to *x*.[25]

Businesses frequently engage in these kinds of transactions. The tie-line built by American Electric Power Company and Commonwealth Edison is one example. Bridges built by interconnecting railroads are another example. Chandler found[26]

> Where roads terminated at a river's edge, the two roads often formed a joint enterprise to build and maintain the connecting bridge. Similar joint enterprises were formed to build belt lines and facilities connecting the lines of different roads terminating in the same cities. By 1870 the Hudson, the Delaware, the Potomac, the Ohio, the Mississippi, and the Missouri had been crossed by railroad tracks, often in several places.

Condominiums are still another example. Condominium owners have separate ownership over their apartments which are "linked" together by commonly owned elevators, plumbing systems, halls, and lobbies. During the early era of telephone competition, the independents sometimes formed a joint venture to provide long-distance service between separately owned independent exchanges and sometimes contracted with a separately owned long-distance company to provide interconnections between separately owned independent exchanges.[27,28]

The following example illustrates another application of the same idea. Company *A* provides service *a* separately. Company *B* provides service *b* separately. Company *A* earns $50 profit. Company *B* earns $200 profit. Company *A* hires a traffic manager to coordinate *a* with *b*. Company *B* hires a traffic manager to coordinate *b* with *a*. Each traffic manager costs $25. After paying the traffic managers, *A*'s profits rise to $75 and *B*'s profits rise to $375. Their total profits rise from $250 to $450 as a result of coordination. If *A* and *B* merged, the merged company could coordinate *a* and *b* more closely with one traffic manager than could the unmerged companies with two traffic managers. Because of superior coordination and lower coordination costs, the profits of the merged company rise to $600, $150 greater than the total profits of the unmerged companies. One might conclude from this that physical coordination can be achieved most cheaply by joint ownership. There are two fallacies in this argument.

First, this situation is really a simple variation of the case of a shared factor of production. *A* and *B* could form a joint venture, consisting of one traffic manager, to coordinate *a* and *b;* *A* could hire a traffic manager to coordinate both *a* and *b* and sell his coordination plan to *B;* or *C* could coordinate *a* and *b* and sell his coordination plan to both A and B. For example, condominium owners set up a joint venture—the condominium association—to provide jointly used services. Railroads in the 19th century set up cooperative ventures called fast freight lines to provide railroad cars that could go "from their point of pick-up to their destination without breaking bulk, thus eliminating delays at transfer points."[29] By 1891, there were more than 20 fast freight line offices in Chicago alone.[30] Independent telephone companies set up associations for coordinating toll traffic and divvying up revenues.

Second, this situation assumes that centralized coordination by a single system planner is more efficient than decentralized coordination by several system planners, each optimizing coordination from the vantage point of her own system. This assumption probably does not apply to many complex systems. For example, the decentralized optimizing actions of businesses and individuals coordinate the production and distribution of goods and services in market systems. Socialist economies have discovered the problems from having a single "traffic manager." As Hayek pointed out, "Even a single controlling mind, in possession of all the data for some small, self-contained economic system would not—every time some small adjustment in the allocation of resources had to be made—go explicitly through all the relations between ends and means which might possibly be affected."[31] Although it is theoretically possible that nondecomposable complex systems exist—that is, systems that are so integrated that only one single controlling mind could operate the system—there are strong economic incentives to devise systems that are decomposable.

Nobel Laureate Herbert Simon has observed

> The fact . . . that many complex systems have a nearly decomposable hierarchic structure is a major facilitating factor enabling us to understand, to describe, and even to "see" such systems and their parts. Or perhaps the proposition should be put the other way around. If there are important systems in the world that are complex without being hierarchic, they may to a considerable extent escape our observation and our understanding. Analysis of their behavior would involve such detailed knowledge and calculation of the interactions of their elementary parts that it would be beyond our capacities of memory or computation.

Aaron Kershenbaum, testifying for the Government in *US* v. *AT&T*, said that the telephone system is a decomposable system. He argued that the national telephone network "is far too large to even seriously consider doing detailed planning of [it] as a single piece. It is decomposed in many ways." He noted that the "overall network problem" is broken down into a series of "subproblems."[32]

Nondecomposable systems are undoubtedly more complex than decomposable ones. The coordination of people's activities is probably more difficult for nondecomposable systems. Common ownership, however, does not make this coordination any less difficult. The next section shows that, if *A* buys *B*, so that there is joint ownership, then *A* will have exactly the same difficulties coordinating the manager of *B* (who is now his employee) that *A* had in coordinating its plans with the owner of *B*.

Standardization

Clothing manufacturers could realize scale economies if they made clothing in only one size, shape, and color. Retail clothing stores could reduce inventories if they stocked uniform clothing. They do not do so because consumers desire, and are willing to pay for, variety. Automobile manufacturers could achieve greater scale economies if they made a standardized car. But consumers shunned the "no-frills Model T" in favor of variety. Many other industries make products in numerous shapes and sizes in order to satisfy diverse consumers with varied tastes. Consumers would incur added expense and inconvenience, however, if they had to buy a different bulb for every lamp, a separate socket for each appliance, a different television set for each television station, or a different record player for every record label. Separately owned manufacturers make these products mutually compatible because consumers desire standardized products that are interconnectible and interchangeable. Many manufacturers realize increased sales and profits by making standardized products. By balancing the benefits of variety against the benefits of standardization, the market system produces an optimal mix of standardized and nonstandardized commodities.

It is easy to see how the price system and self-interest will lead to the optimal product variety and standardization. Consumers reveal the benefits they receive from variety and standardization through the prices they are willing to pay for particular bundles of goods and services. Standardization and variety may be associated with different costs of production (e.g., clothing retailers can keep smaller inventories if all their clothes are of the same size, color, and style). These considerations are expressed in the marketplace by the prices sellers require consumers to pay for particular goods and services. Competition promotes the production of bundles of goods whose costs are less than the benefits that consumers derive from them. The difference between the consumer's benefit and the producer's cost is the producer's profit per unit. When this difference is positive the producer has an incentive to produce more units. Equilibrium is reached at that product variety that just balances the relative costs and benefits of each product so that no net social or private benefit would be derived from increasing or decreasing production or variety.

When several products are highly complementary or interdependent—as are bulbs and sockets, records and record players, and electrical appliances and electrical sockets—consumers generally desire standardization. As we show below, the market system encourages standardization in these circumstances. There are two cases to consider. In the first case, there is an optimal configuration for a set of products. Suppose consumers connect products a, b, and c together in order to obtain some service like light, music, or refrigeration. They are willing to pay $10 for a unit each of a, b, and c when these products are provided in some nonoptimal configuration and $15 for a unit each of a, b, and c when these products are provided in the optimal configuration. Many manufacturers make a, b, and c. It costs them $9 for a unit each of a, b, and c when these products are provided in some nonoptimal configuration and $10 when these products are provided in the optimal configuration. Obviously, the benefits from changing to the optimal configuration ($15 − $10 = $5) exceed the costs from changing to the optimal configuration ($10 − $9 = $1). An *a* manufacturer could switch to the optimal configuration with the reasonable expectation that a *b* and a *c* manufacturer will follow. There are strong profit incentives for them to do so.[33]

The introduction of the $33^1/_3$ rpm long-playing microgroove record by Columbia Record Company illustrates these principles. Columbia did not manufacture record-playing equipment. Consequently, in order to launch the new record, they first commissioned Philco to make the necessary turntable attachments. They then[34]

> . . . freely furnished information to all of the several cooperating manufacturers who agreed to produce radio–phonographs and record players of their own for playing the new discs. This offered a new and golden opportunity for the various companies to develop and introduce record players which would play both the new discs and the older 78 rpm-records. Especially, this seemed . . . a wonderful opportunity to the smaller manufacturers, for the largest one, RCA Victor, had said officially that it was going to have nothing to do with Columbia's LP.

The $33^1/_3$ rpm record became the industry standard for long-playing records. RCA, which was vertically integrated into the production of record players, was unable to force the industry to adopt its 45 rpm record as a standard.

In the second case, consumers want the compatible products although there is no unique optimal configuration for these products. For example, consumers may not care whether the separation between the prongs of electrical connectors is one or two inches but do care that the plugs and outlets are compatible with each other. If the first manufacturer of an electrical connector chose a one-inch separation, the second manufacturer of an electrical connector would find it profitable to have a one-inch separation: She would sell fewer units if she chose a different separation because consumers would be unable to use her connector (an electrical socket perhaps) with the first manufacturer's connector (an extension cord perhaps). Because consumers desire compatible connecting devices, manufacturers have enormous incentives to make compatible devices.

The standardization of railroad gauges illustrates these principles.[35] The early railroads were designed to serve local needs and to divert trade from competing sources of supply. For example, the line between Boston and Worcester "was designed primarily to secure for Boston the trade of the Worcester area and to divert it from the Blackstone Canal which led to Providence," according to Taylor and Neu's history of this country's railroad network.[36] Railroad gauges— the inside distance between the tracks—were a weapon used to thwart the interconnection of railroad systems and thereby to inhibit transportation from competing sources of supply. Taylor and Neu claim ". . . mercantile interests stimulated the deliberate adoption of divergent gauges."[37] But divergent gauges made long-haul rail transportation extremely expensive since shippers had to transfer passengers and goods from one rail car to another whenever there was a break in the tracks. Taylor and Neu concluded that[38]

> Satisfying as such traffic breaks might be to local interests, they placed a heavy tax on through traffic. As this class of freight became more important, the need to eliminate physical obstacles became imperative, and uniformity of gauge was forced upon the railroads of the United States and Canada.

By adopting the standard gauge, railroads could get more through traffic and thereby make greater sales and profits.

By providing convenient forums for trading technological information, professional associations make market coordination easier. Alfred Chandler argues that "middle managers were the persons who devised the organizational procedures and worked out the technological standardization necessary to achieve a national railroad system. Constant consultation and cooperation on complex problems brought these managers a sense of professionalism that had never existed before in American business."[39] Standards evolved partly through deliberations in professional associations. Chandler says[40]

> The middle managers who met regularly to discuss common problems in performing their difficult functions soon set up permanent quasi-professional associations. . . . By the early 1880's such associations had been formed for nearly every major railroad activity. They included the American Society of Railroad Superintendents, American Railway Master Mechanics Association, Master Car Builders Association [and many others].

Meetings of these associations discussed national standardization of procedures and equipment. Professional papers were presented at the meetings and published in journals.[41] The independent telephone associations also provided forums for discussing the standardization of equipment and procedures used by the independent telephone companies and for disseminating information on recommended standards to both independent telephone companies and telephone manufacturers.[42]

Horizontal agreements between firms on standards provide less restrictive methods than horizontal integration across firms for achieving desirable standardization.[43] Trade associations can readily dissolve agreements on standardization when market or technological conditions change. Manufacturers who discover superior configurations can defect from the trade association and test the market. In discussing the standardization of toll line equipment a member of the independent telephone movement noted in this regard, "King George scoffed at the thought of harmony among the colonies just as the 'dictator' in Boston [AT&T was then a Boston-based company] scoffs at the possibility of harmony among the independent telephone companies; but history demonstrated that the path of progress lies in the direction of harmony without a dictator."[44]

The market's ability to encourage standardization does not imply that there are never conflicts over which standard to adopt. Consumers and producers may require experience with several alternative configurations in order to determine the best standard. H. P. Clausen noted that "those who have closely studied telephone practice since 1894 have observed anything but a tendency to standardize the apparatus until it was found that a certain degree of superiority was obtained in a certain piece of apparatus, and not until such excellence in a certain device was universally acknowledged did the manufacturers gradually re-design their apparatus so as to comply with the acknowledged superior form of construction."[45] American railroads tried several different gauges before adopting the 4'8"-gauge as their standard. Taylor and Neu found that "At the very time that old established roads were changing from broad to standard gauge, a number of new roads were being built to a narrow, usually 3-foot gauge."[46] Narrow gauge roads were supposedly cheaper to build, less expensive to maintain, better adapted to mountainous regions, and provided cheaper feeder lines than standard gauge roads. Proponents of the narrow gauge maintained that the costs of transfer would represent but a small fraction of the amount ultimately saved in construction, equipment, and maintenance.[47] By the middle 1880s, this contention was disproved by experience. A central authority could have imposed the 4'8" gauge *ab initio* and prevented the costs of this failed experiment. But a central authority could have imposed the narrow gauge and could thereby have prevented railroads from discovering the superiority of the 4'8" gauge.

In a world of uncertainty, innovations are made by entrepreneurs willing to risk their capital for their ideas. The best test of whether an alternative configuration is desirable is to produce the configuration and allow the market to reveal which configuration consumers prefer. For example, RCA claimed that the 45 rpm record, which it introduced along with a compatible turntable in 1948, was the optimal record speed.[48] Columbia, its chief competitor, almost simultaneously introduced the $33^{1}/_{3}$ rpm record while independent record player manufacturers introduced turntables to play this record. The $33^{1}/_{3}$ rpm record quickly demonstrated its superiority. A history of the *War of the Speeds* found[49]

> Several million dollars were spent by RCA Victor the first year to "put over" the 45's and every device in the books was used to persuade other record manufacturers to go along. The acceptance of the LP [the $33^{1}/_{3}$ rpm] record by other manufacturers had been almost 100%. Not all of the originally issued LP records issued by Columbia and other companies had been completely successful. However, comparisons of the LP's with the standard disc of the same performances as to reproduction quality, as published in the Saturday Review, revealed that for the most part, the LP's very quickly assumed a position of general superiority as to a quieter playing surface, frequency range, and clarity.[50]

The FCC recognized the advantages of market coordination when it refused to impose technical standards for direct satellite systems, in part because ". . . by allowing experimentation with a variety of technical characteristics, the Commission will, in fact, provide entrepreneurs the best possible opportunity to provide the services most valued by consumers [and because] members of the industry are likely to have more resources and better information than the Commission with which to address the technical details of system design, and to have very strong incentives to make correct technical judgements regarding [direct broadcast satellite] system performance parameters."[51] Recognizing these same advantages, four committees appointed by the Secretary of Commerce since World War II have reaffirmed the national commitment to voluntary standard setting by businesses.[52]

Knowledge as a Public Good

To conclude this section, we briefly address AT&T's argument that extensive horizontal and vertical integration is necessary for the effective performance of research and development. Prices are the driving force behind market coordination. When the resources required to produce a good become dearer, producers raise prices and consumers reduce consumption. When the resources required to produce a good become more abundant, producers expand production thereby bidding price down and inducing consumers to increase consumption. The invisible hand works well for resources over which individuals have well-defined property rights—so-called private goods. These individuals have vested interests in the profits or losses realized from their property. The invisible hand may work poorly for resources over which individuals do not have well-defined property rights—so-called public goods. No one has a vested interest in the efficient use of these goods.

Clean air is an example of a public good. Everyone values clean air. But because no one has property rights over air, no one can capture profits from producing it. Consequently, the market system allocates insufficient resources towards "clean air production." Government regulation of pollution is the often-recommended solution to this problem.

In theory, consumers could form a corporation to reduce pollution. Through the corporation, they could contract with polluting firms to reduce pollution. These contracts could provide a monetary reward to polluting firms for reducing pollution. The extent to which pollutors would reduce pollution depends on (a) the costs of reducing pollution and (b) the size of the reward offered. The size of the reward offered by the corporation depends on the aggregate value consumers who belong to the corporation place on clean air. These contracts between consumers and pollutors effectively replace the market system.

The *free-rider problem* could make this corporation unworkable. Any consumer could refuse to contribute to the corporation's expenses and yet still enjoy all the benefits of the clean air fostered by the corporation. Selfish consumers would thereby obtain a free ride from less selfish consumers. Government regulation solves this problem because the government can coerce all consumers to "join" and "contribute to" the corporation, namely the government itself.[53]

Knowledge produced by a firm's research efforts has the characteristics of a public good. An innovation produced by one firm may directly or indirectly benefit many firms. To the extent that the firm can patent this innovation, it can assert property rights over the innovation and capture the profits generated by the innovation over the life of the patent. The market provides incentives to produce patentable innovations. If the firm cannot patent this innovation, it may not be able to capture much of the value generated by the innovation. The market provides limited incentives to produce unpatentable innovations.

Common ownership over all firms producing intermediate and final products affected by an innovation is one solution to this problem. But there are alternative and less restrictive solutions. If there are a few firms that are going to use an innovation, they can contract in advance to share the costs of producing an innovation and to share the benefits from the innovation. When few firms benefit from an innovation, a joint venture can easily "police" free riders. Unlike the corporation to reduce pollution, where millions of people have to contract to form a joint venture and where policing is difficult because of the large numbers involved, a joint venture to finance research is feasible when few beneficiaries are involved.

Joint research and development ventures are not uncommon. Considerable data are available on joint ventures involving European firms because the European Economic Community must approve these ventures. In approving a joint company to market and distribute reprocessing services for oxide fuels, the Commission found that "[b]y combining research efforts, the agreement will help progress to be made as quickly as possible in a new branch of technology where further developments are still awaited on a number of points." In a joint venture to make lenses and lens controls used in television cameras

> The parties coordinate their activities by continuing contact between their respective teams and share research and development programs . . . by joint agreement as appropriate. . . . "Joint products," that is to say products obtained as a result of development work in which both parties have had a considerable share, are protected by jointly owned patents . . . to be applied for and maintained . . . by agreement and jointly in the names of both companies which share the costs equally.

Recent guidelines promulgated by the Justice Department permit joint ventures when single firms would not undertake the venture because of the costs or risks; when the parties to ventures have different uses for a new product developed by the venture; and when the venture would strengthen a weak company and thereby increase competition. These guidelines frown on joint ventures that would commit several firms to the same approach to a problem.[54] According to a recent article,[55]

> . . . in the late 1960's the Justice Department brought suit against four auto companies that proposed a research venture on emission control devices, mainly

because the agreement would not have allowed any company to use the resulting technology without the consent of the others. Such a venture could actually impede the development of new technology. By contrast, early this year the Justice Department approved a "cooperative research program" within the industry as a positive basic research program that no company could have conducted individually and that would enhance the science base of the whole industry.

There are some situations where integration enhances incentives for innovation because many firms (or divisions) benefit from the innovation, thereby creating a severe free-rider problem. In these situations, where there are a large number of beneficiaries, each of whom feels he can free ride on the investment of others, joint ventures may be infeasible. Thus integration may have some benefits. But, as shown in the next section, monitoring and coordination problems associated with joint control can easily outweigh these benefits. Integration creates large, difficult-to-manage organizations. Where market forces work poorly, an integrated company will also work poorly: The difficulties associated with the management of a large research enterprise may be as large as those faced by the participants of a joint venture.

Transactions Costs and Integration

We have shown how the market system coordinates the provision of complex goods and services despite the presence of what would appear, superficially, to be overwhelming problems. We do not mean to imply, however, that the market system is a frictionless machine which never sputters nor malfunctions. Indeed the market is a delicate system. Profit opportunities may go for the asking. Businesses make mistakes. Sometimes information diffuses too slowly through the marketplace. The paths taken by the market system to achieve the optimal kind and degree of coordination are, at times, long and tortuous. For example, society would have realized savings if every railroad had adopted the 4'8" gauge from the beginning. They did not do so because the advantages of a national rail system were not widely recognized, railroads used gauges strategically, to thwart interconnections, and engineers disagreed over which gauge was best. Record consumers and turntable producers faced some uncertainty during the brief War of the Speeds.

In advocating common ownership over the means of production, proponents of central planning often point to these bugs in the market system. But they fail to show that common ownership corrects these deficiencies. For example, a central authority could have imposed the 4'8" railroad gauge and the $33^1/_3$ rpm record speed thereby saving society the costs of discovering the best standard. It also could have imposed a 3' gauge and a 45 rpm speed thereby preventing society from discovering the best product.[56] Coordination by *fiat* looks advantageous only in hindsight. A central authority cannot gather, let alone synthesize, the many pieces of information needed to coordinate economic activity optimally.

Its engineers may overlook preferable alternatives in selecting standards. Its salesmen cannot possibly discern every nuance of consumer preferences. Its managers cannot react quickly to changes in technology and tastes.

That the market system has bugs is undeniable. That integration eliminates these bugs is wishful thinking. The recently advanced transaction-cost theory of integration illustrates this point. According to Teece,[57]

> The notion of frictionless market exchange is a fiction. . . . Information is never perfect. Uncertainty creates a large number of possible contingencies which could arise in any given situation and it is generally costly if not impossible to know and to specify in advance responses to all conceivable contingencies. Common ownership can eliminate these transactions costs by eliminating the need for such contracts. With common ownership [integration], the uncertainty can be handled in an adaptive sequential fashion, and need not be accomodated ahead of time.

This section argues that common ownership does not eliminate these transaction problems.

Proponents of the transaction-cost theory have identified transactions which, they believe, ordinary market mechanism cannot accomplish inexpensively. Williamson, for example, focuses on markets in which there are a small number of buyers and sellers and in which the product traded is "technically complex" and "periodic redesign and/or volume changes are made in response to changing environmental conditions."[58] He concludes that long-term contracts are susceptible to haggling in these kinds of markets. Short-term contracts may limit haggling opportunities. But these are not feasible when "(1) efficient supply requires investment in special-purpose, long-life equipment or (2) the winner of the original contract acquires a cost advantage. . . ."[59] Under these circumstances, one of the parties to a short-term contract can renege on the contract terms and try to extort better terms from the other party. When either party can misrepresent the facts surrounding a transaction, contracts are even poorer trading mechanisms. Williamson believes integration may eliminate these contractual problems more efficiently than alternatives like arbitration, more precise contracts, and litigation.

The problems identified by Williamson are special cases of a more general economic phenomenon called "*ex post* opportunistic haggling" or the "appropriable quasi-rent" problem.[60] The phenomenon is simple despite these long names. Two parties enter into an agreement which creates some valuable good or service. The agreement specifies the costs incurred and the benefits received by each party. In carrying out her part of the agreement, the first party incurs an unrecoverable cost. After she has incurred this "sunk" cost, she would be willing to accept a smaller share of the benefits created by the agreement to backing out of the agreement altogether, although she would not have entered the agreement originally and incurred a sunk cost in return for this same share.

This situation enables the second party to the contract to haggle with the first party after the first party has incurred her sunk cost. Such behavior is called *ex post* opportunistic haggling because the opportunities for haggling come after the agreement has been made and unrecoverable costs have been incurred.[61]

In order to understand why integration fails to solve these contractual problems and why the possibility of contractural problems fails to justify integration, it is helpful to understand the role of contracts in market systems. As Richard Posner has noted, "There are many contingencies that may prevent the process of exchange from operating to reallocate resources to higher valued users, especially when the exchange is carried out over a period of time or when the performance of both parties involves a complicated undertaking. To minimize breakdowns in the process of exchange is the basic function of the law of contracts."[62] Contracts therefore improve coordination by creating "prices" which are specifically tied to the contingencies detailed in the contract.

For example, you might need a hotel room on a business trip to a distant city. If you wait until the evening before you arrive in this city, you will probably have considerable trouble finding a decent and affordable room. Therefore, you will make your reservation for a hotel room before leaving for your destination. In effect, you will make a contract for a hotel room prior to your trip. The hotel could try to renege on its contract and refuse to give you the room unless you pay more than was originally agreed. We rarely see hotels engaging in this sort of *ex post* opportunistic behavior for two obvious reasons: (1) its reputation could be ruined and (2) it could become legally responsible for damages suffered through its violation of the original contract.

Someone might argue that, because you have trouble finding a hotel room on short notice, your employer should merge with a Washington hotel. Obviously, this argument is silly. A failure to contract leads to chaos. A failure by employees of different divisions of an integrated company to coordinate their activities also leads to chaos. Advocates of the transaction-cost theory of integration appear to believe contractual problems evaporate upon merger. Indeed, transactions within integrated businesses are afflicted by the same contractual problems as transactions between unintegrated businesses. A close examination of how contractual problems arise and how transactions take place in integrated firms establishes this conclusion.

Parties to a transaction usually trade contracts rather than commodities when either or both parties have to make unrecoverable investments in order to provide the future delivery of some commodity. They trade contracts in two types of situations. The first type concerns specialized locations. Examples include the placement of electronic cables adjacent to input or output devices such as computers or telephones; bank vaults near jewelry stores; restaurants in airports; and pipelines between oil wells and refineries. The second type concerns specialized design. Examples include tooth dentures; automobile parts for specific automobile bodies; the frames and housing of household electronic equipment; computer software; specialized integrated circuits; made-to-order office furnishings; and tailor-made clothing. Many commodities have unusual features designed for particular users.

In these examples, either or both parties make unrecoverable investments. If an airport closed down, its restaurants would attract few customers. If either the refinery or the oil well closed down, the pipeline connecting them would prove useless. If you tire of your tailor-made suit, you will find few buyers interested in it. In each of these cases, a manufacturer has to make a specialized item and then rely on a unique buyer to purchase this item. Clearly, the manufacturer will seek contractual assurance of payment from the buyer. A bond (or deposit) may provide this assurance.

When he needs a specialized piece of equipment, a buyer may solicit bids from potential manufacturers. He will choose the best price–quality combination from these bids. To the extent he requires an extremely specialized item requiring manufacturers to make a large irreversible investment, he will receive high bids and have to pay a high price. Therefore, he must weigh the added benefits of specialized equipment against the increased cost of specialization. He will demand a specialized component only when its benefits exceeds its costs. For example, in evaluating different radio components, a radio manufacturer will compare the prices of standard components with the bids he receives for specialized components. He will choose the component that leads to the greatest increase in profit. He need not integrate into the production of radio parts in order to either obtain specialized equipment or to know how specialized a component should be. The bids he receives from parts manufacturers will reveal the costs of the specialized items and allow him to select the most cost-effective device.

We do not mean to imply contracts always work perfectly. Many transactions are so complicated that the traders could not possibly draft a contract dealing with every contingency that might arise between making and consummating an agreement. The seller's costs or the buyer's needs may change over time. The traders may disagree over the meaning of the contract's language. A dishonest buyer may renege on his commitment. These problems are important. They make contracting more problematic than purchasing a commodity for immediate delivery.

There are two types of contractual problems that we will describe in more detail. First, there are temporal difficulties. Conditions inevitably change over time, especially when items take a long time to make and technological change is rapid. For example, when interest rates rise construction delays may become more costly and a different size plant may become optimal. The parties to such transactions can, however, make their contracts contingent on interest rates or construction costs. Bank credit agreements sometimes guarantee the borrower a credit line at a variable interest rate. The rate is set several points above the prime rate. These contingent agreements enable borrowers to make long-range plans knowing that they have a guaranteed line of credit and insures banks against fluctuating interest rates.

It is easy to write a contract contingent on interest rate changes when both parties can readily verify when interest rates have in fact changed. It is more difficult to write a contract contingent on some event that is difficult to specify or verify. For example, a computer user's software needs depend on the internal workings of her computer and on the tasks she wants performed. She may sign a contract to develop software for her system. Because of business changes, the tasks she wants performed may change or, because of technological advances, she may want to replace her computer with a more powerful machine. Consequently, she may wish to change her software specifications after she has signed a contract but before her supplier has finished the software. She and her supplier may not have been able to agree upon a sufficiently precise *reopener clause* which would have covered these changes in circumstances. Similarly, a newspaper might contract with a printer to print the evening paper. The newspaper would, presumably, require the printer to have the paper ready at some specified time in order for the newspaper to meet its delivery schedule. But, because of a major news event like a Presidential assassination, the newspaper may want a late *extra* edition. Unless this event were specified in the initial contract, the printer could refuse to oblige or insist on an exorbitant price.

Second, there are enforcement problems. Well-specified contractual provisions are easily enforced when the parties can post a bond with a mutually agreed-upon arbitrator. A party who failed to carry out his part of the contract would lose his bond. For this mechanism to work, the parties must (*a*) post bonds that are large enough that neither party would forego the bond and breach the contract and (*b*) find a mutually acceptable arbitrator. They may be unable to do so in many circumstances.

Integrating the printing company and the newspaper firm removes the printing plant manager's incentives to charge the newspaper for special services—when the conglomerate's headquarters does not charge the newspaper for printing services. Integrating the radio manufacturer with a parts supplier could enable the radio manufacturer to demand substantial changes on previously agreed upon designs—when the conglomerate's headquarters lets the radio manufacturer run the company. But these would be disastrous methods for running multidivisional companies.

Multidivisional companies do not, in fact, use such methods. Most have decentralized accounting systems designed to make each division accountable to corporate headquarters. By assigning *transfer prices* to resources exchanged between divisions, these systems try to allocate resources to their most profitable uses. According to a recent study by the Financial Executives Research Foundation (FERF),[63]

> Nearly all large U.S. manufacturing corporations are decentralized in the sense that corporate managers have segmented their enterprises into business units in order to hold a subordinate manager accountable for the performance of each line of business. . . . [T]he subordinate manager is responsible for the financial performance, including the profitability of "his" or "her" business.[64]

Most of the companies surveyed by this study used transfer prices to exchange resources across divisions.[65]

In a market economy, prices ration goods so that people who consume these goods bear the marginal cost of producing them and place the highest value on them. In a conglomerate, transfer prices and related incentives encourage resources to be allocated so that each division bears the marginal cost of employing them and creates the greatest profit from them. Conglomerates try to mimic markets. They do this by creating *profit centers,* which are essentially small firms managed by an "entrepreneur."[66] According to the FERF study, "Decentralized firms use rewards . . . to 'turn on' their profit center managers, motivating them to initiate actions that will result in high performance both for the profit centers and for the firm as a whole."[67] Most rewards are explicitly linked to the financial performances of the profit center.[68] Headquarters further encourages managerial initiative by giving profit center managers a vested interest in their businesses. The FERF study noted, "Just as ownership of assets is a primary source of power in a society built on the concept of private property, so physical custody of resources is a primary source of authority in a centralized firm."[69]

With these insights into the operation of multidivisional companies, let us return to the radio example. When (*a*) the costs of and technology for producing radio parts, (*b*) the technology for producing radios, and (*c*) the quality of parts desired by the radio division are fairly constant, headquarters could easily coordinate the activities of the radio and radio parts divisions. It could simply tell the parts division what to produce and the radio division which parts to use. Deviations from its corporate plans would be recognizable as incompetence or neglect, because in such circumstances managers do not need flexibility.

Situations in which changes or revisions do not occur are ones in which central headquarters can easily coordinate the divisions. These are, however, exactly the situations in which no common ownership is necessary. These are the situations in which two separately owned companies could have written a contract to describe their mutual obligations. It is easy to coordinate activities that require little coordination. But contractual problems arise between arms-length companies when the benefits and costs of traded goods change unpredictably over time. For example, if the technology for producing radios is volatile, the radio company may not precisely understand its needs when it signs a supply contract. Yet the parts supplier may have to sink money immediately into building a machine for making the required parts. The radio manufacturer may then insist on costly midstream modifications to this machine. These contractual problems make coordination between separately owned companies more difficult.

But the integrated firm faces these same problems. It has two divisions. Each division has a manager. Each manager has an incentive scheme which encourages him to act in the conglomerate's financial interest. There is a division of labor between the divisions. The radio division's personnel are knowledgeable about the products desired by consumers. The parts division's personnel are knowledgeable about the costs of innovation and of producing small parts. Given this decentralized knowledge, corporate headquarters will coordinate its divisions by trying to mimic the price system. By allowing the parts division to charge the radio division for the cost of the parts and forcing the radio division to pay for the parts, headquarters gives the radio division incentives to use the most profitable number and type of parts. The radio division manager is rewarded according to the resources realized by her division less the cost of the parts used.

To take another example, suppose the radio division wants the parts division to make costly midstream modifications to a machine. We'll consider two cases. In the first case, headquarters can easily judge the benefits and costs of midstream modifications. It has the knowledge and expertise required to resolve disputes that may arise between the two divisions fairly. Because there is no reason to believe integration itself creates this knowledge and expertise, two arms-length companies could obtain reliable arbitration under these circumstances. If the parts and radio divisions disagree over the cost of the modification, headquarters acts like an arbitrator by judging these claims. An arbitrator could perform the same task for two unintegrated companies.

In the second and more complex case, headquarters cannot easily judge the benefits and costs of midstream modifications. It lacks the knowledge and expertise. This second case applies to many technologically complex companies which are subject to rapid, unpredictable technological change. These companies have to decentralize knowledge and control. Their divisions face the same contractual problems as do arms-length traders in technologically volatile industries.

In these circumstances, headquarters will set budget and performance measures for its parts and radio divisions. When information is decentralized, the best performance measure is net profit contributed by each division. The FERF study supports this conclusion.[70]

> For nearly two thirds of our profit center managers, the size of their annual bonus is determined by a defined formula rather than by reliance on a potentially subjective judgement by their superior. More surprisingly, for nearly half of those managers, their bonus was determined *solely* by the financial performance of their profit center.

Production targets are poor substitutes for profit measures. In an attempt to meet his target, a manager will overspend company resources. Headquarters could set a production target and a budget limit. This method would work when headquarters has sufficient information and expertise to compute the optimal level and cost of production. But such is not the case with complex decentralized companies.[71]

Given that companies set up profit centers in order to decentralize information and control, it is quite obvious that the contractual difficulties that would arise between separately owned companies will arise between commonly owned companies. Common ownership creates neither new information nor expertise. Nor does it create new auditing opportunities. Nonintegrated companies permit arbitrators to audit their operations when there is a dispute between them. For example, when it provides a line of credit, a bank gets the right to audit the borrower's books.

Under common ownership, headquarters uses profit centers to establish divisional efficiency. Profit centers, however, may create their own difficulties. Consider the example where the radio parts division builds a specialized machine to produce parts for the radio division. The radio division determines the value of the parts. The parts division determines the cost of the machine needed to produce them. In simple cases, headquarters will weigh benefits and costs in deciding whether to proceed with production. In complex cases, headquarters will rely on the profit-center approach. It will tell the radio division to proceed if the project increases the radio division's profits and it will tell the parts division to proceed if the project increases the parts division's profits.

In order to compute divisional profit, headquarters must establish a transfer price for the parts exchanged between divisions. If the divisions were separately owned companies, the radio company would solicit bids; the best bid would determine the price of the part. The integrated company may face a serious problem. Its radio division is stuck with its parts division. If it relies solely on its parts division for parts, and sets a transfer price that enables the parts division to recover its costs plus a normal profit, its parts division would have few if any incentives to economize on the construction of the machine.

The FERF study demonstrates these conflicts. It found that negotiations between profit center managers usually determined transfer prices.[72] It notes an example.[73]

> The managers of the product profit centers are not required to purchase components from the manufacturing profit centers, but when they do the price is negotiated between the two profit center managers just as it would be negotiated between an independent buyer and seller. Nevertheless, interdivisional squabbles over transfer prices are frequent and sometimes bitter.

Headquarters faces these same incentive problems in resolving disputes between its divisional managers. For example, suppose its radio division asks for a costly midstream design change. If it allows the parts division to charge the radio division for any and all costs incurred as a result of the design change, the parts division will have no incentives to economize on the midstream modifications. If it allows the radio division to demand any midstream modification desired without paying the costs of these changes, the radio division would have no incentive to economize on the midstream modifications demanded.

For both integrated and nonintegrated companies, prices play a crucial role in coordinating economic activity. For both, prices that relay economically efficient incentives are important. For both, complex and changeable technology leads to contractual disputes. For both, there are wastes if midstream modifications are priced incorrectly.

Because information is dispersed and because individuals require rewards for their efforts, market systems in which property ownership is dispersed among numerous self-interested businesses provide economic coordination most efficiently. Unless they become profit center managers, entrepreneurs lose these incentives when they become the conglomerate's employees. It is strange that the transactions cost literature assumes that the mere act of integration transforms selfish humans into selfless ones whose only goal is their company's welfare. Common ownership, unfortunately, eradicates neither indolence nor dishonesty.

Appendix

AT&T claimed it received various benefits from its relationship with Western Electric which it would not receive from an arms-length profit-maximizing supplier. These benefits include

Complete subservience to demands by the Bell operating companies. Western is obligated to provide the Bell operating companies with equipment to meet service needs whereas the Bell Operating companies have no obligation to purchase from Western. Western is required to organize based on the Bell operating companies' (BOCs) needs and to gear up to meet forecasted needs regardless of the impact of schedules, inventories, or the likelihood of making existing products obsolete.[74]

Allegedly free insurance against equipment shortages or surpluses due to changes in demand, cost, or technology. Western makes capital commitments for production of equipment without reciprocal commitments from Bell operating companies and thereby bears the risks of changes in demand. A profit-maximizing supplier would not bear risks to the extent Western does.[75]

Replacement parts for old equipment. Western continues to provide replacement parts for equipment long after other manufacturers would have discontinued production of replacement parts.[76]

AT&T's claims raise two questions. First, is AT&T able to receive the benefits listed above without an offsetting cost—in other words, is AT&T really able to receive a free lunch from Western? Second, does the master–slave relationship between AT&T and Western reduce the social cost of providing telecommunications services? Consider the first question. Western cannot provide AT&T with the benefits listed above without reducing its rate of return below that received by competitive manufacturers facing comparable risks (granting, for the sake of argument, AT&T's assertion that Western is effectively regulated). Each benefit entails either a cost, which, unless recouped through higher prices, must reduce Western's rate of return, or a risk, for which competitive manufacturers would require a premium, realized through higher prices, over the riskless rate of return. It is useful to identify the costs and risks associated with AT&T's alleged benefits.

Because of complete subservience to the Bell operating companies, Western must maintain excessively large inventories to meet BOC demand for equipment; introduce new technology regardless of the impact on revenues; and must make potentially unprofitable capital investments. These obligations reduce Western revenues. Because of "free insurance," Western incurs the risk of demand, cost, and technology changes instead of the BOCs. Competitive firms require a higher rate of return and consequently higher market prices for incurring risk. A competitive supplier continues to provide replacement parts for old equipment until the cost of doing so exceeds expected revenues. Western, according to AT&T, apparently provides replacement parts even when the cost of doing so exceeds expected revenues.

In order for Western to earn a competitive rate of return (relative to manufacturers facing comparable risks) and to provide AT&T with the benefits listed above, Western prices must exceed prices charged by an arms-length competitive supplier by enough to offset the cost to Western of providing these so-called benefits and to conpensate Western for bearing risks. Consequently, AT&T pays Western for the benefits listed above in the form of higher equipment prices. AT&T does not receive a free lunch from Western.

The intuition behind the argument presented above is this. Somebody has to pay for the costs and risks involved in providing telecommunications services. AT&T does not eliminate these costs and risks by shifting them to Western. Either Western pays for them by taking losses—which would be passed on to AT&T stockholders because AT&T owns all of Western—or Western recovers them by charging higher equipment prices.

The fact that Western may charge AT&T higher equipment prices to offset the cost of some of the benefits it provides AT&T, does not imply that the net cost of these benefits to society is zero. Indeed, the master–slave relationship between AT&T and Western probably results in inefficiencies that raise the social cost of providing telecommunications services. An efficient equipment market would not generally impose all the costs and risks associated with changes in demand, cost, and technology on the supply side of the market.

For example, AT&T and the Bell operating companies presumably have superior information on likely demand changes than Western, which is not responsible for delivering telecommunications services. Imposing the costs and risks of demand changes solely on Western removes incentives to forecast future capacity requirements from AT&T and the BOCs accurately. This incentive structure, if it really exists, makes little business sense and would not occur in an efficient equipment market.

AT&T apparently argued that Western not only bears all the risks of changes in demand, cost, and technology, but that AT&T forced Western to make capital commitments for equipment production and to enlarge inventories irrespective of the risks involved. An efficient equipment market would select an optimal tradeoff between risk and the expected rate of return on inventories and capital commitments. It would ignore risks only under extremely unusual circumstances.

NOTES

1. AT&T's witness Schwartz said "a centralized authority must exist to gain the cooperation of all companies, to set priorities, to allocate resources and to resolve conflicts among companies." See D-T-125 in *US* v. *AT&T*, p. 121. Schwartz is a business consultant with Booz-Allen and Hamilton. Ian Ross, President of Bell Labs, claimed, "It is the Bell System's ability to engage in centralized planning, operation and management of this network as an integrated whole, and careful attention to the standardization of operating plans and methods and to the technical compatibility of facilities, which have made the network as effective and efficient as it is today." See D-T-195 in *US* v. *AT&T*, p. 15.

2. In their testimony, economists David Teece, Almarin Phillips, and William Nordhaus made these claims repeatedly.

3. It also claimed it received various benefits from its relationship with Western Electric that it would not receive from an arms-length, profit-maximizing supplier. Schwartz cited instances where Western Electric's interests were "subordinated in favor of the BOCs and Long Lines" and where the Bell System minimized risks to the operating companies and Long Lines by having Western Electric incur substantial start up costs without purchase commitments from the operating companies. See D-T-124 in *US* v. *AT&T*, pp. 45–47. Procknow, the President of Western Electric, said his company was subservient to the other portions of the Bell System. An appendix addresses these claims.

4. As of 1979, based on data reported in the "Fortune 500 Industrials," *Fortune Magazine*, May 5, 1980, and in FCC, *Statistics of Communications Common Carriers*, 1979. Public utilities are not listed on the Fortune 500. The comparison is nevertheless useful.

5. *Ibid.*, based on sales.

6. *International Monetary Fund Statistics*, October 1981.

7. Assuming that the operating companies retain only local exchange revenues, and based on 1979 data. There are plans to form seven regional operating companies from the divested portions of the 22 Bell operating companies.

8. Based on 1979 sales from sources cited in Note 4, above.

9. See, for example, Abram Bergson, *Planning and Productivity Under Soviet Socialism* (Pittsburgh: Carnegie–Mellon University, 1968).

10. Schwartz's testimony in *US* v. *AT&T* [D-T-125 at 17–18].

11. F. A. Hayek, "The Use of Knowledge in Society," *American Economic Review*, September 1945, pp. 519–530.

12. *Ibid.*, p. 524.

13. *Ibid.*, p. 525.

14. *Ibid.*, p. 526.

15. *Ibid.*, p. 525.

16. Abram Bergson, "The Current Soviet Planning Reforms," in Alexander Bolinky *et al.*, *Planning and the Market in the USSR: The 1960's* (New Brunswick, NJ: Rutgers, 1967), pp. 43–64.

17. Adam Smith, *Wealth of Nations* (New York: Modern Library, 1937), p. 423.

18. The physical coordination of people and materials differs from the economic coordination of the production and distribution of goods and services. Physical coordination requires physical intervention. The price system achieves economic coordination. For example, attaching airplane seats to airplane frames requires physical coordination; ensuring that the supply of passenger seats equals the demand for passenger seats requires economic coordination. We show below that the profit system encourages physical coordination without common ownership of the resources being coordinated.

19. Tax laws allow the owners of equipment to deduct the depreciation of the equipment. This depreciation deduction is of value if the company has offsetting income. Thus, companies with high income will buy equipment and lease it to companies with low income. This is an artifact of the tax laws, and probably explains a good deal of why some firms own and others lease heavy equipment.

20. George Rogers Taylor and Irene D. Neu, *The American Railroad Network* (Cambridge: Harvard University Press, 1956), p. 16.

21. Alfred Chandler, *The Visible Hand: The Managerial Revolution in American Business* (Cambridge: Harvard University Press, 1977), p. 123. Although Chandler's main argument is that administrative coordination was superceding market coordination, his factual account of physical coordination by separately owned businesses supports our thesis that the market provides strong incentives for physical coordination without common ownership.

22. Transshipment involves physically moving produce and passengers from one railroad to another. In the early days of the railroads, roads did not interconnect so that produce would have to be carted from one railroad terminal to another railroad terminal.

23. T. J. Nagel, "Interconnection and Reliability," in Harry Trebing, ed., *Energy and Communications in Transition* (East Lansing, MI: Institute for Public Utilities, Michigan State University, 1981), pp. 275–281.

24. *Ibid.*, p. 281.

25. Some economists have argued that market transactions are prohibitively expensive under certain circumstances thereby necessitating common ownership. In the second section, we show that common ownership may not alleviate transactions problems.

26. Chandler, *op. cit.*, p. 124.

27. For example, the Long-Distance Telephone Company of Virginia sought "arrangements with the numerous local telephone companies whereby the long-distance circuit can be strung on the poles already erected." This company intended to give subscribers to local telephone companies throughout the state "the benefit of the long-distance facilities by having the long-distance circuit 'tapped' in every county at the county exchange. . . ." *The Telephone Magazine*, March 1901, p. 132. Many independents formed toll associations which often had clearing houses for dividing up toll revenues. See, for example, the report of the Special Tariff Committee for the New York State Independent Telephone Association in *The Telephone Magazine*, July 1905.

28. Ian Ross stated that the independent telephone companies have profit incentives to cooperate with the Bell System on managing the telecommunications network. Because these companies "participate in the settlements in which they receive a portion of . . . the long haul revenues . . . , they have an interest in seeing that [the] network will continue to carry the maximum of traffic with which it is capable." See Ross Tr. 11501-2 in *US v. AT&T*. The settlement procedure thereby divides the profit created from physical coordination among the firms that participate in this coordination.

29. Taylor and Neu, *op. cit.*, p. 68.

30. The cooperative fast freight lines replaced independent fast freight lines which contracted with each railroad separately. According to Taylor and Neu, "Each cooperative fast freight line had a Board of Directors composed of a representative from each cooperating road. The function of the Board was to decide the policy for the line, and the relative weights of each representative's vote was decided on the basis of the length of his road or on the basis of the proportion of line business which it carried. This board had no jurisdiction over freight rates; such rates continued to be set by the individual railroads or by agreement among all the roads on the line. A general office was maintained by each cooperative line, and was presided over by a general manager. It employed a staff of clerks and functioned as a clearing house." *Op. cit.*, p. 72.

31. Hayek, *op. cit.*, p. 525.

32. Kershenbaum Tr. 11098. For the opposite view see Ronald Skoog, *Design and Cost Characteristics of Telecommunications Networks* (Holmdel, NJ: Bell Laboratories, 1980).

33. The allocation of profits across manufacturers of different products depends on the demand and supply of these products.

34. Oliver Reed and Walter L. Welch, *From Tin Foil to Stereo* (New York: Bobbs-Merrill, 1959), p. 348.

35. Taylor and Neu, *op. cit.* There is some evidence that there was no compelling reason to adopt a particular gauge so long as everyone used the same gauge. See pp. 12–13.

36. *Ibid.*, p. 4.

37. *Ibid.*, p. 13.

38. *Ibid.*, p. 52.

39. Chandler, *op. cit.*, pp. 123–124.

40. *Ibid.*, pp. 130–131.

41. *Ibid.*, pp. 131–132.

42. See H. P. Clausen, "Standardization Possibilities," *Telephony*, January 1908, pp. 32–34; James B. Hoge, "Necessity of State and National Organizations," *The Telephone Magazine*, June 1905, pp. 373–375; and W. J. Stanton, "Standardization of Toll Line Equipment," *Telephony*, May 1907, p. 306.

43. Recognizing the value of horizontal agreements, the Telecommunications Competition and Deregulation Act of 1981 provided that "it shall be lawful for telecommunications carriers jointly to meet for the purposes of planning or agreeing to . . . the planning, development, construction, management, or coordination of any network of telecommunications services or facilities . . . ."

44. W. J. Stanton, *op. cit.*, p. 306. For a discussion of the desirability of flexible standards also see H. P. Clausen, *op. cit.*, p. 33.

45. H. P. Clausen, *op. cit.*, p. 33.

46. Taylor and Neu, *op. cit.*, p. 63.

47. Taylor and Neu, *op. cit.*, p. 62.

48. Reed and Welch, *op. cit.*, p. 341.

49. Reed and Welch, *op. cit.*, p. 343.

50. Nevertheless, the 45s were superior for short selections and were produced for juke boxes in considerable numbers.

51. 86FCC2d719 at 747–748 (1981).

52. Congressional Research Service, *Voluntary Industrial Standards in the United States*, Report to the Subcommittee on Scientific Research and Development of the Committee on Science and Astronautics, U. S. House of Representatives, 93rd Congress, 2nd Sess., July 1974.

53. Unfortunately, government intervention may create its own problems. Once consumers are forced to join a coalition for reducing pollution, what forces ensure that the coalition—the government—acts in the consumers' interest? Tenuous ones at best.

54. The article quotes Joel Davidow of the Antitrust Division, U.S. Department of Justice to this effect.

55. *Science*, January 2, 1981.

56. David Teece mentioned examples where "the process by which a common standard [was] selected and adopted [was] quite slow." But, he failed to recognize that this process helps society choose the best kind and degree of standardization. He also failed to recognize that monopolists, which are no less fallible than competitive firms, may impose poor standards. See D-T-362 in *US* v. *AT&T*, p. 7.

57. David Teece, D-T-362 in *US* v. *AT&T*, p. 7.

58. Oliver Williamson, "The Vertical Integration of Production: Market Failure Considerations," *American Economic Review*, May 1971, pp. 114–115.

59. *Ibid.*, p. 116.

60. B. Klein, W. Crawford, and A. Alchian, "Vertical Integration, Appropriable Rents, and the Competitive Contracting Process," *Journal of Law and Economics*, October 1978, pp. 297–326.

61. The basic idea behind the transaction-cost theory is that, after consummating a contract involving a specialized investment, one of the parties to the contract could haggle with the other party and thereby receive a greater share of the value created by the contract. If the extorted party could have anticipated that the extorting party would engage in opportunistic haggling, he would not have entered into the contract in the first place. The extorted party must be unable to anticipate opportunistic haggling by the extorted party for the opportunistic-haggling scenario to be plausible. An obvious method to anticipate opportunistic tendencies on the part of a potential partner to a contract is to examine his behavior on previous contracts. Companies surely avoid doing business with other companies who have developed a reputation for engaging in opportunistic haggling. Consequently, companies that expect to remain in business for a long time—such as GM, Western Electric, and AT&T—have particularly strong incentives to develop a reputation for fair dealing. Companies that engage in multiple transactions involving specialized investments also have particularly strong incentives to develop such a reputation. These incentives for fair dealing would make the devil an honest man.

62. Richard Posner, *Economic Analysis of Law* (Boston: Little, Brown, 1972), p. 42.

63. Richard F. Vancil, *Decentralization: Managerial Ambiguity by Design*. A research study prepared for the Financial Executives Research Foundation (Homewood, IL: Dow Jones-Irwin, 1978), p. 1. Hereafter, referred to as the FERF study.

64. The study found that "decentralization is *the* most common organizational mode in U.S. manufacturing corporations today, not only in large diversified firms, but also in relatively homogenous businesses with sales of $100 million or less." *Ibid.*, p. 3.

65. *Ibid.*, p. 15.

66. A recent survey of Fortune 1000 companies found that 95.8% of the 620 respondents used profit centers or investment centers (which the authors distinguish from profit centers but which are the same for our purposes). See James S. Reece and William R. Coal, "Measuring Investment Center Performance," *Harvard Business Review*, May–June 1978, p. 29.

67. FERF Study *op. cit.*, p. 13.

68. FERF Study, *op. cit.*, p. 95.

69. FERF Study, *op. cit.*, p. 54. The study goes on to observe, "A profit center manager . . . doesn't want you to forget that he has 'his' inventory, 'his' lab, or 'his' plant."

70. FERF Study, *op. cit.*, p. 95.

71. Like most large companies, AT&T uses objective rating systems for awarding salary increases and promotions. Wallace Bunn, President of South Central Bell, testified that, "My compensation and my opportunity for advancement . . . is geared to the effectiveness with which I manage the service job, and the effectiveness with which I perform the earnings job. That is the thing that motivates me." See Bunn Tr. 17499 in *US* v. *AT&T*. He goes on to say that AT&T has a written salary administration policy based on specific, quantitative measurements.

72. FERF Study, *op. cit.*, p. 15.

73. FERT Study, *op. cit.*, pp. 48–49.

74. Procknow, O'Neill, Eichols, and Schwartz, testifying for AT&T, made these claims.

75. For example, Schwartz said the "Bell System minimized risk to the operating companies and Long Lines" because "Western Electric authorized considerable development and start-up costs and began manufacture without order commitments from Long Lines and the operating companies." See D-T-125 in *US* v. *AT&T*, p. 47. Also see p. 112.

76. For example, Procknow said, "Western Electric cannot discontinue the manufacture of a product without AT&T approval, which is not granted unless AT&T determines that there is no substantial Bell operating company or Long Lines need for the product. . . . Each of these efforts [to continue manufacture] diverted scarce resources which, if Western were an unaffiliated manufacturer, I would have preferred to use to develop new products." See D-T-195 in *US* v. *AT&T*, pp. 11–12.

Chapter 6

Natural Monopoly

David S. Evans and James J. Heckman

In *US* v. *AT&T*, AT&T submitted considerable evidence that the telephone industry is a natural monopoly.[1] It claimed that "a single interactive and interdependent network [is] the most efficient means for providing all of this Nation's telecommunications services. . . . [S]uch a network can be planned, constructed, and managed most efficiently by an integrated enterprise that owns the major piece-parts of the facilities network. . . ."[2] This claim was supported by James Rosse, an economist, who testified that "telecommunications services in general, and Bell System offerings in particular, possess economies of scale and scope to such an extent that no competitor with less volume or growth could match Bell's costs for providing a comparable set of services."[3]

The natural monopoly evidence was used to support three arguments. First, that AT&T's competitors in intercity service were less efficient than AT&T and were able to survive in the marketplace only because rate averaging made AT&T's rates exceed its costs for competitive services. Second, in lowering rates in response to competition, AT&T was merely exploiting its inherent scale economies and was not engaging in predatory pricing. Third, breaking up the Bell System would eliminate the efficiencies realized from common ownership and control over the telephone system.

This chapter reviews the empirical evidence presented by AT&T concerning the production characteristics of the telecommunications industry and examines the relevance of this evidence for determining whether a single firm can provide any or all telecommunications services more efficiently than several firms. It has four sections. The first section shows how certain cost characteristics of the technology used by firms in an industry help reveal whether this industry is a natural monopoly. The second section describes econometric methods for determining these cost characteristics from cost and production data. Together, these two sections provide a useful framework for evaluating evidence that the telecommunications industry is a natural monopoly. The third section critiques the case made by AT&T, primarily through testimony submitted by James Rosse, that a single firm that owns the major parts of the telecommunications network can provide telecommunications services more efficiently than several firms that own different parts of the telecommunications network. The fourth section reports the results of our own study of whether the telecommunications industry is a natural monopoly.

Theory of Natural Monopoly

A firm transforms inputs such as labor and capital into products. The technology used by a firm consists of the various techniques used to accomplish this transformation. It includes methods for managing the firm and methods for physically manipulating inputs. The cost function, which we use extensively in this chapter, is a convenient device for summarizing the nature of the technology used by the firm.[4] It describes how production costs vary with production levels and input prices when the firm uses the most efficient combination of inputs and production techniques. It can be written as $C(Q_1, Q_2, r, w)$, for the case of a firm that produces two products with capital and labor, where Q_1 represents the level of output one, Q_2 represents the level of output two, r is the price of capital services, and w is the price of labor services. When all firms use the same technology, certain properties of the cost function provide valuable information concerning the most efficient industry structure for meeting the demand for the outputs produced by this technology.[5] In order to describe these properties, and their usefulness for determining whether an industry is a natural monopoly, we first consider the case of a single-product technology and then consider the more complicated, but more realistic, case of a multiproduct technology.[6]

Single-Product Technology

An industry is a natural monopoly if a single firm can meet market demand more efficiently than several firms. When all firms use the same technology, the concept of subadditivity helps us determine whether an industry is a natural monopoly. For convenience, we suppress the factor prices r and w in writing the cost function. Firm M produces Q units of output for $C(Q)$ dollars. When a is some number between zero and one, firm A produces aQ units of output and firm B produces $(1 - a)Q$ units of output for a total of

$$aQ + (1 - a)Q = Q \qquad (6.1)$$

units of output. Firm A spends $C(aQ)$ dollars and firm B spends $C[(1 - a)Q]$ dollars for a total of $C(aQ) + C[(1 - a)Q]$ dollars. If

$$C(aQ) + C[(1 - a)Q] \text{ exceeds } C(Q) \qquad (6.2)$$

for all possible values of a, so that a single firm can produce Q more cheaply than two firms, then the cost function is said to be subadditive at the level of output Q.[7]

In the single-product case, it is easy to determine whether the cost function is subadditive by examining the shape of the average cost function $C(Q)/Q = AC(Q)$. Figure 6.1 depicts two possible average cost curves denoted by AC and two demand curves denoted by D. The average cost curve on the left is constantly declining. A single firm, constrained to break even by its regulators, would produce Q_M units at average cost AC_M.[8] Dividing Q_M units of output between two firms would always raise cost. Let firm A produce Q_A and firm B produce Q_B, with $Q_A + Q_B = Q_M$. Firm A's average cost AC_A and firm B's average cost AC_B exceed firm M's average cost AC_M.

The average cost curve on the right declines through Q_A and then rises. A single firm, constrained to break even by its regulators, would produce Q_M units at average cost AC_M. Dividing Q_M between two firms would enable one firm to produce at an average cost lower than AC_M and force one firm to produce at an average cost higher than AC_M. Together, these firms would produce Q_M at an average cost higher than AC_M. For example, the average cost AC_A for Q_A is slightly lower than the average cost AC_M for Q_M, but the average cost AC_B for Q_B is substantially higher than the average cost AC_M for Q_M. At output level Q_M, the cost functions shown in the diagram are therefore both subadditive and the industries with these cost functions are both natural monopolies.

In analyzing natural monopoly and subadditivity, it is useful to distinguish between local and global concepts. Local concepts apply at a particular level of

Figure 6.1 Two subadditive cost functions.

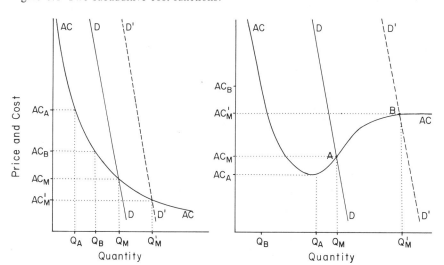

output. Global concepts apply to all levels of output. We say that the cost function is globally subadditive if it is subadditive at all levels of output. There is a global natural monopoly if the cost function is globally subadditive. Figure 6.1 demonstrates the distinction between local and global natural monopoly. Suppose demand shifts from D to D'. The cost function on the left-hand side is subadditive at the new equilibrium output Q_M', as the reader may readily verify. In fact, this cost curve is subadditive for all levels of demand and is therefore globally subadditive. The cost function on the right-hand side is not subadditive at the new equilibrium output Q_M'. As drawn on the right-hand side, $Q_M' = 2Q_A$. The average cost of a single firm producing Q_M' is AC_M'. The average cost of two firms producing Q_M' is AC_A when each firm produces half of the total quantity (equal to Q_A). The two-firm, average cost AC_A is substantially lower than the single-firm, average cost AC_M'. Therefore, the cost function on the right-hand side is not globally subadditive. The distinction between local and global subadditivity and local and global natural monopoly is important. As the right-hand side of the diagram demonstrates, a natural monopoly that is only local may collapse as demand increases. As the left-hand side of the diagram demonstrates, a natural monopoly that is global remains a natural monopoly as demand increases.

The older industrial organization often used the degree of scale economies to determine whether an industry was a natural monopoly. There are scale economies when a proportionate increase in output leads to a less than proportionate increase in cost. There are scale diseconomies when a proportionate increase in output leads to a more than proportionate increase in cost.[9] It is well known that there are scale economies when the average cost curve slopes downward and scale diseconomies when the average cost curve slopes upward. Thus, the average cost curve on the left-hand side of Figure 6.1 has scale economies at all levels of output while the average cost curve on the right-hand side of Figure 6.1 has scale economies between zero and Q_A and scale diseconomies thereafter. A cost function that has scale economies at some output Q is subadditive at output Q, as the left-hand side of Figure 6.1 demonstrates. A cost function that is subadditive at some output Q does not necessarily have scale economies at output Q, as the right-hand side of Figure 6.1 demonstrates. At Q_M, the cost function on the right-hand side of Figure 6.1 has scale diseconomies but is subadditive.

Researchers are often tempted to conclude, from evidence that a firm has scale economies (or subadditive costs), that the industry in which this firm operates is a natural monopoly and that, therefore, single-firm production is more efficient than multifirm production. This conclusion is usually based on several assumptions which these researchers seldom verify and which may not hold in many practical circumstances.

First, it assumes that all firms use the same technology and therefore have the same costs for all levels of production. This may not be the case. Firms may be able to choose between alternative technologies which exhibit varying degrees of scale economies.[10] Figure 6.2 shows a situation with two alternative tech-

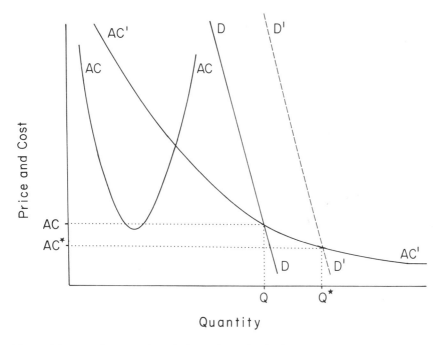

Figure 6.2 Natural monopoly and alternative technologies.

nologies T and T' described by average cost curves AC and AC'. At the level of demand Q, the average cost of a single firm producing Q with technology T' is greater than the average cost of three firms producing Q with technology T. At the level of demand Q^*, the average cost of a single firm producing Q^* with technology T' is less than the average cost of two or more firms producing Q^* with technology T or T'. In general, whether an industry is a natural monopoly depends on the relative costs of alternative technologies and the level of market demand.

Second, the theory ignores whether common ownership over the means of production is necessary for an industry to realize economies from a technology with a subadditive cost function. The Bell System is a case in point. Even if the technology for providing telephone service has a subadditive cost function, it is conceivable that several firms with separate ownership over different portions of the telephone network might be able to provide telephone service at least as efficiently as a single firm. Multiple ownership might reduce managerial diseconomies and promote innovation. Whether common ownership is better than multiple ownership depends on the relative cost of coordinating within a firm and coordinating between firms.[11]

Third, researchers who use scale economy evidence (or evidence of subadditive costs) to make policy recommendations concerning whether an industry should be subject to regulations that protect an incumbent monopoly from competition often ignore the facts that (*a*) regulations are costly to administer and may induce the regulated firm to act inefficiently, (*b*) competition by inefficient producers may keep the regulated natural monopoly in line and thereby serve as a substitute for direct regulation, and (*c*) entry barriers erected by regulators may stifle the development of more efficient technologies.[12]

These pitfalls in inferring that an industry is a natural monopoly from evidence of subadditive costs also apply to multiproduct industries, to which we now turn.

Multiproduct Technology

Most industries produce several products. As Baumol, Panzar, and Willig have observed[13]

> . . . although much of received theory focuses on single-product firms, virtually all firms in reality produce and sell more than one good or service. This multiplicity of outputs can take the form of a variety of physically-dissimilar offerings, a wide variety of offerings of similar outputs (such as shoes of different sizes) adpated to the demands of individual consumers, or just physically-similar outputs sold at various places or time.

With several important qualifications and additions, the single-product cost concepts discussed above have analogous multiproduct concepts.

For simplicity, we shall consider an industry that produces two products.[14] Firm A produces aQ_1 units of the first product and bQ_2 units of the second product, where a and b are between zero and one. Firm B produces $(1 - a)Q_1$ units of the first product and $(1 - b)Q_2$ units of the second product. Firm A spends $C_A = C(aQ_1, bQ_2)$ dollars. Firm B spends $C_B = C[(1 - a)Q_1,(1 - b)Q_2]$ dollars. Together, firms A and B produce Q_1 units of the first product and Q_2 units of the second product for $C_M = C(Q_1, Q_2)$ dollars. The cost function is subadditive at Q_1 and Q_2 if $C_A + C_B$ exceeds C_M for all possible values of a and b (i.e., for any possible division of production between firms A and B). In this case, a single firm can produce both products, in the amounts Q_1 and Q_2, more cheaply than two separate firms can. Therefore, there is a natural monopoly over both products when the market demands Q_1 units of product one and Q_2 units of product two. The cost function is globally subadditive, and there is a global natural monopoly, when the cost function is subadditive at all levels of demand for both products.

In the multiproduct case, there are several scale economy concepts. Product-specific scale economies is the most useful concept. Suppose a firm originally produces only the first product in the amount Q_1, spending $C(Q_1, 0)$ dollars. It subsequently produces the second product as well, in the amount Q_2, spending $C(Q_1, Q_2)$ dollars. The average incremental cost of producing Q_2 is

$$AIC_2 = \frac{C(Q_1, Q_2) - C(Q_1, 0)}{Q_2} \qquad (6.3)$$

If average incremental costs decline with a small increase in the output of product two, product two is said to have product-specific scale economies at output level Q_2. If average incremental costs increase with a small increase in the output of product two, product two has product-specific diseconomies of scale at output level Q_2. Product two has global, product-specific scale economies if it has product-specific scale economies at all levels of output.

Single-product scale economies imply single-product natural monopoly. But product-specific scale economies in both products do not necessarily imply multiproduct natural monopoly. If common production of both products yields no synergies, firm A could specialize in product one and firm B could specialize in product two with no loss of efficiency.[15] If common production of both products yields substantial synergies which peak and then decline at small output levels and if product-specific scale economies are not extensive, several multiproduct firms could produce these products more cheaply than either one multiproduct firm or two single-product monopolies.[16] Moreover, a multiproduct natural monopoly may not necessarily have product-specific scale economies in either product. If common production of both products yields enormous synergies which increase as production of both outputs increases, a multiproduct natural monopoly could exist even though there are product-specific scale economies in neither product. Thus, product-specific scale economies by themselves provide little useful information about the existence of natural monopoly.

An additional concept, however, is useful for determinining when there is a natural monopoly. There are economies of scope when the cost of producing product one and product two separately exceeds the cost of producing product one and product two jointly. Formally, there are economies of scope at output levels Q_1 and Q_2 when

$$C(Q_1, 0) + C(0, Q_2) \text{ exceeds } C(Q_1, Q_2) \qquad (6.4)$$

There are global economies of scope when there are economies of scope at all levels of output. Local (global) economies of scope together with local (global), product-specific scale economies imply local (global) natural monopoly, although natural monopoly could exist in other circumstances as well. Empirically, economies of scope and product-specific scale economies are hard to measure because econometricians seldom observe separate production of either product.

Natural monopoly is, as we hope this discussion has shown, a more subtle concept in the multiproduct case than in the single-product case. Moreover, in order to determine whether an industry is a natural monopoly, we have to know much more about the cost function in the multiproduct case than in the single-product case. For these reasons, empirical workers are tempted to treat multiproduct industries as single-product industries and to use evidence on the degree of scale economies to determine whether an industry is a natural monopoly. Indeed, the major Bell System studies relied upon by Rosse in his testimony in US v. AT&T treated AT&T as a single-product firm and used scale economy estimates to conclude that AT&T is a natural monopoly. But, as Baumol, Panzar, and Willig have observed,[17]

> We can see why analysts have attempted to use the analytically and statistically
> tractable concept of scale economies as a surrogate test of natural monopoly.
> Unfortunately, . . . such traditional tests simply can not do the job.[18]

The fact that the cost function for some aggregate measure of output exhibits
scale economies is consistent with the absence of natural monopoly over some
and possibly all of the products that comprise this aggregate. A simple example
illustrates this fact. Consider the production of apples and oranges.[19] Suppose
orange production exhibits scale economies and apple production exhibits scale
diseconomies. The average cost curves are drawn in Figure 6.3. Suppose we try
to ascertain whether there is a natural monopoly over apple and orange production
by determining whether some aggregate measure of apple and orange production
exhibits scale economies. It is easy to verify that the cost function for an aggregate
measure of apple and orange production, where the aggregate measure is formed
by taking a fixed, price-weighted average of apple and orange output, lies between
the cost functions for the separate outputs, As drawn, the aggregate cost function
exhibits scale economies even though apple production exhibits scale diseconomies. Therefore, we are led to conclude from the aggregate evidence that apple
and orange production is a natural monopoly even though apple production is
clearly not a natural monopoly.

When the shares of apples and oranges in total revenues change over time,
we obtain even stranger results. Suppose oranges exhibit diseconomies of scale
and apples exhibit constant returns to scale.[20] Also suppose that the demand for
apples rises and the demand for oranges fall over time. Under these circum-

Figure 6.3 Scale economies for apple and orange production.

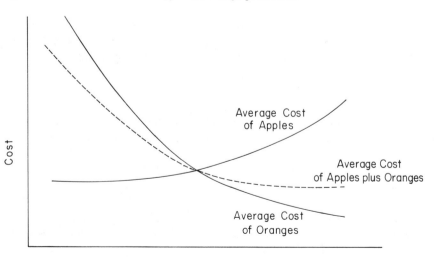

Table 6.1 Calculation of Aggregate Scale Economies for Apple
            and Orange Production[a]

	Output			Input of people		
	Oranges	Apples	Total	Oranges	Apples	Total
Period 1	2	2	4	2	2	4
Period 2	1	4	5	$\frac{1}{2}$	4	$4\frac{1}{2}$
Change	-1	2	1	$-1\frac{1}{2}$	2	$\frac{1}{2}$
Percentage change	-50%	100%	25%	-75%	100%	12.5%

$$\text{Aggregate elasticity} = \frac{25\%}{12.5\%} = 2.00$$

$$\text{Orange elasticity} = \frac{-50\%}{-75\%} = .67$$

$$\text{Apple elasticity} = \frac{100\%}{100\%} = 1.00$$

[a]We assume oranges and apples sell for $1 each and that labor is the only input into apple and orange production.

stances, the unit cost of producing apples and oranges may appear to fall with increases in output despite the fact that neither apples nor oranges exhibits scale economies. Table 6.1 describes a simple situation where this perverse result occurs. Orange demand decreases while apple demand increases from periods one to two. Because oranges exhibit diseconomies of scale, the reduction in orange demand dramatically decreases the personhours required for orange production. Because apples exhibit constant returns to scale, increased apple demand leads to a small proportionate increase in personhours required. Assuming the prices of apples and oranges are $1 each, the net result is that aggregate output increases by 25% from 4 to 5 whereas total input increases by only 12.5% from 4.0 to 4.5. The scale elasticity, calculated by dividing the percentage of increase in output by the percentage of increase in inputs, is 2.0, indicating large scale economies.[21] But neither apple production nor orange production exhibits scale economies. Indeed, orange production exhibits strong diseconomies of scale. A large aggregate scale elasticity, therefore, does not necessarily indicate that any or all of the products that comprise the aggregate have scale economies.[22]

Estimating the Cost Function

The previous section showed that the cost function can provide considerable information concerning the technology used by a firm. Given sufficient data, econometricians can estimate the cost function. From the estimated cost function, they can determine whether a single-product firm has scale economies and whether a multiproduct firm has subadditive costs, product-specific scale economies, or economies of scope. In this section, we describe econometric methods for estimating cost functions.

Specifying a Cost Function to Estimate

Let us begin with a simple example. A firm uses labor and capital to produce widgets. The prices of capital and labor have remained constant over a long period of time. The technology for producing widgets has also remained constant over a long period of time. Therefore, we can ignore factor price changes and technological change and concentrate on the relationship between costs and output. For example, assume that a widget company has provided data on costs *per annum* and output *per annum* for the last 30 years. These points are plotted in Figure 6.4. We can draw a curve through these points. The cost function is simply the mathematical expression for this curve.

Figure 6.4 Estimation of alternative cost functions.

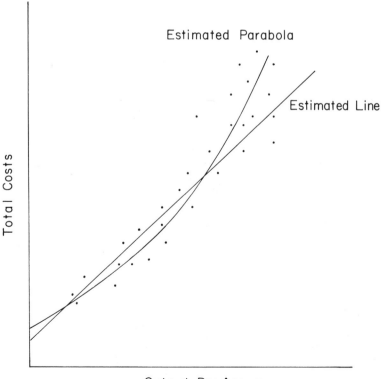

We could try to fit a straight line to the data by estimating the following function

$$C = a + bQ \tag{6.5}$$

where C is cost *per annum*. Regression analysis enables us to estimate the values of a and b that make this straight line fit the data as closely as possible. Visual inspection of the data shows that a straight line provides a rather poor description of the data. The data seem to follow a curve. We could fit a parabola to the data by estimating the following function

$$C = a + bQ + cQ^2 \tag{6.6}$$

Regression analysis enables us to estimate the values of a, b, and c that make this parabola fit the data as closely as possible.

How can we decide whether a straight line or a parabola better describes the data?[23] We can test the hypothesis that the data are more consistent with a parabola than with a straight line by the following procedure. Notice that equation (6.5) is a special case of equation (6.6) with $c = 0$. We can estimate equation (6.6) and test whether the value of c is significantly different from zero. If c is not significantly different from zero, we can infer that equation (6.5) describes the data adequately.

We could use an even fancier function such as

$$\log (C) = a + b[\log(Q)] + c[\log(Q)]^2 \tag{6.7}$$

Equation (6.7) is a special case of the translog cost function.[24] Several economists have argued that the translog cost function provides a general approximation to any valid cost function.[25] In estimating cost functions, most econometricians working in the past 10 years have used the translog cost function.

Recently, several econometricians have proposed more general cost functions such as

$$g(C) = a + b[g(Q)] + c[g(Q)]^2 \tag{6.8}$$

where $g(\cdot)$ is some function. Instead of relating logarithmic values of the variables as in (6.7) or arithmetic values of the variables as in (6.6), we can relate some function of the variables as in (6.8). By letting[26]

$$g(x) = \frac{x^\lambda - 1}{\lambda}$$

where λ is a parameter to be estimated, we can test whether the data are more consistent with equation (6.7), in which case λ should be insignificantly different from zero, with equation (6.8), in which case λ should be insignificantly different from one, or with some other specification, perhaps where $\lambda = .75$. For example, we could estimate

$$\frac{C^\lambda - 1}{\lambda} = a + b\left(\frac{Q^\lambda - 1}{\lambda}\right) + c\left(\frac{Q^\lambda - 1}{\lambda}\right)^2 \tag{6.9}$$

There are standard statistical techniques for estimating the parameters a, b, c, and λ from the data.[27] Equations (6.5), (6.6), and (6.7) are special cases of equation (6.9). By estimating equation (6.9), we can test which equation is more consistent with the data. To the extent the data permit, it is important to test different functional specifications before imposing a particular functional specification on the data.

Let us turn to a slightly more complex example. A firm uses capital and labor to produce widgets and gizmos.[28] We could estimate the following multiproduct translog cost function

$$\log(C) = b_1 [\log(Q_1)] + b_2[\log(Q_2)] + c_1[\log(Q_1)]^2 \\ + c_2[\log(Q_2)]^2 + d[\log(Q_1)\log(Q_2)] \qquad (6.10)$$

We could use this function to determine whether the firm has a subadditive cost function. Analyzing the cost characteristics of a multiproduct firm, however, is much more difficult than analyzing the cost characteristics of a single-product firm. Therefore, we would like to treat the widget-gizmo firm as a single-product firm and estimate

$$\log(C) = a + b[\log(Q)] + c[\log(Q)]^2 \qquad (6.11)$$

where $Q = f(Q_1, Q_2)$. The function $f(\cdot)$ represents some method for aggregating Q_1 and Q_2 into a single measure of output.[29]

If Q_1 and Q_2 are almost identical products produced with roughly the same technology, we can legitimately create an aggregate measure of output by somehow "adding" Q_1 and Q_2 together. We would not hesitate to add red Rabbits and blue Rabbits together. In many cases, however, there is no obvious reason for believing that firms produce homogenous products with a common technology. If we are unsure whether widgets and gizmos are similar products produced by the same technology, we can estimate equation (6.10) and perform a statistical test of whether Q_1 and Q_2 can be replaced by some aggregate measure $f(Q_1, Q_2)$.[30]

Estimation Problems

In practice, we are likely to encounter a number of problems in estimating a cost function. In order to illustrate these problems, let us consider the single-product widget firm. We now assume, as is in fact the case, that factor prices and technology change over time. We want to estimate the following translog cost function

$$\log(C) = a_0 + a_1[\log(Q)] + a_2[\log(Q)]^2 + a_3[\log(r)] \\ + a_4[\log(r)]^2 + a_5[\log(w)] + a_6[\log(w)]^2 + a_7[\log(w)][\log(r)] \\ + a_8[\log(r)][\log(Q)] + a_9[\log(w)][\log(Q)] + a_{10}[\log(T)] \qquad (6.12)$$

where T is some measure of technological change.[31] We have 30 yearly observations on C, Q, r, w, and T from which we want to estimate 11 parameters $(a_0, a_1, \ldots, a_{10})$.

We are trying to obtain a great deal of information—estimates of 11 parameters—from rather little data—30 yearly observations. Statisticians sometimes measure the informational demands placed on their data by the difference between the number of observations and the number of parameters. This difference is called the *degrees of freedom* available from the data. If the degrees of freedom were less than zero—that is, if the number of parameters to be estimated exceeded the number of observations available—it would be impossible to estimate the parameters from the available data. The parameters can be estimated more precisely—that is, estimated with a smaller degree of error—as the number of degrees of freedom increases. When the data suffer from a high degree of *multicollinearity,* a term we shall explain below, the data must provide many more degrees of freedom in order for statistical methods to provide reliable results. Our widget data provides us with $19 (30 - 11 = 19)$ degrees of freedom. Because these data are yearly observations for a single company, they are probably highly time trended and thus afflicted by a high degree of multicollinearity. Consequently, as a practical matter, we probably cannot obtain reliable estimates of the cost function parameters from the data available to us.

Before discussing solutions to this difficulty, it is useful to describe the multicollinearity problem. Let us consider regression analysis in more detail. Suppose we wish to estimate the quantitative relationship between a dependent variable y and two independent variables x_1 and x_2. There are two extreme cases that could characterize the relationship between x_1 and x_2. The first case is where x_1 and x_2 are uncorrelated with one another; the value of x_1 provides no information whatsoever about the value of x_2. The second case is where x_1 and x_2 are perfectly correlated; the value of x_1 tells us exactly what the value of x_2 is. The second case is known as perfect multicollinearity between x_1 and x_2. As the correlation between x_1 and x_2 increases, and as we go from a situation of no multicollinearity to a situation of perfect multicollinearity, our ability to disentangle the separate effects of x_1 and x_2 decreases and our estimates of their separate effects become less precise. Multicollinearity is a characteristic of the data. The data are simply not as informative as we would like them to be. The only way to obtain more precise estimates in the face of data afflicted by extreme multicollinearity is to obtain more data or less collinear data. That is, the only way to alleviate the multicollinearity problem is to obtain more information.[32]

We have 30 observations on the widget firm. There is no way to obtain additional observations. The data exhibit a high degree of multicollinearity. Given this state of affairs, we cannot reliably estimate the cost function. The economic theory of production, however, provides us with some additional pieces of information. First, the firm's demand functions for capital and labor depend on the same parameters as the firm's cost function. Second, the cost function is *homogeneous of degree one* which means that, for a given level of output, doubling input prices must double costs. Homogeneity follows from the fact that the firm has no incentive to substitute one factor for another factor when all factor prices double. Third, the cost function exhibits a mathematical property called *symmetry*.[33] These implications of producer theory provide a great deal of information which, in principle, we can add to the statistical analysis.[34] Indeed, by incorporating this information, the degrees of freedom available for estimating equation (6.12) increase from 19 to 53. We can incorporate this information by forcing our estimated parameters to be consistent with economic theory even though the parameters that would be estimated from the data alone may not be consistent with economic theory.

Generally, econometricians are rather hesitant to force their estimates to be consistent with economic theory. They prefer to let the data speak for themselves. If the data are consistent with economic theory, then the data provide confirmation that this theory is correct. If the data are not consistent with the theory, then there are two possibilities, which require further analysis. First, the theory could be flawed. Second, the theory may be correct but the analysis itself may be flawed. Perhaps the econometrician has used an incorrect functional specification, has made a computational error, or has ignored some basic detail. Consequently, before imposing economic assumptions upon the data, it is important to check first that the data are consistent with these economic assumptions. There are standard statistical techniques for doing so. Unfortunately, most of the studies relied upon by Rosse fail to test these crucial assumptions.

A Critique of AT&T's Case That the Telephone Industry Is a Natural Monopoly

This section critiques the case made by AT&T, primarily through testimony submitted by James Rosse, that the telephone industry is a natural monopoly. Rosse reviewed numerous studies concerning the technology used by the Bell System and Bell Canada. These studies included engineering studies that looked at the cost of constructing telecommunications networks; productivity studies that related changes in the level of output to changes in the levels of inputs; and econometric studies that analyzed the economically relevant aspects of the technology used by the firm. We believe engineering and productivity studies have little bearing on the question of natural monopoly. Before turning to the econometric evidence, we briefly discuss our grounds for dismissing the engineering and productivity evidence.

There are two basic problems with engineering studies. First, these studies ignore the costs of managing the firm and thereby ignore the possibility that managerial diseconomies may outweigh engineering economies. Many technologies exhibit engineering scale economies. Yet few industries are considered natural monopolies. Bureaucratic inefficiencies generally swamp engineering economies from large-scale operation. The furniture business, an example used by Rosse to illustrate scale economies, demonstrates this point. Rosse argued that a bigger furniture dealer could buy a single semi or large truck instead of a number of separate vans and thereby economize on the number of drivers required and reduce the unit costs of delivering furniture. But few economists would argue that the furniture business is a natural monopoly. Managerial costs and other costs of running a furniture business presumably outweigh engineering economies in the furniture distribution system. Second, these studies are often based on theories concerning the techniques for physically manipulating inputs rather than on hard data concerning how inputs are actually physically manipulated. Ronald Skoog, an engineer testifying for AT&T, claimed that a single firm can realize economies in planning, constructing, and operating the telephone network by using techniques that minimize costs over the whole network. Aaron Kerschenbaum, an engineer testifying for the Government, claimed that it is impractical to treat the network as a whole and that, in fact, the Bell System decentralizes the planning, construction, and operation of the network. Thus, the economies which Skoog argues are available in theory may not be available in practice. This discrepancy between theory and practice is endemic in engineering studies. Although engineering studies may be useful to businessmen choosing between alternative technologies, they are of little use for determining whether an industry is a natural monopoly.

Most productivity studies, including the major ones relied upon by Rosse, do not separate out technological change, which several firms may be able to realize as easily as a single firm, from scale economies which only a single firm may be able to realize. High productivity growth, therefore, may result from either scale economies or technological change or both. We are not aware of any studies that determine the correlation between high productivity growth and scale economies across industries. Microelectronics manufacturing has a high rate of productivity growth but has numerous competing producers with mimimal scale economies. Automobile manufacturing has an average rate of productivity growth but a relatively high degree of scale economies.[35] Therefore, productivity studies shed little light on whether an industry is a natural monopoly or on whether a particular product line has scale economies.

Rosse reviewed several econometric studies concerning the cost characteristics of the Bell System and Bell Canada. He noted that most econometric studies have some limitations. He then concluded[36]

> For this reason, an analyst, when seeking to validate a conclusion, looks for evidence that persists across a variety of specifications and statistical techniques. The econometric studies that I have reviewed employ a diversity of production specifications and use both standard and novel statistical methods. Nevertheless, the finding that economies of scale exist within the Bell System persists across all reasonable variations. Furthermore, the econometric analyses of Bell Canada reach similar and consistent results. Accordingly, the econometric studies strongly confirm the conclusion of the engineering studies that substantial economies of scale exist both within the Bell System as a whole as well as in individual components such as interexchange service.

It is useful to divide the studies reviewed by Rosse into two groups: (1) the Vinod studies which estimated production functions with Bell System data; (2)

the cost function studies which estimated cost functions with either Bell Canada data or Bell System data. We begin by discussing the Vinod studies.[37]

Vinod Studies

H. Vinod has prepared two major studies of scale economies in the Bell System. Both studies were submitted as Bell exhibits in FCC Docket 20003 as well as in *US* v. *AT&T.*

"Application of New Ridge Regression Methods to a Study of Bell System Scale Economies," *Journal of the American Statistical Association,* December 1976, pp. 835–841. An earlier version with the same title appeared as Bell Exhibit No. 42 in FCC Docket 20003.

"Bell System Scale Economies and the Economics of Joint Production," which appeared as Bell Exhibit No. 59 in FCC Docket 20003. The statistical methods upon which this paper is based appear in his "Canonical Ridge and the Econometrics of Joint Production," *Journal of Econometrics,* 4 (1976), pp. 147–166.

These studies suffer from numerous problems which make them unreliable sources of evidence on Bell System scale economies.[38] Their most serious problem, upon which we concentrate exclusively, is that they use a statistical technique called *ridge regression.* Ridge regression makes arbitrary and unjustified adjustments to the data and produces estimated values of underlying economic parameters which, on average, do not equal the true values of these parameters. Few econometricians or statisticians recommend ridge regression or publish studies that use it.

Let us describe the relationship between standard regression and ridge regression. Standard regression estimates are the most precise, unbiased estimates obtainable from a given set of data. But, if we are willing to consider estimates that are potentially biased, regression estimates are not necessarily the most precise, potentially biased estimates obtainable from a given set of data.[39] We can, in theory, trade bias off for precision. The rationale for doing so is that an estimator that is unbiased but highly imprecise will frequently lead to estimates that are widely off the mark. Some statisticians have suggested balancing precision and bias by calculating an estimator that minimizes the sum of squared bias and variance.[40] Such estimates are called minimum mean square error (MSE) estimates. Although they are biased, in theory MSE estimates are more precise than standard regression estimates.

Hoerl and Kennard have suggested a particular MSE estimator as a solution to the multicollinearity problem discussed earlier.[41] They show that, by making certain systematic perturbations to the data, it is possible to obtain an estimate with smaller mean square error than the standard regression estimate.[42] The perturbations necessary for this result depend on the value of an unknown parameter. The literature on ridge regression since Hoerl and Kennard's 1970 paper has concentrated on devising largely *ad hoc* methods for deciding exactly how much to perturb the data.[43]

Ridge regression has come under sharp criticism from both econometricians and statisticians. We would like to highlight some of these criticisms. The first point to note about ridge regression is that it is a special case of minimum mean square error estimation. Kmenta has noted that such estimation, whatever its theoretical appeal, is useless from a practical standpoint.[44]

> Unfortunately, in practice the formula for an estimator that would give the minimum value of mean square error very frequently includes the true value of the parameters to be estimated. This obviously makes the formula quite useless; it would be like a recipe for a cake that starts with "take a cake. . . . " Our definition of an estimator specifically excluded such formulas as not being worthy of the name "estimator".

The ridge regression estimator contains an unknown factor k. Most econometricians view k as a fudge factor. By choosing a suitable k, it is possible to obtain any estimate desired. Refer to Table 6.2 which reproduces Vinod's Table 3 from Bell Exhibit 42. Column 1 gives values of m, which is simply a modified version of k. Column 6 gives values of Vinod's estimate of the scale elasticity. The estimated value of the scale elasticity changes as factor m changes. With a suitable choice of m, one could conclude that Bell has diseconomies of scale (an elasticity

Table 6.2 Sensitivity of Scale Elasticity Estimate to the k Factor[a]

1	2	3	4	5	6	7
					Scale elasticity	
m	RSS	R^2	$\hat{a}_1^*$	$\hat{a}_2^*$	$\hat{a}_1^* + \hat{a}_2^*$	$\hat{a}_3^*$
0.0000	0.1316	0.9976	0.7215	0.3194	1.0409	0.0033
0.2500	0.0140	0.9975	0.6403	0.4201	1.0604	0.0045
0.5000	0.0164	0.9970	0.5617	0.5207	1.0824	0.0057
0.7500	0.0205	0.9963	0.4487	0.6207	1.1094	0.0068
1.0000	0.2714	0.9951	0.4285	0.7174	1.1459	0.0074
1.2500	0.4286	0.9923	0.3849	0.8021	1.1870	0.0074
1.5000	0.0857	0.9923	0.3849	0.8021	1.1870	0.0074
1.7500	0.1948	0.9651	0.3127	0.8723	1.1850	0.0063
2.0000	0.4508	0.9191	0.2682	0.8207	1.0890	0.0054
2.2500	0.9866	0.8230	0.2132	0.6958	0.9090	0.0043
2.5000	1.9511	0.6500	0.1485	0.5056	0.6540	0.0030
2.7500	3.4584	0.3797	0.0766	0.2681	0.3441	0.0015
3.0000	5.5704	0.0000	0.0000	0.0000	0.0000	0.0000

Source: Table 3 from Bell Exhibit 42 in FCC Docket 20003.

[a]RSS denotes the residual sum of squares, R^2 denotes the multiple correlation coefficient.

less than one), constant returns to scale (an elasticity equal to one), or scale economies (an elasticity greater than one).

How does Vinod select his m and thus one of the many possible scale elasticity estimates obtainable from ridge-regression methods? He uses a device called a ridge trace, a technique that has come under repeated criticism. Thisted, a supporter of ridge regression in some limited circumstances, has noted that " . . . the ridge trace carries little information of use in selecting k."[45] The authors of a recently published econometrics textbook have commented that " . . . the use of the ridge trace as a way of obtaining improved estimators may be rejected. It can illustrate the sensitivity of coefficients to small data perturbations, but that is all."[46] Vinod selects a scale elasticity estimate that is least likely to change as a result of small, arbitrary changes in the data. There is no compelling reason for this choice, which results in a biased estimate with unknown precision. Vinod follows this same procedure in Bell Exhibit 59.

Most econometricians, Vinod included, recognize that ridge regression introduces so-called prior information. Prior information is information known to the analyst but not contained in the data. In his comments on Smith and Campbell's critique of ridge regression, Vinod stated, "Certain prior knowledge about intrinsic measurement errors can be cleverly incorporated in ridge regression by adding p fictitious observations to the data."[47] There is nothing wrong with bringing prior information to bear on an estimation problem. Bayesian methods for doing so are fairly well regarded by most econometricians. But, econometricians who use Bayesian methods state their prior information explicitly. They might say, for example, that the elasticity of scale is 1.00 plus or minus .25 and then use sample information—data—to revise this prior belief. Vinod, and most other people who have used ridge regression, never state what their prior information is and never write as if they held any particular prior beliefs. It can be shown, however, that ridge regression, as usually practiced, is equivalent to Bayesian estimation where the analyst believes *a priori* that every parameter but the constant term equals zero. In estimating a production function, as Vinod does in Bell Exhibit 42, this prior information means that ridge regression assumes that output is constant regardless of the level of inputs employed. Generally, such prior beliefs are absurd.[48]

As Smith and Campbell remarked, "[Ridge regression] is often motivated by *a priori* information that it does not accurately describe and can degenerate into *ad hoc* manipulations."[49] Van Nostrand commented, "Like [Smith and Campbell] I am dubious about the mechanical use of ridge regression when there is no source of external information."[50] In a similar vein, Thisted has said, " . . . ridge estimators contain auxiliary information the appropriateness of which is seldom assessed."[51] Finally, Smith and Campbell respond to Vinod's reference to ridge as a method for "cleverly incorporating prior information about intrinsic measurement error"[52]

> There is also considerable irony in Vinod's argument that ridge regression cleverly incorporates prior knowledge about intrinsic measurement error. As

noted in the oral presentation, ridge regression is analytically equivalent to the old tongue-in-cheek remedy for multicollinearity, in which a zero-mean random number generator is used to add measurement error without changing their covariances. The one-liner for this joke is "Use less precise data to get more precise estimates."

The second point to note about ridge regression is that it is rarely used. Vinod and other advocates of ridge regression claim this technique is valuable for alleviating multicollinearity problems. He says " . . . ridge regression offers new hope for avoiding most serious ill effects of multicollinearity on ordinary least squares regression coefficients."[53] He uses ridge regression to "solve" multicollinearity problems in both of the Bell Exhibits he prepared. Multicollinearity is, indeed, a common problem with economic studies. Many economic series move in tandem over time. Consequently, it is difficult to disentangle the relationship between several independent variables and a dependent variable. The data, too often, are uninformative. Because multicollinearity is such a common problem in economic studies and because ridge regression methods have been available for almost 13 years now, it is surprising to discover that there are virutally no published economic studies in major journals that use ridge regression to solve multicollinearity problems and thereby provide "superior" estimates of underlying economic parameters. Vinod's 1978 survey paper includes what appears to be a fairly comprehensive bibliography of ridge-regression papers.[54] Most of these papers were not published in economic journals. Most look at the theoretical properties of the ridge estimator but do not actually apply ridge regression to economic data. Because of the controversy surrounding ridge regression, few articles that use this technique to estimate economic parameters have been published in top economic journals. In his oral testimony in *US* v. *AT&T*, Vinod said "[Ridge regression] is a relatively new technique and even people like Kmenta, the textbook people, change their mind over the years. It takes some educating."[55] Given the dearth of econometric studies that have used ridge regression, the *ad hoc* data manipulation upon which ridge regression relies, and the fact that ridge regression has been available for almost 13 years, we doubt that many economists will change their opinion about ridge regression in the coming years.

The third point to note is that econometricians cannot test statistical hypotheses with ridge-regression procedures. In standard regression analysis, the researcher reports not only his estimate of the parameter in question but also the standard error of the estimate. With both the estimate and the standard error, it is possible to test whether he has obtained so precise an estimate that other alternative values of the parameter in question are unlikely. Let us give an example. Suppose we run a regression and obtain an estimated scale elasticity of 1.5. If we are advancing the proposition that there are scale economies, we should test this hypothesis against some alternative hypothesis. The obvious alternative hypothesis in this example is that there are constant returns to scale, that is, the scale elasticity equals one. Basic statistics tells us how to do this. If $1.5 - 1.0$ is less than roughly twice the standard error of the estimated scale elasticity, then we cannot reject the hypothesis that the true scale elasticity is 1.0 and that we have obtained an estimate of 1.5 simply as a matter of chance. In other words, our data do not provide an estimate that is precise enough to reject the hypothesis of constant returns to scale.[56] Generally, the researcher who advances a hypothesis has the burden of proof for showing that an alternative hypothesis is not also consistent with the data.[57] In the example above, we failed to satisfy our burden of proof for our hypothesis that there are scale economies because the alternative hypothesis, that there are constant returns to scale, is also consistent with the data. Because neither Vinod nor anyone else to our knowledge has developed a valid statistical formula for the standard error of a ridge-regression estimate, Vinod was unable to report standard errors or report tests of statistical significance in the ridge regression studies he submitted in FCC Docket 20003 and in *US* v. *AT&T*.

The Vinod studies fall far short of providing reliable evidence concerning the cost and production characteristics of the telephone industry. These studies rely on an arbitrary, and highly controversial, statistical technique. It is surprising that AT&T has relied so heavily on these studies, both in FCC Docket 20003 and in *US* v. *AT&T* to support its arguments that the telephone industry is a natural monopoly.

Cost Function Studies

Several economists have used data on either Bell Canada or the Bell System to estimate cost functions using some of the methods discussed in the second section of this chapter. We begin by discussing the two major Bell Canada studies. Both studies assumed that Bell Canada provides three services—local, toll, and private line, with three inputs—labor, capital, and materials. Both studies used Bell Canada data for the period 1952–1976 to estimate multiproduct cost functions. Smith and Corbo estimated a multiproduct translog cost function and found that there were aggregate scale economies.[58,59] Their study did not address the question of whether Bell Canada is a natural monopoly. Fuss and Waverman estimated a cost function similar to the translog cost function but with the quantity variables entered as $Q_i^* = (Q_i^\lambda - 1)/\lambda$ rather than as $\log(Q)$.[60] Because $Q_i^* = \log(Q)$ when $\lambda = 0$, their cost function contains the translog cost function as a special case. They found that λ was significantly different from zero and were therefore able to reject the cost function used by Smith and Corbo.[61] Moreover, they found that their more general cost function showed that (*a*) Bell Canada did not have aggregate scale economies and (*b*) Bell Canada did not have a natural monopoly over local, toll, and private line services. They concluded[62]

> There is little evidence that Bell Canada is a natural monopoly with respect to all of its principal service offerings. In particular, tests of overall economies of scale and tests of economies of scale and scope with respect to private line services fail to reject the hypothesis that private line services can be provided on a competitive basis without efficiency loss.

Therefore, the most recent and most comprehensive study of Bell Canada rejects the hypothesis that the telecommunications industry is a natural monopoly.[63]

With the exception of our own study, which had not been completed at the time AT&T presented its case that it was a natural monopoly, there was one major study of the Bell System's cost characteristics.[64] This study was prepared by Laurits Christensen who presented testimony on behalf of AT&T. His study used data on the Bell System for the period 1947–1977. He estimated numerous translog cost function specifications.[65] Under every specification he estimated, he found statistically significant scale economies. There are two serious flaws in his study.

First, he assumed that the Bell System produces a single output and aggregated diverse telecommunications outputs in order to measure this single output.[66] Using standard statistical techniques, we tested whether this assumption is correct and found that it is not. Moreover, many of the single-output cost functions he estimated violated basic tenets of producer theory. For example, the specification most consistent with the data showed that the demand curves for capital and labor sloped upward. By relaxing his assumption that there is a single telecommunications output, we found that the estimated cost functions were much more in accord with producer theory.[67] In particular, the demand curves for capital, labor, and materials sloped downward as one would expect.

Second, the fact that his aggregate output measure exhibits scale economies provides little information concerning whether intercity service exhibits scale economies or whether the telephone industry is a natural monopoly. His conclusion that evidence of aggregate scale economies "is consistent with the view that a proliferation of suppliers of telecommunications services would result in a large sacrifice of efficiency" is disingenuous.[68] Evidence of aggregate scale economies is consistent with the opposite view as well.[69]

Conclusions

Having examined the major Bell System econometric studies relied upon by Rosse, we conclude that (*a*) the Vinod studies rely on an inappropriate statistical technique; (*b*) the Christensen study uses an invalid and irrelevant aggregate measure of telecommunications outputs; and (*c*) none of these studies provide credible evidence concerning whether a single firm can provide any or all telecommunications services more efficiently than several firms. Unlike the Bell System studies, the Bell Canada studies reviewed by Rosse used generally accepted statistical techniques and estimated multiproduct cost functions. Multiproduct cost functions, unlike the single-product functions estimated by Christensen, provide information concerning whether particular lines of service have scale economies and whether a single firm realizes efficiencies from jointly operating these services. The most comprehensive of the Bell Canada studies—the Fuss and Waverman study—rejects the hypothesis that there are scale economies in private line service and that there are economies of scope between private line service and other telecommunications services. The most reasonable conclusion one can make from the studies reviewed by Rosse is that, to the extent these studies shed any light at all on the issue of whether a single firm can operate any or all telecommunications services more efficiently than several firms can, they suggest that the provision of telephone service by a single firm may not be the most efficient arrangement available to society.[70] This conclusion is supported by the findings of our own study of the Bell System, to which we now turn.

Multiproduct Cost Function Estimates for the Bell System

Using the same data used by Christensen, we assumed the Bell System provides local and intercity service and we estimated numerous two-product specifications of the Bell System cost function.[71] For most of these specifications, we found that the implied demand curves for capital, labor, and materials were downward sloping. Therefore, these specifications were generally consistent with producer theory.

We then used these specifications to examine the cost characteristics of the Bell System. We found that product-specific scale economy estimates and economies of scope estimates varied considerably between specifications. These estimates were particularly sensitive to the proxy we used for technological change and the manner in which we allowed technological change to affect the cost function. Given that the variables used in the cost function are highly time trended and that local and intercity calls have increased in tandem over the period for which we have data, we were not surprised by these results. Because local and intercity calls are highly collinear, statistical methods cannot disentangle the impact of local calls on cost from the impact of intercity calls on cost.

Although multicollinearity may make particular parameter estimates imprecise, multicollinearity does not necessarily make mathematical combinations of these parameters imprecise. Let us give an example. Suppose we estimate the following production function

$$\log(Q) = a_0 + a_1\log(L) + a_2\log(K) \qquad (6.13)$$

where Q is the quantity of output, L is the quantity of labor, and K is the quantity of capital. The scale elasticity is simply

$$E = a_1 + a_2 \qquad (6.14)$$

Suppose that $\log(L) = \log(K)$. Because these variables are perfectly collinear, we cannot estimate either a_1 or a_2 precisely. But we can estimate $a_1 + a_2$ precisely. Given $\log(L) = \log(K)$, we can write equation (6.13) as

$$\log(Q) = a_0 + a_1\log(L) + a_2\log(K)$$

$$= a_0 + (a_1 + a_2)\log(L) \qquad (6.15)$$

$$= a_0 + E[\log(L)]$$

By estimating equation (6.15), we can generally obtain a precise estimate of the scale elasticity E even though we cannot obtain a precise estimate of a_1 or a_2 separately.

Fortunately, the test of whether the Bell System can supply all intercity and local telephone services most efficiently depends on a linear combination of many parameters of the cost function. Suppose the Bell System provides Q_L local calls and Q_T toll calls. Its costs are $C = C(Q_L, Q_T)$. Suppose we divide this output between two firms. Firm A provides aQ_L local calls and bQ_T toll calls for a cost of $C_A = C(aQ_L, bQ_T)$. Firm B provides $(1 - a)Q_L$ local calls and $(1 - b)Q_T$ toll calls for a cost of $C_B = C[(1 - a)Q_L, (1 - b)Q_T]$. The numbers a and b are between zero and one. By changing a and b, we change the allocation of output between firm A and firm B while holding the total output of both services constant. The Bell System has a subadditive cost function at the output level Q_L and Q_T if $C_A + C_B$ exceeds C.

For every two-product specification we estimated, we tested and rejected the hypothesis that the Bell System had a natural monopoly at the output levels produced between 1958 and 1977. Dividing output between two firms never resulted in a statistically significant increase in costs and often resulted in a decrease in costs. On the basis of these tests, we have concluded that the Bell System did not have a natural monopoly over telephone services for the levels of demand experienced between 1958 and 1977. Several telecommunications firms could have met demand for local and intercity services more cheaply than the Bell System did over this period. Our conclusion is consistent with the Fuss and Waverman conclusion discussed in the previous section.

Nevertheless, our study shares two flaws with other studies of the cost characteristics of the telephone industry. First, it was necessary to impose the homogeneity and symmetry constraints implied by producer theory in order to increase the degrees of freedom available for estimating the cost function. These constraints were rejected resoundingly by the data.[72] Second, the data available for estimating cost functions for the telephone industry are rather poor for determining whether this industry is a natural monopoly.[73] Because all of the Bell Canada and Bell System studies are afflicted with these problems, none of these studies can provide decisive evidence that the telephone industry is or is not a natural monopoly.[74]

The most reasonable conclusion that can be made from the available evidence on the cost characteristics of the telephone industry is that there is weak evidence that this industry is not a natural monopoly. The most comprehensive analysis of Bell Canada rejects the hypothesis that the telephone industry is a natural monopoly. The most comprehenisve analysis of the Bell System also rejects the hypothesis that the telephone industry is a natural monopoly. In the FCC hearings concerning the desirability of competition in the telephone industry and in the *US* v. *AT&T* case, AT&T bore the burden of proof for establishing that the telephone industry is a natural monopoly. Although more sophisticated studies with better data may reverse our finding that the telephone industry is not a natural monopoly or may find that particular telephone services can be provided most efficiently by a single firm, it is clear that the evidence presented by AT&T fell far short of meeting AT&T's burden of proof.

NOTES

1. Two engineers, George Mandanis and Ronald Skoog, presented evidence that telecommunications technologies exhibit natural monopoly characteristics. Two economists, H. Vinod and Laurits Christensen presented econometric evidence that the telecommunications industry has scale economies and, by implication, is a natural monopoly. Christensen also presented evidence that the telephone industry has experienced a faster rate of productivity growth than most other industries. James Rosse, an economist, reviewed available studies concerning whether the telephone industry has economies of scale and scope and testified that the overwhelming body of evidence supports the conclusion that the telephone industry is a natural monopoly. Because Rosse presented the most detailed and comprehensive treatment of the natural monopoly question, we focus on his testimony in this chapter.

2. AT&T, *Defendants' Third Statement of Contentions and Proof*, in *US* v. *AT&T*, p. 35.

3. James Rosse, written testimony in *US* v. *AT&T*, p. 37.

4. The production function also conveniently characterizes the nature of the technology used by an industry. It expresses the relationship between the quantity of inputs used and the quantity of outputs obtained. Until the early 1970s, most economists used the production function to describe the technology used by the firm. Since then, economists have increasingly used the cost function to describe the technology used by the firm. The cost function contains the same economic data as the production function. In theory, these functions are merely different methods for summarizing the same information.

5. Industry structure refers to the number of firms that operate in an industry and the allocation of production across these firms.

6. We focus on whether an industry consisting of a single firm can meet market demand more efficiently than an industry consisting of several firms. We assume that, whatever the industry structure, society prevents firms from restricting output below the socially desirable level and from thereby earning monopoly profits.

7. We have compared single-firm production to two-firm production for simplicity. More generally, the cost function is subadditive at output Q if the cost of having a single firm produce Q is less than the cost of having two or more firms produce Q for all possible allocations of output Q across these firms. Formally, $C(Q)$ is subadditive if $C(Q) < \sum_{i=1}^{n} C(Q_i)$, where Q_i is the output of the ith firm, $\sum_{i=1}^{n} Q_i = Q$, and $Q_i > 0$ for at least two firms. Conversely, $C(Q)$ is said to be superadditive.

8. In the single-product case, the intersection of the demand curve and the average cost curve gives the so-called Ramsey optimum where social surplus is maximized subject to the constraint that the public utility exactly breaks even.

9. The degree of scale economies is often measured by the scale elasticity which equals the inverse of the percentage change in cost due to a percentage change in output. There are scale economies (diseconomies) when the scale elasticity is greater than (less than) one. Some of the older industrial organization literature defined the scale elasticity as the percentage change in output resulting from a common percentage change in all inputs. In general, these measures are not equal because, in order to produce a given percentage change in output, it might be inefficient to change inputs by a common percentage.

10. Firms may, in fact, choose technologies that have extreme scale economies in order to deter entry. See Avinash Dixit, "Recent Developments in Oligopoly Theory," *American Economic Review*, May 1982, pp. 12–17 and the references cited therein for a discussion of this possibility.

11. For further discussion of this question, see Chapter 5 of this volume.

12. On the other hand, if the industry is really a natural monopoly, regulators may face insurmountable obstacles in trying to maintain uneconomic and basically self-destructive competition.

13. William Baumol, John Panzar, and Robert Willig, *Contestable Markets and the Theory of Industry Structure* (San Diego: Harcourt, Brace, Jovanovich, 1982), p. 3.

14. No new concepts arise with more than two products.

15. There are synergies, or cost complementarities, if the production of one product facilitates the production of another product. Firm *A* might have a natural monopoly over product one and firm *B* would have a natural monopoly over product two in the absence of synergies.

16. For example, there may be engineering economies in producing two products but managerial diseconomies in coordinating these two products might swamp engineering economies at some relatively low level of production.

17. Baumol, Panzar, and Willig, *op. cit.*

18. Analysts treat multiproduct industries as single-product industries by forming an aggregate measure of the diverse outputs produced by the industry. Usually, this aggregate measure is formed by adding the revenues generated by each product and then deflating this sum by a suitable price index. The resulting quantity is a price-weighted index of output. See Note 29 for further details.

19. We assume that there are no synergies in apple and orange production.

20. Constant returns to scale means that there are neither scale economies nor scale diseconomies so that the scale elasticity equals one.

21. Because there is only one input, this measure of scale elasticity is appropriate here and is equal to the measure obtained by taking the inverse of the percentage change in cost due to a percentage change in output. See Note 9.

22. Moreover, as we discussed earlier, scale economies in both outputs does not imply that one firm is the most efficient producer of both outputs or that two firms, each specializing in the production of one of the outputs, can produce more cheaply than several firms each producing both outputs.

23. If there were two or more right-hand variables we would be unable to plot the cost function in a two-dimensional graph and make a visual determination.

24. The general translog cost function includes the squares and cross products of the independent variables (in logarithmic form) on the right-hand side.

25. Their argument is based on a mathematical theorem concerning the approximation of functions. Suppose we have a function $F(x)$. Over a small portion of the curve described by $F(x)$, $F(x)$ can be written as the sum of several linear terms plus an approximation error. The translog cost function, which is the sum of several linear terms, is a simple application of this theorem. Recently, the translog cost function has fallen into some disrepute for several reasons. First, it provides a local but not a global approximation to the true cost function. Second, it may provide an approximation over a portion of the true cost curve but may not provide a particularly good approximation. Third, there is some limited evidence that it tends to overstate scale economies. For further discussion, see Chapter 10 of this volume.

26. This particular expression for $g(x)$ is called the Box–Cox transformation. See G. E. P. Box and D. R. Cox, "An Analysis of Transformations," *Journal of the Royal Statistical Society*, Series B, 26, 1964, pp. 211–243.

27. The parameters appear linearly in equations (6.5), (6.6), and (6.7). Multiple regression is the standard statistical technique for estimating parameters in these circumstances. The parameters appear nonlinearly in equation (6.9). Maximum likelihood methods can estimate parameters under these circumstances. Multiple regression analysis can be viewed as a special case of maximum likelihood analysis.

28. We continue to ignore factor price changes and technological change.

29. For example, $Q = Q_1 + Q_2$, $Q = p_1 Q_1 + p_2 Q_2$ where p_1 and p_2 are the real prices of outputs one and two, respectively, or $\log(Q) = w_1 \log(Q_1) + w_2 \log(Q_2)$, where w_1 and w_2 are the revenue shares for outputs one and two, respectively.

30. The statistical test involves determining whether certain relationships hold among the estimated parameters of the multiproduct cost function. For further information, see Michael Denny and Cheryl Pinto, "An Aggregate Model with Multiproduct Technology," in M. Fuss and D. McFadden,

eds., *Production Economics: A Dual Approach to Theory and Applications* (Amsterdam: North Holland, 1978), pp. 255–258.

31. Technological progress implies that the cost of producing output at given factor prices decreases over time, presumably because the factors of production have improved in quality or because the techniques for combining these factors of production have improved. Unfortunately, it is extremely difficult to measure technological change. In cost studies, econometricians usually include a proxy for technological change as an independent variable. There are various ways for including the technological change variables, which we shall not discuss here. The manner in which we have entered the technological change variable in equation (6.12) is the simplest way and makes the least demands on the degrees of freedom. This method assumes that technological change has resulted in a proportional shift of the cost curve over time.

32. Some statisticians have proposed a method called *ridge regression* to alleviate the multicollinearity problem. This method can be shown to introduce pseudo-information into the data. Because this information is usually nonsensical and because ridge regression cannot be used to test statistical hypotheses—a main objective of most econometric studies—few econometricians use this method. We return to this subject below because several of the studies of Bell System scale economies relied upon by economists testifying for AT&T in *US* v. *AT&T* used ridge regression quite extensively.

33. Symmetry means that the second-order derivatives of cost with respect to the independent variables are invariant with respect to the order of differentiation.

34. We have eleven parameters. These parameters appear in three equations—the cost function, the capital demand function, and the labor demand function. The second and third implications enable us to write these eleven parameters as combinations of only seven parameters. We can then use standard statistical techniques for estimating a system of equations.

35. Christensen testified that the Bell System has a higher rate of productivity growth than independent telephone companies and the US economy as a whole. Even if high productivity growth necessarily implied large scale economies, high productivity growth would not imply the existence of a natural monopoly over a group of services or the existence of scale economies for particular services for the reasons discussed earlier.

36. James Rosse, *op. cit.*, p. 75.

37. For other reviews of the econometric evidence on scale economies in the telecommunications industry see Melvyn Fuss, "A Survey of Recent Results in the Analysis of Production Conditions in Telecommunications" (Toronto: Institute for Policy Analysis, University of Toronto, June 1981); Melvyn Fuss and Leonard Waverman, *The Regulation of Telecommunications in Canada,* Technical Report No. 7, Economic Council of Canada, March 1981, pp. 11–27; and John Meyer *et al., The Economics of Competition in the Telecommunications Industry* (Cambridge, MA: Oelgeschlager, Gunn, and Hain, 1981), pp. 111–153 and 317–327. A large number of studies have been performed concerning the cost and production characteristics of the telephone industry. Many of these studies are seriously flawed because they rely on needlessly restrictive functional forms or because they ignore the impact of technological change on cost. Some of these studies supplant earlier studies by the same authors. Our review concentrates on the most recent and comprehensive studies.

38. These studies supplant a preliminary study by Vinod which does not use ridge regression but fails to disentangle the impact of technological change from the impact of scale economies. See H. D. Vinod, "Nonhomogenous Production Functions and Applications to Telecommunications," *Bell Journal of Economics and Management Science,* Autumn 1972, pp. 531–543. Recently, Vinod published another study that uses ridge regression. See Baldev Raj and H. D. Vinod, "Bell System Scale Economies from a Randomly Varying Parameter Model," *Journal of Economics and Business,* February 1982, pp. 247–252. The first Vinod study cited in the text determines whether there are scale economies for an aggregate measure of output. As discussed earlier, the presence of aggregate scale economies is consistent with there being scale diseconomies in any or all services. Therefore, this study provides little relevant information. The second study cited in the text examines whether there are scale economies in local and intercity services separately. But this study relies on a statistical technique that has even less to recommend it than ridge regression. See Note 44 for further details.

39. In statistical parlance, an estimator Z of some parameter z is unbiased if the average value of this estimator equals the true value of the parameter for repeated applications of this estimator

to different sets of data and biased otherwise. In principle, a biased estimator need not be based on subjective judgement although in practice these estimators often depend upon subjective judgements made by the analyst. Regression estimates are the most precise unbiased estimates of estimates that are linear functions of the data.

40. High variance means low precision. Formally, this estimate minimizes the distance between the true but unknown parameters and the estimated parameters and therefore makes the estimated parameters closer to the true parameters on average.

41. A. E. Hoerl and R. W. Kennard, "Ridge Regression: Biased Estimation for Nonorthogonal Problems," *Technometrics,* February 1970, pp. 55–67.

42. The regression estimator of the vector of parameters of the standard regression equation is $B = (X'X)^{-1}X'y$ where X is the matrix of observations for the independent variables and y is the vector of observations for the dependent variable, and B is the estimate of the true vector of parameters b. Ridge regression uses $B_R = (X'X + kI)^{-1}X'y$ where I is the identity matrix of order corresponding to $X'X$ and k is the "ridge fudge factor." The matrix kI perturbs the data matrix $X'X$.

43. Given the optimal k, the ridge regression estimator has a lower mean square error than the standard regression estimator. But k is unknown. The fact that the researcher has to guess about the value of k means that the ridge regression estimator does not necessarily have a lower mean square error than the standard regression estimator. Therefore, there is no theoretical ground for choosing ridge regression over standard regression even if minimum mean square error is a sensible criterion for selecting an estimator. Moreover, the method Vinod used in Bell Exhibit 59, Docket 20003 to estimate a joint production function for the Bell System is completely *ad hoc*. According to Vinod, in reference to the canonical ridge estimator he used in Exhibit 59, "[I have not] succeeded in proving an extension of the 'existence theorem.' This paper offers the empirical possibility that our approach will work. . . . Further, our data processing is not claimed to be optimal." See H. D. Vinod, "Canonical Ridge and Econometrics of Joint Production," *Journal of Econometrics,* May 1976, p. 149. This oblique statement means that Vinod has not proved that canonical ridge estimators minimize mean square error even when the analyst knows the true value of k. In short, even in theory, canonical ridge has no desirable statistical property.

44. J. Kmenta, *Elements of Econometrics* (New York: MacMillan, 1971), p. 158. In his oral testimony, Vinod claimed that Kmenta has written a paper that "strongly recommends ridge regression." This statement is disingenuous. Kmenta has written a paper that suggests that ridge regression may be helpful in forecasting economic time series but that states quite emphatically that ridge regression should not be used for testing economic hypotheses. He states, " . . . [the ridge regression] procedure is not suited for testing hypotheses. This makes ridge regression uninteresting for many econometric problems. It would seem, though, that [ridge regression] may become a powerful tool in forecasting." See K. Lin and J. Kmenta, "Some New Results on Ridge Regression Estimation" (Ann Arbor, Michigan: Department of Economics, University of Michigan, September 1980), p. 23.

45. R. Thisted, "Comment," on Smith and Campbell's "A Critique of Ridge Regression Methods," *Journal of the American Statistical Society,* March 1980, p. 85.

46. G. Judge, W. Griffiths, R. Hill, and T. Lee, *The Theory and Practice of Econometrics* (New York: Wiley, 1980), p. 475.

47. H. D. Vinod, "Comments," on Smith and Campbell, *op. cit.,* p. 97.

48. Ridge regression forms what is known as a *matrix-weighted average* of a vector containing zeroes and a vector containing the standard regression estimates. Vinod appears to believe that this average means that the ridge regression estimator biases the estimated coefficients towards zero and therefore biases his production function estimates of the scale elasticity towards a smaller number. See Vinod Tr. 18222 in *US* v. *AT&T*. His belief has no mathematical basis. In his analysis of multicollinearity published in 1973, Leamer noted that the individual regression parameters or the sum of the individual regression parameters obtained from a matrix-weighted average do not necessarily fall between zero—the prior estimate—and the standard regression estimate. See Edward Leamer, "Multicollinearity: A Bayesian Interpretation," *Review of Economics and Statistics,* August 1973, pp. 371–380. As Vinod shows in Bell Exhibit 42, using ridge regression increases the scale elasticity from a number approximately equal to one (indicating constant returns to scale) to a number well in excess of one (indicating scale economies).

49. Smith and Campbell, *op. cit.*, p. 81.

50. Van Nostrand, "Comment," on Smith and Campbell, *op. cit.*, p. 92.

51. Thisted, "Comment," on Smith and Campbell, *op. cit.*, p. 86.

52. Campbell and Smith, "Rejoinder," in Smith and Campbell, *op. cit.*, pp. 102–103.

53. H. D. Vinod, "A Survey of Ridge Regression and Related Techniques for Improvements Over Ordinary Least Squares," *Review of Economics and Statistics*, February 1978, p. 121.

54. *Ibid.*

55. Kmenta does not support the use of ridge regression for testing economic hypotheses. See Note 44.

56. There are several equivalent ways of stating this test. We can say that our estimate of 1.5 is not significantly different from 1.0. Generally, this means that there is more than 1 chance out of 20 that we could obtain an estimate of 1.5 or higher even though the true scale elasticity is only 1.0. We can also say that the 95% confidence interval includes parameter values less than 1.0. A confidence interval of .7 to 2.3 means that there is less than 1 chance out of 20 that the true parameter lies outside of this interval.

57. In recent years, standard errors and tests of statistical significance have gained some legal recognition because of several landmark cases involving the discriminatory selection of juries and in employment discrimination cases. In order to establish a *prima facie* case of discrimination, the plaintiff must prove not only that there is an average pay disparity between black and white employees but also that this disparity is statistically significant in the sense that there would be less than 1 chance out of 20 of observing a disparity this large if there were in fact no disparity. See M. Finkelstein, "The Judicial Reception of Multiple Regression Studies in Race and Sex Discrimination Cases," *Columbia Law Review*, May 1980, pp. 737–753, and the references cited therein.

58. J. B. Smith and V. Corbo, "Economies of Scale and Economies of Scope in Bell Canada," Working Paper, Department of Economics, Concordia University, March 1979. Also see J. Breslaw and J. B. Smith, "Efficiency, Equity, and Regulation: An Econometric Model of Bell Canada," Report to the Department of Communications, Interim Report, December 1979, and Final Report, March 1980. See Fuss and Waverman, *op cit.*, for a critique of the Breslaw and Smith study which suffered from some serious data and methodological problems.

59. In a multiproduct cost function study, aggregate scale economies are present if a proportionate change in all outputs leads to a less than proportionate change in costs.

60. The Fuss and Waverman study, *op. cit.*, supplants earlier studies by Denny, Fuss, and Waverman and Fuss and Waverman.

61. See our discussion of the Box–Cox transformation and tests between alternative functional forms in the previous section.

62. Fuss and Waverman, *op. cit.*, p. *viii*.

63. Rosse appears to reject the Fuss and Waverman study for two reasons. First, in his oral testimony he stated that Fuss and Waverman relied "on a very novel method of transformation for representing the production process, the so called Boss–Cox [sic] transformation." (See the previous section for a discussion of the Box–Cox transformation and its usefulness for testing between alternative functional forms.) The Box–Cox transformation is not a novel statistical technique. It was introduced by G. E. P. Box and D. R. Cox, *op. cit.*, in a classic article published in 1964. One of us used this transformation in a 1974 paper to test whether, in a regression of earnings on education, earnings should be measured arithmetically or logarithmically. See J. Heckman and S. Polachek, "Empirical Evidence on the Function Form of the Earnings–School Relationship," *Journal of the American Statistical Association*, June 1974. In a recent paper, Caves, Christensen, and Trethway proposed the same function as was used by Fuss and Waverman. See D. W. Caves, L. R. Christensen, and M. W. Trethway, "Flexible Cost Functions for Multiproduct Firms," *Review of Economics and Statistics*, August 1980, pp. 474–481. The important point to note about this functional form is that it includes several commonly used functional forms as special cases and therefore enables us to test which of these special cases is most consistent with the data. It is unclear from Rosse's testimony whether he believes Fuss and Waverman's study is flawed because they rely on a "novel technique" or whether he believes

this "novel technique" is the source of some flaw. Second, in his written testimony, Rosse rejected Fuss and Waverman's latest study because "their estimates have several problems that lead me to give their conclusions little weight. The first of these problems (upward sloping demand curve for capital) they recognize and attempt to repair. But, their 'corrected estimates' still violate the economic conditions for a cost function; this violation is not recognized in their paper." He does not specify what economic condition for a cost function is violated by their estimated cost function. In this oral testimony, he said he rejected their study because it implied an upward sloping demand curve for capital, a problem that he said they corrected in his written testimony. In any event, most of the Bell System studies he relied upon have equally if not more serious problems. For example, the Christensen specifications that are most consistent with the data have upward sloping demand curves for both capital and labor. Although we believe Rosse's criticisms are far from sufficient to reject the Fuss and Waverman study in favor of other studies which reach contrary conclusions, we do not believe the Fuss and Waverman study is flawless or the last word on the cost characteristics of Bell Canada. In obtaining their estimates, Fuss and Waverman assumed that Bell Canada can set prices at the profit-maximizing level on message toll and private line services and that rate-of-return regulation is not operative in these areas. Smith and Corbo, *op. cit.*, made this same assumption. The assumption increases the amount of information available for estimating scale economies. Kiss, Karabadjian, and LeFebvre dispute this assumption and show that relaxing this assumption reverses Fuss and Waverman's conclusions concerning economies of scale and scope. See F. Kiss, S. Karabadjian, and B. LeFebvre, "Economies of Scale and Scope in Bell Canada," paper presented at the Telecommunications in Canada Conference, March 4–6, 1981. It is possible to perform a statistical test of whether this assumption is correct. Unfortunately, none of these authors did so.

64. Laurits Christensen, Diane Cummings, and Philip Schoech, "Econometric Estimation of Scale Economies in Telecommunications," SSRI Working Paper No. 8124 (Madison WI: Social Systems Research Institute, University of Wisconsin at Madison, September 1981). This study uses data and methods and reaches conclusions similar to M. I. Nadiri and Mark A. Shankerman, "The Structure of Production, Technological Change, and the Rate of Growth of Total Factor Productivity, in the Bell System," in T. Cowing and R. Stevenson, eds., *Productivity Measurement in Regulated Industries* (New York: Academic Press, 1980). Rosse relies on another Christensen study which is not publicly available to our knowledge. "Multiproduct Cost Function Analysis of the Bell System, 1947–1967 [sic]" was cited by Rosse on p. 33 of his written testimony but has not been published to our knowledge. Moreover, Christensen neither referred to nor discussed this study in his own testimony on the cost characteristics of the Bell System.

65. The specifications differed from one another in the number of squared and cross terms entered on the right-hand side of the cost equation and in the procedure used to estimate the equation.

66. Christensen's "output variable for the Bell System was based on its five principal sources of revenue: local, intrastate, and interstate services, directory advertising, and miscellaneous. The revenue categories are deflated by appropriate price indexes and then combined into a translog index of aggregate output." See Christensen, Cummings, and Schoech, *op. cit.*, p. 8.

67. Chapter 10 of this volume reports our study of the Bell System's cost characteristics in more detail.

68. Christensen, Cummings, and Schoech, *op. cit.*, p. 1.

69. Likewise, Nadiri and Shankerman's finding of aggregate scale economies provides little relevant evidence concerning the natural monopoly question, although these authors make no pretense that it does.

70. In his testimony, Rosse argued that the conclusion that the Bell System has scale economies is robust because so many studies using different methods found scale economies. This argument is without merit. Most of the Bell System studies estimated single-product cost or production functions using post-World War II Bell System data. It is not surprising that these studies all find that the Bell System has aggregate scale economies. Moreover, as discussed above, this common conclusion is irrelevant for most of the legal and policy question concerning the telecommunications industry. The important point to note concerning the studies reviewed by Rosse is that the most comprehensive, relevant, and professionally executed of these studies controverts Rosse's conclusion.

71. Christensen has data on local, intrastate, and interstate service revenues, directory advertising revenues, and miscellaneous service revenues. We created two measures of output by (1) combining interstate and intrastate revenues into a measure of intercity revenues and (2) absorbing directory advertising and miscellaneous revenues into local revenues. We also estimated numerous three-product specifications of the Bell System cost function, using local, intercity, and combined directory advertising and miscellaneous revenues. These results were highly unstable.

72. The Christensen, Fuss and Waverman, and Smith and Corbo studies also impose the homogeneity and symmetry restrictions in order to obtain cost function estimates. These restrictions are rejected by the data for the Christensen study and, we conjecture, for the Fuss and Waverman and Smith and Corbo studies as well.

73. See Appendix A to Chapter 10 in this volume for a discussion of the kinds of data that would be useful for determining whether the telephone industry is a natural monopoly and the reasons why such data were not available.

74. Moreover, evidence that Bell Canada or the Bell System has a subadditive cost function would not necessarily prove that the telephone industry is a natural monopoly and that competition should be prohibited for the three reasons discussed in the first section of this chapter.

Chapter 7

The Impact of Divestiture on the Cost
of Capital to the Bell System*

David S. Evans and Michael Rothschild

Under its historic settlement with the Justice Department, AT&T will divest its 22 local operating companies.[1] According to several financial analysts, divestiture will raise the cost of debt and equity capital to the local operating companies. The *Washington Post* reported that many local phone companies "were AAA rated because they were under the umbrella of AT&T, one of the nation's most solid companies."[2] According to a senior vice president of Moody's, "The absence of the AT&T umbrella and divergences in local rate regulations will lead to greater differences in credit quality than has been seen in the past."[3] Indeed, in opposing the remedy sought by the Justice Department, AT&T claimed that, "because they are part of the integrated System, all Bell operating units have access to capital funds at a lower cost than if they were unaffiliated companies."[4]

As we shall show, the Bell umbrella is less fact than folklore. If the umbrella lowered the cost of capital to the Bell operating companies, these companies would have lower average capital costs than independent telephone companies. The evidence shows these differences are small and statistically insignificant. In setting telephone company rates, state public utility commissions estimate the cost of capital for companies under review. Based on 1974–1979 data from these commissions, Bell operating companies have almost the same average cost of capital as independent telephone companies.

Moreover, divestiture will probably not change the cost of capital to operating companies for three reasons. First, the cost of capital to the divested operating companies will be at least as low as the cost of capital to the independent companies. Since the Bell operating companies have roughly the same cost of capital as the independents, the cost of capital to the divested operating companies

---

*We would like to thank William Brock for many helpful comments and suggestions and John Bender and S. Ramachandran for excellent research assistance.

will be roughly the same as the cost of capital to the Bell operating companies. Second, the bond markets will continue to treat the operating companies more favorably than the independents after divestiture. The favorable interest rates presently paid by the Bell operating companies are due, not to the Bell umbrella, but to financial characteristics that these companies will retain after divestiture. High ratings and low interest rates go to large companies with low debt–equity ratios. Moody's has awarded Aaa ratings to the bonds issued by AT&T and 21 of its 22 operating companies.[5] Of the independent telephone companies we examined, most have A ratings and all have an Aa rating or less. But the divested operating companies will be much larger and have substantially lower debt–equity ratios than the independent telephone companies.[6] Consequently, the rating firms will probably award higher ratings to the divested operating companies than to the independents. Because bond ratings largely determine interest rates, the operating companies will continue to pay substantially lower interest rates than the independents will.[7]

Third, modern financial theory suggests that divestiture will not change the cost of capital to the Bell System.[8] The cost of capital is determined by the value that investors attach to the projects undertaken by the Bell System. Divestiture will change neither the returns earned by these investments nor the returns earned by the companies making these investments. It will only change the manner in which the claims to these returns are packaged and sold to investors. Investors have had to buy these investments in a single package, namely as a share in AT&T and its associated companies. For example, in order to purchase equity in Illinois Bell, investors have had to purchase equity in AT&T and, therefore, in all of the companies that comprise AT&T. Divestiture will enable investors to purchase equity in one or more operating companies without also purchasing equity in the remaining operating companies, AT&T, and Western Electric. According to modern financial theory, these kinds of packaging considerations do not affect the market value of investment opportunities. Unbundling the parts of the Bell System will not decrease their attraction to investors. Investors who wish to hold the present Bell System portfolio of telephone investments will be able to do so by holding stock in every part of the Bell System. Should many small investors desire this portfolio, a mutual fund could provide it to them.[9] Thus, divestiture will not change the market value of Bell System securities and therefore will not change the cost of capital to the Bell System.

This chapter is divided into three sections. In the next section, we show that the difference in the cost of capital between the operating companies and the independents is small and statistically insignificant. We then show that, although the Bell operating companies have lower debt capital costs than the independent telephone companies, the divested operating companies will probably also have lower debt capital costs than the independent telephone companies. Finally, we show why financial theory predicts that divestiture will not change the cost of capital to the Bell System.

Cost of Capital to Bell and Non-Bell Telephone Companies

State public utility commissions estimate the cost of capital to telephone companies seeking tariff increases. Their estimates provide evidence concerning whether common ownership of the operating companies and the resulting Bell umbrella lowered the cost of capital to the Bell operating companies. Furthermore, because they determine the rates telephone companies charge, their estimates are relevant for determining whether divestiture will raise telephone rates by raising capital costs.[10] Rate hearings take the following form. The commission determines the utility's operating expenses and the value of its capital stock. It then determines the rate of return the utility needs to attract capital. It sets the rate of return equal to the cost of capital. Finally, it sets tariffs it believes will enable the utility to earn this rate of return.[11]

Algebraically, the rate-setting procedure sets the firm's revenues equal to its costs

$$R = O + \rho K \qquad (7.1)$$

where R is the firm's revenue, O is the firm's operating expenses, K is the value of the firm's capital stock, and ρ is the rate of return on (or cost of) capital. The commissions' estimates of the cost of capital, ρ, largely determine the tariffs charged by telephone companies. Although it is easy to quarrel with the procedures used by particular commissions to calculate the cost of capital, the commissions' estimates of the cost of capital ultimately determine local and intrastate telephone rates. Even if other procedures found Bell operating companies had lower capital costs than independent telephone companies, these procedures could lead to lower tariffs only if the commissions' procedures also found lower capital costs.

If the Bell operating companies had lower capital costs than the independents did, the commissions would have allowed lower rates of return for the Bell operating companies than for the independents. In order to determine whether this was the case, we used multiple linear regression to estimate the difference between the allowed rates of return for the Bell operating companies and for the independents that received tariff increases during the years for which we had data.[12] We estimated[13]

$$\rho_{it} = a_0 + a_1 T_{i,75} + a_2 T_{i,76} + a_3 T_{i,77} + a_4 T_{i,78} + a_5 T_{i,79} + bd_i + \varepsilon_{it} \qquad (7.2)$$

where T_{it} equals one, if the observation on company i is for test year t, and zero otherwise; d_i equals one if company i is an independent and zero if company i is a Bell operating company; ρ_{it} is the allowed rate of return for company i in test year t; ε_{it} is an error term with mean zero and constant variance across companies and test years; and the b and the a are parameters to be estimated.[14] The parameter b equals the difference between the allowed rate of return for the independents and the allowed rate of return for the Bell operating companies. We found that this difference, although positive, was not significantly different from zero at the 5% level of statistical significance.[15] Table 7.1 reports the estimated regression equation.[16]

The Bell operating companies and the independents had almost the same cost of capital even though the Bell operating companies had lower costs of both debt and equity capital than the independents had. Two facts explain this paradox. First, the Bell operating companies financed a higher proportion of their investments from equity than from debt than did the independents. From 1971 to 1979, the average ratio of debt to total assets was 0.325 for the Bell operating companies and 0.400 for the independents.[17] Because debt capital is less expensive than equity capital, the cost of capital is greater the greater the fraction of equity capital. Thus, a firm with a low debt ratio may have a lower cost of capital than a firm with a high debt ratio even though the firm with a low debt ratio has higher costs for both debt and equity capital than the firm with a high debt ratio.[18]

Second, the fact that Bell operating companies paid lower interest rates on debt than the independents does not necessarily guarantee that the Bell operating companies' cost of debt capital is less than the independents' cost of debt capital. To commissions and to telephone ratepayers, the cost of debt capital depends on the interest rates paid on outstanding debt rather than on the present market interest rate for new debt. Present interest rates may differ from past interest rates. Consequently, the cost of debt capital depends on past interest rates and the fraction of debt outstanding at those interest rates. Suppose a company has $100 million of bonds outstanding and that half of these bonds were floated when interest rates were 8% and half when interest rates were 12%. It will have interest payments of $10,000,000 (= .08 × $50,000,000 + .12 × $50,000,000) and its average cost of debt capital will be 10%. Another firm, also with $100

Table 7.1 Rate of Return Estimates for Bell Operating Companies and Independent Telephone Companies

Variable	Parameter estimate	Standard error
Constant	8.844[a]	(.102)
T_{75}	.019	(.150)
T_{76}	.127	(.149)
T_{77}	.204	(.157)
T_{78}	.335[a]	(.146)
T_{79}	1.220[a]	(.140)
D	.099	(.101)
R^2		.343
Error sum of squares		.665
Number of observations		196

[a]Statistically significant at the 5% level.

million of bonds outstanding, faced the same interest rates but floated 75% of its debt at the 8% rate and 25% of its debt at the 12% rate. Its total interest payments are only $9,000,000 and its average cost of debt capital will be 9%. Even if the second firm had faced higher interest rates of 9% and 13%, its average cost of debt capital would only be 10%. This example shows that a company's cost of debt capital reflects interest rates faced by it, as well as its skill (or luck) at floating bonds when interest rates were relatively low.

We computed an implied rate of return on debt for the Bell operating companies and the independents for which data were available from the Argus Research Corporation. If s is the rate of return on equity, r the rate of return on debt, and α the share of equity in the capital structure, then the average rate of return ρ may be written as

$$\rho = \alpha s + (1 - \alpha)r \qquad (7.3)$$

Solving equation (7.3) for the rate of return on debt r we obtain

$$r = \frac{(\rho - \alpha s)}{1 - \alpha} \qquad (7.4)$$

This value is reported in Table 7.2 along with the rate of return on equity, the share of equity in the capital stock, and the total rate of return.[19] It is clear from columns 5 and 6 that Bell and non-Bell companies had very similar rates of return on (and costs of) debt capital. Because Bell operating companies faced lower market rates than the independents at every point in time, it is clear that, in retrospect, the independents played the bond market more intelligently than the Bell operating companies.[20]

The estimated difference between the cost of capital for the Bell operating companies and the independents may be biased for two reasons. First, the independents' success at forecasting interest rates during the 1970s may have resulted from luck rather than skill and therefore may not enable these companies to achieve lower capital costs in the future. If so, we should purge these fleeting successes (or random fluctuations) from the cost of capital estimates made by the commissions. We can do so by assuming that the cost of debt capital equals the current market interest rate. As shown in the appendix, this correction raises the difference between the cost of capital for the independents and the Bell operating companies by an average of 0.51 basis points.[21]

Second, because it ignores the effect of taxes, the allowed rate-of-return formula (7.1) overestimates the cost of capital ultimately borne by ratepayers. Debt interest payments are deductible from corporation income taxes whereas dividend payments are not. As shown in the appendix, when this deduction is taken into account the cost of capital is

$$\hat{\rho} = \rho - (1 - \alpha)r\,\tau \qquad (7.5)$$

where τ is the corporate income tax rate and ρ is the cost of capital allowed by the public utility commission. This correction is larger for the independents than

Table 7.2   Asset-Weighted Rates of Return and Shares of Equity for Bell and Non-Bell
            Companies (1974–1979)

	1 Bell	2 Non-Bell	3 Bell	4 Non-Bell
Test year	Rate of return		Rate of return on equity	
1974	9.16	8.95	11.62	12.86
1975	9.03	8.97	11.67	11.87
1976	8.90	9.03	11.46	12.04
1977	9.10	9.29	11.86	12.14
1978	9.18	9.06	11.78	13.24
1979	10.01	9.76	12.59	13.78

Source: Argus Research Corporation, *Argus Reports,* 1974–1979.

While the regression reported in Table 7.1 used 196 observations, this table is based on only 169 observations (121 Bell and 48 non-Bell) because some *Argus Reports* omitted either the rate of return on equity or the share of equity and because some companies had more than one decision in a given year.

for the Bell operating companies because the independents have a larger fraction of debt and thus a relatively larger adjustment than the Bell operating companies. As shown in the appendix, this correction lowers the difference between the cost of capital for the independents and the Bell operating companies by an average of 0.34 basis points. The corrections for interest rate fluctuations and taxes roughly cancel each other out and lead to no net change in the estimated difference in the cost of capital between the independents and Bell operating companies.

Our finding that the Bell operating companies have virtually the same costs of capital as the independents is rather surprising. After all, even ignoring the Bell umbrella the investments in the Bell operating companies were surely less risky and therefore should have commanded lower interest rates than investments in the independents. The Bell operating companies apparently have failed to realize lower capital costs than the independents because they have chosen lower debt–equity ratio than the independents. There are several explanations for their choice, which we explain in more detail in the last section of this chapter.

Cost of Debt Capital to Bell and Non-Bell Telephone Companies

Bonds issued by the Bell operating companies pay lower interest rates and receive higher ratings than the bonds issued by the independents. All but one of the 22 Bell operating companies had Aaa bond ratings in 1979.[22] Of 22 independents for which we were able to obtain data, 15 had an A rating, six had an Aa bond rating, and one had a Baa bond rating, according to Moody's.[23]

Are the lower interest rates and higher credit ratings received by Bell bonds

Table 7.2 *(continued)*

5 Bell	6 Non-Bell	7 Bell	8 Non-Bell	9 Bell	10 Non-Bell
Implied rate of return on debt		Shares of equity in total captial		Number of observations	
6.79	6.33	.49	.40	26	15
6.61	6.94	.48	.41	15	10
6.44	6.63	.50	.39	21	8
6.30	7.17	.50	.43	18	5
6.73	6.44	.48	.38	12	1
7.66	6.91	.49	.41	29	9

a result of the Bell umbrella—which divestiture will eliminate—or a result of financial characteristics—which will largely remain after divestiture—of the individual Bell operating companies? Although it is hard to answer this question with great certainty, the evidence presented below suggests that divestiture will not have a significant effect on the bond ratings, and thus the interest rates paid, by the Bell operating companies.

Bond rating agencies award high ratings to companies that are large and have low debt ratios. The divested operating companies will be large and probably have lower debt ratios than independents whose bonds are rated.[24] Therefore, the divested operating companies will probably have higher bond ratings than the independents whose bonds are now rated. Bonds issued by the independents are usually rated A. Therefore, bonds issued by the divested operating companies will probably have ratings of Aa or Aaa. Because bond rates and bond ratings are highly correlated and apparently determined by the same variables, divestiture will not significantly raise the cost of debt capital to the operating companies.[25] We now turn to the evidence upon which we have based this conclusion.

In 1979, Robert S. Kaplan and Gabriel Urwitz published a comprehensive review and extension of attempts to determine the bond ratings awarded by Moody's and Standard and Poor's rating services.[26] They developed a simple model to predict the ratings received by new industrial bonds.[27] Different variants of their model correctly predicted the ratings of newly issued bonds at least 60% of the time. Their preferred model correctly predicted the ratings 69% of the time. Predicted ratings were never more than one rating away from the actual rating. That is, if the model predicted an Aa rating, the actual rating was always

Aaa, Aa, or A. The same factors that determine industrial bond ratings probably determine utility bond ratings.[28] To the best of our knowledge, Moody's and Standard and Poor's rating services use the same procedure and look at the same data to rate these two kinds of bonds.[29]

The model developed by Kaplan and Urwitz used six variables to predict bond ratings.[30] The first three were financial ratios: (1) CFBIT/INT, the ratio of cash flow before interest and taxes divided by interest charges, which measures the firm's ability to meet its debt obligations; (2) LTD/NA, the ratio of long-term debt to net assets, which measures the probability of default—as the ratio of debt to net assets increases, the probability of default increases and the ability of the company to satisfy its creditors decreases; (3) NI/TA, the ratio of net income to total assets, which measures the firm's profitability. As expected, Kaplan and Urwitz found that bond ratings are higher the larger the first and third variables and the smaller the second variable. The fourth variable, TA, is the total assets owned by the firm. Kaplan and Urwitz found that larger firms had higher ratings than smaller firms.

The first four rows of Table 7.3 present data on these variables for the 22 Bell operating companies and for 22 independent telephone companies. Columns 1 and 2 report average values for these variables for the Bell operating companies

Table 7.3 Effects of Financial Variables on Bond Ratings[a] (1971–1979)

Financial variables	1 Average Bell value[b]	2 Average non-Bell value[b]	3 Difference in average (1)–(3)	4 Sign of effect on bond rating
1. Interest coverage (CFBIT/INT)	3.190 (0.287)	3.290 (0.268)	.620[c] (0.084)	+
2. Debt–equity ratio (LTD/TA)	.323 (0.030)	.398 (0.027)	−0.075[c] (0.009)	−
3. Rate of return (NI/TA)	0.068 (0.004)	0.070 (0.006)	−0.002 (0.002)	+
4. Total assets (TA)	3271 (2634)	480 (420)	2791[c] (569)	+
5. Coefficient of variation in net income (CVNI)	32.791 (4.837)	31.411 (6.395)	1.380 (1.709)	−
6. Coefficient of variation in net assets (CVNA)	25.607 (5.119)	23.775 (6.284)	1.832 (1.729)	−

Source: FCC, Statistics of Communications Common Carriers, 1971–1979.

[a]Data were available for 22 Bell operating companies and 22 independent telephone companies. See Table 7.8 for a list of the independent companies.
[b]Standard deviations are reported in parentheses.
[c]Significantly different from zero at the 5% level. See Note 33 for an explanation of the statistical tests used.

and the independents, respectively.[31] Kaplan and Urwitz also included two measures that reflect the systematic risk of the firm's common stock. They tried several alternative measures. Unfortunately, the two measures that yielded the best predictions are hard to calculate for the Bell operating companies. However, they tried two other variables which are easier to calculate and which appear to proxy well for their preferred measure of systematic risk. These are the coefficients of variation of net income (CVNI) and of total assets (CVTA).[32]

Column 4 lists the direction of each variable's effect on bond ratings. Column 3 contains the amount by which the average Bell value of this variable exceeds the average non-Bell value of this variable. Thus, a plus sign (+) in column 4 and a positive number in column 3 indicate that this variable predicts a higher Bell than non-Bell bond rating; a negative sign (−) in column 4 and a negative number in column 3 should be interpreted in the same way. A combination of a positive number in column 3 and a negative sign in column 4 or a positive sign in column 4 and a negative number in column 3 indicates that this variable predicts a lower Bell than non-Bell bond rating.

It is natural to ask whether the differences reported in column 3 are significant, both statistically and practically. The first question is easy to answer. The differences in the means reported for CFBIT/INT, LTD/TA, and TA in Table 7.3 are statistically significant. The probability that they could have arisen by chance is extremely small. The differences in the means of the other variables are statistically insignificant.[33]

Practical significance is, as always, more difficult to assess. When we naively extrapolate the Kaplan–Urwitz model, we find that the most important variable is total asset size. But the differences between the sizes of the Bell and non-Bell companies are too large to make naive extrapolation believable. Taken together, the other variables account for between one-third and one-half of a bond rating.[34] Aside from total assets, the only important variables are the debt ratio and the stability variables. Naive extrapolation, however, is not a particularly attractive technique for estimating practical significance. Although we believe the Kaplan–Urwitz model does explain bond ratings for telephone companies, we do not believe that the values of the coefficients are necessarily the same for telephone companies as for manufacturing companies.

The most sensible and conservative conclusion we can make from this analysis is that large firms with low debt ratios get high bond ratings.[35] The Bell operating companies are larger, and have lower debt ratios than the independents. The divested operating companies, even in the absence of mergers to establish large regional operating companies, will have the same relatively favorable characteristics and therefore have high bond ratings. Because most independents in our sample have A bond ratings, the local operating companies will probably have Aa or Aaa ratings. This evidence implies that the high bond ratings and favorable rates presently enjoyed by the Bell operating companies will continue after divestiture.[36]

Why Divestiture Will Not Raise the Cost of Capital

Before we set forth our theoretical analysis, we must state exactly what we are asserting and what we are not asserting. We are not asserting that divestiture will either raise or lower the profits of the companies that comprise the Bell System. These profits will fall if the Bell System earns monopoly profits and if divestiture effectively deters anticompetitive behavior. These profits will rise if the Bell System was inefficient and if divestiture eliminates the sources of these inefficiencies. These profits will fall if the Bell System has extensive economies of scale and scope which divestiture may reduce. We are abstracting from any and all nonfinancial consequences of divestiture. We shall show that divestiture, in and of itself, will not make investments by the financial community in the Bell System after divestiture inherently riskier than investments in the Bell System before divestiture and that, therefore, the financial community will advance funds to the divested companies under the same terms and conditions as they would have advanced funds to the undivested companies.

Our analysis proceeds by (a) defining the cost of capital, (b) showing that the cost of capital to a conglomerate consisting of several merged businesses generally equals the average cost of capital to the unmerged businesses, and (c) demonstrating that possible exceptions to this generally accepted proposition fail to apply to the Bell System.

The Cost of Capital and the Value of the Firm

The firm has two financial characteristics. First, it may be characterized by its prospective stream of returns. It has a collection of projects that will generate a stream of profits or losses over time. Because they depend on unpredictable factors such as consumer tastes and technological change, its future profits are uncertain and variable. If it is large, the number and variety of projects it undertakes may be staggering. For example, AT&T supplies many different kinds of telecommunications services to its customers; manufactures telephone equipment; installs, maintains, and expands the local and intercity telecommunications networks; does research on telecommunications technology; and develops the fruits of research into products that may be used commercially. In order to abstract from this variety, we view the firm as a collection of projects that will produce a stream of variable profits in the future.

Second, the firm may be characterized by the value that investors attach to its future projects. It returns profits, when they occur, to investors whose capital enabled it to generate these profits. Investors provide it with capital for investments in exchange for securities which promise a portion of its future earnings. There are many different kinds of securities. We consider the two principal types—bonds and equity (also called common stock). Bondholders give the firm money today in exchange for fixed and regular future payments. If the firm fails to meet its payments, bondholders have a claim on its assets. Stockholders receive profits remaining after bondholders have been paid. The firm may distribute profits—or residual earnings—to stockholders in the form of dividends or it may retain and reinvest these profits. Reinvestment of residual earnings may increase the value of the firm and thereby benefit stockholders whose stock will appreciate in value. The firm does not promise stockholders a minimum return on their investment. If the firm does poorly, they may get nothing. If it does well, they will reap the benefits.

The firm's value is determined by the value that the market places on the firm's future stream of returns. Since payments to stockholders and bondholders together exhaust the firm's profits, its value is just the sum of the market value of its bonds and its equity. Symbolically we have

$$V = E + D \qquad (7.6)$$

where V = value of the firm, E = value of the firm's equity, and D = value of the firm's debt. The values on the right-hand side of equation (7.6) are market values because stocks and bonds are negotiable instruments traded on efficient and well-organized markets. The owners of an ongoing firm need not—and in general will not—be those who put up the capital that launched the firm. Instead, they will be participants in the security market who value the firm's future profits at least as highly as any other potential investor.

The firm's capital structure can be characterized by the relative values of its debt and equity. An implication of equation (7.6) is that the firm's capital structure is just a way of packaging the profits generated by the firm. The arbitrary labeling of some of these profits as returns to bondholders and some as returns to stockholders cannot affect the total value of these profits. This proposition was first proved by Franco Modigliani and Merton Miller.[37] Our argument that divestiture will not raise the cost of capital is merely an application of the Modigliani–Miller theorem that an arbitrary division of the returns from investments between several financial instruments cannot change the total market value of the returns from these investments.[38]

From time to time the firm will consider undertaking a new investment project. A new investment project has two aspects. The first is an initial expenditure. The second is the resulting change in the firm's future earnings. Changes in its future earnings change its value. Its owners will profit from launching a new investment project if the value of the change in its earning stream exceeds the initial cost of the project. This is obvious if its owners personally provide the initial capital for the project. Suppose the project costs P to launch and increases profits by ΔV. If its owners do nothing they have an asset worth V. If they decide to undertake the project their asset will be worth $V + \Delta V$. Because they will have paid P, they will be better off by investing if $V + \Delta V - P \geq V$ or if

$$\Delta V \geq P \qquad (7.7)$$

Generally the firm's owners do not finance an investment out of their own pockets. When many people own the shares of the firm, coordinating the raising of capital in this way would be time consuming and costly. Even if the profitability condition (7.7) held, some owners of equity might be unable (or unwilling) to come up with their fair shares of the required investment. The result (7.7) holds true, however, when the firm issues securities, either stocks or bonds, in exchange for the capital, P, required to undertake the investment, as shown below.

First, consider financing by issuing new stock.[39] In effect, the present owners sell a fraction of the firm's future earnings in return for capital which they invest in order to increase the value of the firm from V to $V + \Delta V$. Those who supply capital worth P will do so only if compensated by an asset of equal value. Thus, if a firm sells sufficient equity to raise P by selling a fraction α of the firm, it must be that

$$\alpha(V + \Delta V) = P \tag{7.8}$$

The left-hand side of equation (7.8) is the value of what those who supplied capital to the firm received in exchange for their capital. If this value were less than P, investors would be unwilling to give up P to obtain it. If this value were more than P, the present owners of the firm could raise the capital they needed by selling a smaller share of the firm. The value of the firm $(V + \Delta V)$ is easily observed and known on a securities market. After a sale is made, the previous owners of the firm own $(1 - \alpha)$ share of a firm worth $(V + \Delta V)$. The ownership claims of individual shareholders are diluted by the same amount. An individual who owned 2% of the firm before the issuing of new equity will afterwards own $2(1 - \alpha)\%$ of the firm. The previous owners are better off and willing to undertake the investment and finance it by issuing new stock if this increases the value of their holding or if

$$(1 - \alpha)(V + \Delta V) \geqslant V \tag{7.9}$$

If we substitute the value of P from equation (7.8) into inequality (7.9) and rearrange we see that inequality (7.9) can hold if and only if

$$\Delta V \geqslant P$$

This is the same conclusion we reached before.

Second, if the owners finance the investment by issuing bonds, they will raise P dollars by making promises to meet a particular schedule of payments to the firm's new bondholders. The market will value these obligations at D; the value of the bonds must just equal the amount of money raised. That is, it must be that $D = P$. If $D > P$, the firm could have raised P by selling bonds worth less—by making less costly obligations. Had it been able to do this, it would have. If $P > D$, then the market will not supply capital to the firm on these terms. After the bonds are sold, and the investment made, the value of the firm to its shareholders will change. The value of the profit stream has increased by ΔV, but from this stream must be subtracted the new payments that will be made to the bondholders. Since the value of these payments is D, the net change is $\Delta V - D$. Thus, the value of the shareholder's claim on the firm's earnings is just $V + \Delta V - D$. Since $D = P$, shareholders will wish to sell bonds to finance the investment if and only if $\Delta V \geq P$.

Thus, a firm's owners will undertake an investment, financing it out of their own pocket or by issuing stocks or bonds, if and only if the investment will change the market value of the firm by as much or more than the cost of the investment. Thus, it is the market's evaluation of the change in the firm's profits that determines whether or not a potential investment will be made. If it appears profitable, it will be made. If it does not appear profitable, it will not be made.[40]

The cost of capital describes the terms under which the capital market provides capital to a firm. We have shown that these terms are completely determined by the market's evaluation of the future profits the firm will earn from its various investment projects. We now ask whether divestiture or merger will affect the market value of (and thus the cost of capital to) a fixed group of investment projects. To do this we address the following question. Suppose that these projects can be organized into firms (recall that a firm is just a collection of investment projects) in two ways. First, in *merger* a single firm owns all the projects. Second, in *divestiture* the projects are split between two firms. Neither the conglomerate nor the two independent firms have any other investment projects. Assume that the organization of the investment projects has no effect on their success or failure. There are no economies or diseconomies of scale or scope and no monopoly profits. Each and every project will yield the same profits whether or not the projects are organized into a conglomerate or split between two independent firms. The question at issue is: Will the market value of the conglomerate firm be the same as or different from the sum of the market value of the two independent firms? We shall show that it must be the same. Our argument proceeds by showing that the value of the firm is simply the sum of the value of its assets. Thus the value of the conglomerate firm is the same as the sum of the values of the two independent firms. We make this argument by first considering a simple example.[41]

Divestiture and the Value of the Firm: An Example

The AB Sales Company has two divisions. Division A has an inventory of 20,000 boxes of cereal and 5000 candy bars. Division B has 12,000 boxes of cereal and 12,000 candy bars. Cereal sells for $.60 per box. Candy bars sell for $.30 cents each. Assuming A and B possess no other assets, then division A has $13,500 worth of assets (20,000 × $.60 + 5000 × $.30) and division B has $10,800 worth of assets. The AB Company as a whole has assets of $24,300 (= $13,500 + $10,800). Tables 7.4 and 7.5 report these facts.

Would a structural remedy that splits AB Company into A Company (corresponding to division A) and B Company (corresponding to division B) reduce the value of the assets owned by AB Company? Obviously not. A's assets would be worth $13,500. B's assets would be worth $10,800. A's and B's assets together would be worth $24,300, the same as the integrated AB company.

Table 7.4 Assets of Divisions of the AB Company

	Cereal boxes	Candy
Division A	20,000	5,000
Division B	12,000	12,000
Total for the AB Company	32,000	17,000
Prices	$.60	$.30

Let us continue this example at a higher level of abstraction before we return to the realities of the business world. Instead of tangible commodities like candy or cereal, AB company may have inventories of investment opportunities which are likely to generate future profits. Division A may be developing a Picturephone, for example, and division B may be investing in Treasury bills. In this case, division A will obtain revenues next year of $20,000 if the Picturephone is successful and $5000 if the Picturephone is not successful, assuming division A obtains the $10,000 to launch the Picturephone. Division B will obtain revenues next year of $12,000 in return for $10,000 invested today. Table 7.6 records the investments and the likely payouts. The reader should note that the entries in the first three rows of Tables 7.4 and 7.6 are the same. In Table 7.6, A has $20,000 if A's project (the Picturephone) succeeds and $5000 if A's project fails. A's project can be viewed as being composed of 20,000 units of one commodity and 3000 units of another. The commodity in the first column of Table 7.6 is "a payment of $1 if the Picturephone succeeds" and the commodity in the second column of Table 7.6 is "a payment of $1 if the Picturephone fails." We call these commodities *basic risky commodities*. They are basic because any investment is composed of various combinations of these commodities. They are risky because their payments depend on future uncertain events.

An investment yields an uncertain stream of future returns. It is useful to characterize this stream of returns as consisting of a combination of various amounts of basic risky commodities. A risky project like the Picturephone consists of units of different combinations of two basic risky commodities, reflecting different payouts depending on the Picturephone's success or failure. A riskless project like investing in Treasury bills consists of equal amounts of two basic risky commodities reflecting equal payments upon the Picturephone's success or failure.

By issuing securities, a company sells claims to the profits from its investment

Table 7.5 Values of Assets of the AB Company

Division A	$135,000.00
Division B	$ 10,800.00
AB Company	$243,000.00

Table 7.6 Prospects of Divisions of the AB Company

	Returns if Picturephone succeeds (in dollars)	Returns if Picturephone fails (in dollars)
Division A	20,000	5,000
Division B	12,000	12,000
Total for the AB Company	32,000	17,000

projects. These securities may be viewed as combinations of basic risky securities that bestow claims over basic risky commodities. For example, AB Company could be viewed as selling claims to 32,000 units of a commodity that pays $1 if the Picturephone succeeds and 17,000 units of a commodity that pays $1 if the Picturephone fails. An AB bond with a face value of $1000 consists of 1000 units of the first basic risky commodity and $1000 units of the second basic risky commodity. Bondholders receive $1000 regardless of whether the Picturephone succeeds or fails.[42] Equity holders in the aggregate own 22,000 units of the first basic risky commodity and 7000 units of the second basic risky commodity. A 1% share of the outstanding stock of AB Company consists of 220 units of the first basic risky commodity and 70 units of the second basic risky commodity.

The most important difference between Tables 7.4 and 7.6 is that the real commodities in Table 7.4 have prices while the basic risky commodities in Table 7.6 do not. We show below that the financial markets place implicit prices on these basic risky commodities by valuing securities these commodities comprise. We then show that divestiture has no impact on the cost of capital by establishing the following propositions: (a) these prices are the same regardless of whether A and B are separate or combined companies; (b) the value of the integrated AB Sales Company is the same as the combined value of the A Company and the B Company; (c) the cost of capital depends intimately on the value of the investment projects for which capital is being raised; and (d) because divestiture has no effect on the value of these investment projects, divestiture has no effect on the cost of capital raised to finance these investment projects. By assumption, the nature of the investment projects undertaken by division A and division B is the same whether A and B are separate companies or divisions of the same company. In either case, investment projects will be undertaken which have, in the aggregate, 32,000 units of the first basic risky commodity and 17,000 units of the second basic risky commodity. If the prices of these commodities are the same before and after the ties are severed between A and B, the aggregate value of these investment projects and the aggregate cost of capital—the return investors will require to pay money today for an uncertain payment a year from today—will remain the same.

The fact that basic risky commodities have implicit prices immediately follows from the fact that stocks and bonds, which are really different combinations of basic risky commodities, have prices and that these prices determine the prices of basic risky commodities. Suppose we do not know the separate prices of cereal and candy (the commodities listed in Table 7.4) but that we do know that an assortment of one candy bar and one box of cereal costs $.90 and that an assortment of 22 candy bars and seven boxes of cereal costs $15.30.[43] Then we could use some high school algebra to determine that the implicit price of candy bars is $.30 and that the implicit price of a box of cereal is $.60. These implicit prices must be the same for each assortment. If they were not, a clever entrepreneur could buy one assortment, create a less expensive version of the other assortment, and make a profit.

We can use the same reasoning to derive the prices of basic risky commodities from stock and bond prices. The investment of AB Company consists of 32,000 units of the first basic risky commodity and 17,000 units of the second basic risky commodity. Suppose AB issues ten bonds; these bonds promise to pay $1000 if sufficient funds are available. AB Company will have sufficient revenues to pay bondholders regardless of whether the Picturephone is a success or not. Suppose each bond is sold for $900. Then we know that an assortment that consists of 1000 units of the first basic risky commodity and 1000 units of the second basic risky commodity yields $900. Suppose, in addition, that AB Company issues stock and that the market price of a 1% equity holding in AB is $153. A 1% equity interest in AB consists of 220 units of the first basic risky commodity and 70 units of the second basic risky commodity. Therefore, an assortment of 220 units of the first basic risky commodity and 70 units of the second basic risky commodity yields $153. The same high school algebra we used above shows that the implicit price of the first basic risky commodity must be $.60 and that the implicit price of the second basic risky commodity must be $.30.[44]

There is economics as well as algebra in the statement that prices of stocks and bonds determine the prices of basic risky commodities. An individual investor can arrange to hold any arbitrary combination of basic risky commodities by buying and selling the appropriate number of stocks and bonds.[45] For example, to buy one unit of basic commodity one and no units of basic commodity two (i.e., to guarantee a return of $1 if the Picturephone is a success and $0 if it is not), it is only necessary to buy shares corresponding to .00667% of AB Company's outstanding equity and sell .004667 bonds short.[46] At the prices we have specified for bonds and equity such a portfolio costs $.0067 × 153 − .0004667 × 900 or $.60. Prices of bonds and stocks not only determine the prices of basic risky commodities; these prices also allow investors to buy basic risky commodities at these prices.

The last part of our argument is that prices of basic risky commodities are the same whether AB Company is a conglomerate or two independent companies. On any market, supply and demand determines prices. On a securities market, supply and demand determines prices of risky opportunities. Divestiture changes neither the supply nor the demand of basic risky commodities and therefore cannot change their equilibrium price.

The total supply of risky assets is just the pattern of returns available to the investment projects of AB Company. Different financial structures, that is, different combinations of debt and equity, simply repackage the same total amounts of basic risky commodities. In looking solely at the financial consequences of divestiture we are assuming that divestiture does not change the projects the firm will undertake. This is just another way of saying that divestiture will not change the supply of risky assets.

Demand for risky assets is determined by the demand by potential investors for basic risky assets. Divestiture by itself should not change investors' attitudes toward risk and therefore should not change the demand for basic risky assets. By holding stocks and bonds in different proportions, investors can obtain any pattern of returns they want.

If investors held the securities of the conglomerate at stock and bond prices that were consistent with a particular pattern for basic risky commodity prices, investors will hold the stocks and bonds of the two separate companies at prices that are consistent with the same pattern of basic risky commodity prices. Suppose, for example, that A and B are separate companies and that A issues three bonds with a face value of $1000 each and raises the rest of its capital by issuing stock. This market will be in equilibrium when the bonds of each company sell for $900, when a stockholding corresponding to a 1% interest in A sells for $108, and a stockholding corresponding to a 1% interest in B sells for $9.00. These are the only prices consistent with prices of the first and second basic risky commodities being $.60 and $.30, respectively. Under either situation— merger or divestiture—there are 32,000 units of one commodity and 17,000 units of the other commodity on the market. Commodity prices of $.60 and $.30, which are consistent with equilibrium in one situation, are consistent with equilibrium in the other situation.

Let us recapitulate. The argument that divestiture will not change the cost of capital to the firm rests on two simple propositions. First, the cost of capital cannot change if the total market value of the securities issued by the present Bell System does not change. Second, because divestiture will change neither the supply of nor the demand for the risky assets represented by the securities, divestiture cannot change the aggregate market value of the securities issued by companies that comprise the Bell System. Once again it is important to stress that only the *financial* consequences of divestiture are at issue here. Whether divestiture, by encouraging competition or restricting the exploitation of economies of scale or scope, would change the nature of the projects that the Bell System invests in is not at issue. By assumption, the supply of risky assets is the same whether the Bell System is one company or several independent companies.

Possible Exceptions to the Theory

It is generally accepted that this relatively abstract model fits the actual business world when three qualifications are made.[47] (1) If a merger creates monopoly power, efficiencies, or economies of scale and scope, then divestiture will change the value of the firm and its cost of capital. (2) Divestiture may change the cost of capital when the effects of taxation are taken into consideration. (3) Divestiture may change the probability and cost of bankruptcy and thereby change the cost of capital. To complete our argument, we now show that these considerations do not alter our conclusion that divestiture will not raise the cost of capital.

Economies of scale and monopoly power. Our analysis abstracted from changes in monopoly power and economies of scale and scope. However, in arguing against breaking up the Bell System, AT&T put forth arguments which suggested that, for purely financial reasons, its present merged status permits it to exploit economies of scale. Financial writers have repeated these arguments. We believe these arguments cannot be taken seriously. There are three claims that we must refute.

The first claim is that AT&T "coordinates the timing of the Bell operating companies' long term debt offerings to permit orderly access to credit markets" and thereby lowers interest rates.[48] After divestiture, supposedly, two telephone operating companies might try to float new bond issues at the same time. Such competition might raise rates. This argument is hard to take seriously for two reasons. First, although AT&T is a huge company, its debt obligations are a small part of the total value of all outstanding credit in the United States. At the end of 1979, the face value of AT&T's outstanding debt was about $41.3 billion, less than 10% of the total value of outstanding corporate and foreign bonds ($455.7 billion) on U.S. credit markets and less than 1% of other outstanding credit.[49] Other credit includes government securities (federal and local), mortgages, consumer credit, bank loans, and other debt. These different debt instruments are close substitutes for one another. All trade is on well-organized markets. Professional traders constantly check the relative prices of these assets and make trades to correct anomalies. It is hard to believe that, with so many close substitutes available, changes in the supply of Bell System debt could affect the overall price of debt (the interest rate) or the price of Bell System debt. Our second reason for believing that the Bell System's ability to coordinate its debt offerings does not lower its cost of capital is that the kind of coordination that AT&T has provided the Bell operating companies is available from underwriters. The Bell operating companies now use underwriters—and will surely continue to do so after divestiture—to advise them about the timing of new debt issues.

The second claim, made by AT&T and some financial analysts, is that the Bell System's financial structure protects investors against risks and thereby makes these securities more valuable. Standard and Poor's says that "the size and structure of the Bell System have the effect of reducing the level of risk which the financial markets attach to the securities of AT&T and the other Bell System companies."[50] Supposedly, the market puts a high value on Bell's "financial umbrella" and thereby lowers the cost of capital to the Bell System. This argument cannot bear close scrutiny, for three reasons. First, it is not clear what the umbrella is or if it protects Bell System bondholders. Bell operating company bonds are secured by the revenues of the individual company. A company that gets into trouble cannot count on AT&T for assistance. AT&T's refusal to subscribe to new issues of equity by its Pacific Telephone subsidiary as long as an adverse regulatory climate prevailed in California illustrates this fact.[51] When a Bell operating company gets into financial trouble, its bond ratings fall as do the price of its bonds. Pacific Telephone's rating fell from Aaa in 1977 to A in 1979.[52] Interest rates on its bonds consequently rose. The market knows that the umbrella is not protection against a bad regulatory climate. Similarly, it is hard to imagine a state public utility commission letting a Bell operating company use its funds to bail out other parts of the Bell System. Second, the umbrella is not a free lunch. Aid to one class of Bell System securityholders must come from the pockets of other Bell System securityholders. In other words, the Bell umbrella is at best a repackaging of claims to the returns of the Bell System. Third, the fact that the umbrella operates uncertainly could actually decrease the market value of Bell System securities.[53] When it operates, the umbrella takes from one class of securityholders and gives to another. It cannot increase the profits that the Bell System pays to those who hold its securities. All it can do is divide up these profits in different ways. Because the umbrella operates uncertainly, the claim that each class of securities has on Bell System profits is uncertain. Investors cannot diversify this uncertainty away costlessly. Thus, this uncertainty reduces the attractiveness of Bell securities to risk-averse investors. Consequently, Bell's uncertain umbrella may actually lower the value of Bell securities. Divestiture may remove this uncertainty and thereby increase the value of Bell System securities.

The final claim is that AT&T can, and has, provided debt or equity capital to the Bell operating telephone companies when temporary conditions produce an unfavorable climate for the company to issue debt in the public market.[54] We do not see how AT&T's munificence could possibly raise the market value of the Bell System's securities. Furthermore, this claim seems to imply that AT&T can predict how the bond market will behave and thus pick the proper time to issue bonds. The evidence is to the contrary. We showed earlier that, relative to the independents, the Bell operating companies predicted interest-rate fluctuations poorly. AT&T's reluctance to issue bonds in the 1960s and early 1970s when interest rates were lower than they are now has increased the Bell System's cost of capital relative to the independents' cost of capital. It was AT&T's ability to make loans to its subsidiaries that, in part, allowed them to make this misjudgement. While divestiture may limit AT&T's ability to speculate on the bond market, the record suggests that past speculations have raised the cost of capital to the Bell System. There is no reason to believe that future speculations would lower the cost of capital to the Bell System.

Bankruptcy and taxes. The possibility of bankruptcy and the realities of the corporate tax system are the two reasons most often put forth as explanations of why the Modigliani–Miller theorem does not apply to the US economy. Both factors constitute possible exceptions to our argument that divestiture should not change the cost of capital.

Bankruptcy and taxes are linked together and we consider them together. Because interest payments are deductible expenses while dividends are not, debt financing is cheaper than equity financing. Consider a firm which raises p dollars to finance a new project. Suppose that, with certainty, the firm will last only a single year and that there is only one source of funding, a banker who is willing to trade $\$p$ this year for $\$p(1 + r)$ next year.[55] If the same amount were raised by equity financing, the firm would have to issue stock paying $\$p(1 + r)$ next year. The banker does not care whether her payment of $\$p(1 + r)$ takes the form of stock or bonds.[56] The firm, however, is not indifferent. When it pays $\$p(1 + r)$ to a bondholder it can deduct $\$pr$ from its taxable income and thereby reduce its tax payments by $\$pr\tau$ where the corporate tax rate is τ. After taxes, debt financing costs $\$p[1 + r(1 - \tau)]$.[57]

Equity financing costs $\$p(1 + r)$ because dividends are not tax deductible. In these circumstances, it is hard to understand why firms issue any equity at all. The only satisfactory answer that anyone has been able to give to this question is that firms and investors are afraid of bankruptcy. The higher the ratio of debt to equity the more likely it is that the firm will default on its bond obligations and be pushed into bankruptcy. If there are real costs to bankruptcy, then the Modigliani–Miller theorem relied upon above does not apply.[58] This theorem does not apply because increasing the debt–equity ratio does more than repackage the stream of returns from the firm's projects. As the debt–equity ratio increases, the probability that bankruptcy costs will have to be paid increases and the stream of returns from the firm's projects thereby decreases. When bankruptcy is possible, the firm has an optimal capital structure. The optimal debt–equity ratio balances the tax advantages of debt against the increased probability of bankruptcy caused by increased debt.

The relevance of this argument to the effect of divestiture on the value of the firm is immediate. Divestiture changes both the probability of bankruptcy and the costs of bankruptcy. It is easy to show that, if divestiture does not change the capital structure, divestiture cannot decrease and will probably increase the probability of bankruptcy.[59] Costs change but not in such a simple way. After divestiture, each company may go bankrupt by itself. Before divestiture, the whole, merged company must go bankrupt. The real costs of AT&T going bankrupt are much greater than the real costs of Illinois Bell going bankrupt. Although divestiture may increase the probability of bankruptcy, it may decrease the expected cost of bankruptcy. As divestiture changes the costs and probabilities of bankruptcy, it changes the firm's optimal capital structure. Because the effects of divestiture on the probability and cost of bankruptcy can go in different directions, it is hard to say whether divestiture will increase or decrease the optimal debt–equity ratio.[60]

While taxes and bankruptcy are avenues through which merger or divestiture can affect the cost of capital, it is difficult to apply these principles to the divestiture of the operating companies from AT&T.[61] If the size and structure of the Bell System were to decrease the cost of capital to its constituent companies, the analysis above suggests it would do so by enabling the Bell operating companies to have higher debt–equity ratios than the independents. But, as we showed above, the Bell operating companies have lower debt–equity ratios than independents. There are three possible explanations for this fact. The first is that the Bell System has not chosen an optimal debt–equity ratio. The second is that, at least for telephone companies, the Modigliani–Miller theorem, which says that there is no optimal debt–equity ratio, holds without qualification. The third is that, because the costs of bankruptcy for the Bell System are so enormous, divestiture, by reducing the cost (although increasing the probability) of bankruptcy, will increase the optimal debt–equity ratio to the Bell System. All of these arguments suggest that divestiture will decrease, or leave unchanged, the cost of capital to the Bell System. The best evidence therefore suggests that the simple theoretical argument that divestiture will simply repackage the claims to the Bell System's profits without changing their value can be applied in this case.

Appendix

Regression equation (7.2) suggested that the cost of capital as measured by state public utility commissions is not significantly different for the Bell operating companies and the independents. This result needs to be qualified in two ways. One correction lowers the relative cost of Bell operating company capital. The other raises it. The net effect is to leave the differential unchanged. The first correction involves the fact that independents have been hurt less by recent changes in interest rates than BOCs. The second correction concerns taxes. Because taxes are part of the rate base and interest payments are deductible, the definition of the cost of capital in equation (7.3) overstates the cost of capital to rate payers.

Arguably, the greater the success which independents have had in predicting interest rates changes in recent times is just a fleeting bit of luck which cannot be expected to continue in the future. Cost of capital estimates should be purged of these random fluctuations in the cost of debt capital. In principle, this is easy to do. As observed in equation (7.3), the cost of capital is $\rho = \alpha s + (1 - \alpha)r$ where s is the cost of debt capital, α the share of equity in the capital stock, and r the historically weighted cost of debt capital. If r is replaced by the current market rate of interest, i, then the resulting cost of capital

$$\rho_1 = \alpha s + (1 - \alpha)i \qquad (7.10)$$

is free of random fluctuations and arguably reflects the correct current cost of capital.

This argument, if correct, suggests that the regression results reported above, and the cost of capital reported in Table 7.2 substantially overstate the cost of capital for Bell operating companies relative to non-Bell companies. It is difficult to estimate the market interest rate, i, for each company and each commission decision. We could not get observations on ρ_1 for each commission decision and rerun regression (7.2) with ρ_1 substituted for ρ. It is possible, however, to check for the order of magnitude of the correction by recalculating the average cost of capital reported in Table 7.2 for test years 1974 to 1979 using an estimate of the current market interest rate in place of the implied cost of debt reported in Table 7.2. We did this by assuming that the cost of debt capital to the Bell operating companies equals the average rate of Aaa public utility bonds issued in the appropriate year, as reported by Moody's. We used the comparable series for A-rated public utility bonds as the cost of debt capital for independents.

As Table 7.7 shows, this adjustment substantially raises the cost of capital for independents relative to that of the Bell operating companies. The first column of Table 7.7 reports the amount by which the Bell operating companies' cost of capital exceeded the independents' cost of capital as reported in Table 7.2. The second column reports this difference when the cost of capital is ρ_1 as given by equation (7.10) (estimated as described above).

We are not sure that this correction should be made. Independents have done

Table 7.7 Cost of Capital Adjusted for Interest Rate Fluctuations and Taxes (1974–1979)[a]

Test year	Unadjusted	Adjusted for interest rate fluctuations	Adjusted for taxes	Adjusted for taxes and interest rate fluctuations
	1	2	3	4
1974	0.21	−0.71	0.37	−0.13
1975	0.06	−0.53	0.36	0.03
1976	−0.07	−0.58	0.29	0.01
1977	−0.20	−0.08	0.26	0.31
1978	0.12	−0.53	0.35	0.00
1979	0.33	−0.65	0.41	−0.12

Source: Table 7.2 and text.

[a]Entries equal the difference between the Bell and non-Bell cost of capital.

better on the bond market than Bell companies in recent years. They floated more of their outstanding debt when interest rates were low than did Bell companies. Their better performance may have been due to blind dumb luck or to their having tried harder and having been shrewder. Furthermore, there is some reason to believe that independents' low cost of debt may lead commissions to estimate higher costs of equity capital. The argument goes as follows. Among the several methods that commissions use to estimate the cost of equity capital is the comparable earnings approach. In this method, the required rate of return on equity to a utility is set equal to the observed rate of return (in terms of dividends and capital gains) of comparable companies. An unregulated company with the debt history of an independent will have a higher rate of return on equity than will an otherwise similar company with the debt history of a Bell operating company. The reason is that the first company will have, in effect, realized more capital gains on the relatively cheap debt it issued than the second company. These capital gains will be reflected in the value of the company's shares. Thus, if when the comparable earnings approach is used, companies are matched on characteristics that are correlated with debt history, a lower historical cost of capital will lead to a higher cost of equity capital. Among the characteristics used in matching companies is the debt–equity ratio and the rate of growth. Together these variables virtually determine the debt history of a company. Thus, to the extent that commissions rely on the comparable earnings approach in setting the cost of capital, the correction made in column 2 of Table 7.7 is overstated.

Because it ignores taxes, the required rate of return as given in equation (7.1) overestimates the cost of capital to rate payers. Adjusting for this bias tends to lower the cost of capital to independents more than it lowers the cost of capital for the Bell operating companies. The adjustment is of comparable magnitude to that discussed above. The net effect of the two adjustments is a wash. We demonstrate this bias, which comes about because interest payments are tax deductible, by going through some simple algebra. Start with equation (7.1) which states that revenues paid by rate payers are the sum of operating expenses and capital costs. Break operating expenses into taxes paid, $\hat{T}$, and other expenses, Q, to get

$$R = Q + \hat{T} + \rho K \qquad (7.11)$$

Recall from equation (7.3) that the rate of return, ρ, is a weighted average of the cost of equity capital r where the weights reflect the share of equity and of debt in the capital structure. That is,

$$\rho = \alpha s + (1 - \alpha)r$$

where

$$K = E + D \qquad (7.12)$$

and

$$\alpha = \frac{E}{K} \tag{7.13}$$

(E is the amount of equity capital, D the amount of debt capital.) However, the total taxes paid $\hat{T}$ are total tax obligations before deductions of interest expenses, T, minus deductions of interest payments of capital gains. That is

$$\hat{T} = T - \tau r D = T - \tau r(1 - \alpha)K \tag{7.14}$$

where τ is the corporate income tax rate. Substituting (7.14) into (7.11) we obtain

$$R = Q + T + [\alpha s + (1 - \alpha)(1 - \tau)r]K \tag{7.15}$$

From the point of view of rate payers—and it is this point of view that is relevant for addressing whether "lower capital costs lead to lower rates"—the cost of capital is not ρ but

$$\rho_2 = \alpha s + (1 - \alpha)(1 - \tau)r = \rho - (1 - \alpha)r\tau. \tag{7.16}$$

To get a quick idea of the order of magnitude of the correction factor $(1 - \alpha)r\tau$, suppose that the corporate tax rate is 46%, that both Bell and non-Bell companies have an interest cost of 7%, and that the share of equity for Bell companies is .5 and the share of equity for non-Bell companies is .4. Then, $\rho - \rho_2$ is 1.61 for Bell companies and 1.93 for non-Bell companies. Using ρ_2 rather than ρ to estimate the cost of capital lowers the cost of capital to non-Bell companies by 32 basis points more than it lowers the cost of capital to Bell companies.

The third column in Table 7.7 reports the amount by which the cost of capital to Bell companies exceeds the cost of capital to non-Bell companies when the tax-adjusted formula (7.9) is used (together with data from Table 7.2) to estimate the cost of capital. This adjustment is roughly as large in magnitude (and opposite in sign) as the adjustment for interest rates.

In the last column of Table 7.7 we combine the two adjustments by calculating differences in the cost of capital for the two types of companies where the cost of capital is given by

$$\rho_3 = \alpha s + (1 - \alpha)(1 - \tau)i. \tag{7.17}$$

The value of the market rate of interest, i, in this formula is the same as that used in calculating column 2 of Table 7.7.

When adjustments for both interest-rate fluctuations and taxes are made, there is no significant difference between the cost of capital, as estimated by the commissions, to Bell and non-Bell companies.

Table 7.8 Independent Telephone Companies for Which Data Were Collected

1. Continental Telephone of California	17. United of Ohio
2. General Telephone of Florida	18. Western Carolina Telephone[a]
3. General Telephone of Indiana	19. Woodbarry Telephone[a]
4. General Telephone of Michigan	20. Carolina Telephone and Telegraph
5. General Telephone of the Northwest	21. Central Telephone
6. General Telephone of Ohio	22. General Telephone of California
7. General Telephone of the Southwest	23. General Telephone of Illinois
8. General Telephone of Upper New York	24. General Telephone of Kentucky
9. Hawaiian Telephone	25. General Telephone of the Midwest
10. Lorain Telephone[a]	26. General Telephone of Pennsylvania
11. Navajo Community Telephone[a]	27. General Telephone of Wisconsin
12. North Carolina Telephone and Telegraph	28. Lincoln Telephone and Telegraph
13. United Inter-Mountain[a]	29. Rochester Telephone
14. United of Indiana	30. United Telephone of Pennsylvania
15. United of the Northwest[a]	31. General Telephone of the Southeast
16. United of the West[a]	

[a]Indicates telephone systems for which bond rating data were not available.

Table 7.9 Bell and Non-Bell Telephone Companies for Which Allowed Rate-of-Return Information Was Available[a]

Bell Telephone Companies	1974	1975	1976	1977	1978	1979
Nevada Bell		X				X
Bell of Pennsylvania		X		X		X
Chesapeake and Potomoc						
District of Columbia	X					X
Maryland	X	X	X	X		X
Virginia	X		X		X	
West Virginia	X		X			X
Cincinnati Bell			X			
Indiana			X			
Kentucky			X			
Ohio		X				
Diamond State	X		X			
Illinois Bell		X		X		
Indiana Bell	X		X	X		
Michigan Bell	X		X			X

(continued)

Source: Argus Research Reports, 1974–1980.

[a]An "X" indicates that a rate of return decision was made for corresponding test year.

Table 7.9 *(continued)*

Bell Telephone Companies	1974	1975	1976	1977	1978	1979
Mountain State						
Arizona	X					X
Colorado	X		X			X
Idaho				X		X
Montana				X	X	
New Mexico				X	X	X
Texas				X		X
Utah	X		X	X		
Wyoming	X	X	X			X
New England Telephone						
Maine	X		X		X	
Massachusetts		X				
New Hampshire	X	X				X
Rhode Island	X	X				
Vermont		X				
New Jersey Bell	X	X			X	
New York Telephone		X		X		
Northwestern Bell						
Iowa	X	X				
Minnesota		X			X	X
Nebraska	X	X		X		X
North Dakota		X		X		
South Dakota		X				X
Ohio Bell	X					
Pacific Northwest Bell						
Idaho	X					
Oregon	X					X
Washington	X					X
Pacific Telephone		X	X			X
South Central Bell						
Alabama	X			X		X
Kentucky	X		X	X	X	X
Louisiana	X	X		X		X
Mississippi			X		X	
Tennessee			X	X		X
Southern Bell						
Florida	X		X			
Georgia	X		X			X
North Carolina	X			X		X
South Carolina	X		X		X	X
Southern New England Telephone	X			X		X
Southwestern Bell						
Arkansas	X			X	X	
Kansas		X		X	X	X
Missouri	X		X			X
Oklahoma	X	X	X			X
Texas			X	X		X

Table 7.9 *(continued)*

Independent Telephone Companies	1974	1975	1976	1977	1978	1979
Wisconsin Telephone	X		X	X		X
Continental Telephone of California		X				
Continental Telephone of Illinois		X				
Continental Telephone of Iowa		X				X
Continental Telephone of Kentucky						X
Florida Telephone	X	X				
General Telephone of Iowa		X				
General Telephone of Kentucky			X	X		X
General Telephone of California	X		X			
General Telephone of Florida	X		X		X	
General Telephone of Indiana	X					X
General Telephone of Michigan	X	X				X
General Telephone of the Northwest						
Oregon	X					
Washington	X					X
General Telephone of the Southeast						
Georgia	X		X			
North Carolina		X				
Tennessee	X					X
West Virginia		X				
Alabama						X
General Telephone of Wisconsin	X			X		X
General Telephone of the Southwest						
Texas						X
General Telephone of Illinois		X				
Hawaiian Telephone			X			
Rochester Telephone	X		X			
United Intermountain Telephone						
Tennessee			X	X		
United Telephone of Ohio		X				
United Telephone of Carolina						
South Carolina						X
United Telephone of Florida	X					
Central Telephone of Florida		X				
Central Telephone of Virginia		X	X			
Central Telephone of North Carolina	X		X			
Carolina Telephone	X			X		
Western Telephone of Ohio	X					
Winter Park Telephone (Florida)	X			X		
Lincoln Telephone of Nebraska				X		X

NOTES

1. The settlement removes AT&T from the local telephone business. It creates 22 independently owned local operating companies which will inherit roughly two-thirds of the assets presently held by the 22 Bell operating companies. It transfers ownership over the intrastate and interstate facilities presently owned by the Bell operating companies to AT&T which will continue to own AT&T Long Lines, Western Electric, and Bell Labs. The 22 local operating companies may be merged to form seven regional operating companies. We use the following terminology throughout this paper. *Bell operating companies* refers to the presently constituted operating companies. *Divested operating companies* refers to the companies divested by AT&T under the consent decree. *Independents* refers to the non-Bell telephone companies. *Bell System* refers to AT&T inclusive of the Bell operating companies both before and after divestiture.

2. *Washington Post*, "S&P Debates Maintaining AT&T Rating," January 13, 1982, p. D-8.

3. *New York Times*, "Credit Quality of Some Units Questioned," January 12, 1982, p. 27.

4. AT&T, *Defendants' Third Statement of Contentions and Proof*, in *US* v. *AT&T*, p. 337.

5. Pacific Telephone and Telegraph Co. has an A rating. Southern New England Telephone Co., the 23rd operating company in which AT&T has minority ownership, has an Aa rating.

6. This statement is true for the 22 operating companies as well as the seven regional operating companies planned by AT&T.

7. Because of the current uncertainty surrounding the details of the settlement, some of the operating companies may have their credit ratings reduced one or two notches (from Moody's Aaa to Moody's Aa or A). But once this uncertainty is resolved, we expect most of these ratings to return to Aaa.

8. We are abstracting from any and all non-financial consequences of the settlement such as the dissipation of monopoly profit or new profit opportunities resulting from AT&T's entry into unregulated markets.

9. Nor will unbundling by itself raise the market value of the Bell System securities. If there is now a demand for the claims on the returns from particular Bell System projects, financial intermediaries could provide the appropriate financial instrument cheaply. The slight increase in the value of AT&T's stock immediately after the settlement was announced was probably due to the modification of the 1956 Consent Decree. This modification allows AT&T to compete in unregulated markets and therefore provides AT&T with new additional investment opportunities.

10. In *Defendants' Note to Contention 396*, Episode 62, in *US* v. *AT&T*, AT&T claimed that "Bell's lower capital costs lead to lower rates."

11. In determining the fair rate of return, the commissions are guided by several Supreme Court decisions. They are required to set rates high enough to give those who provide capital to the telephone companies a fair rate of return. A fair rate of return (*a*) compensates investors for the risks they incur by investing in a regulated industry to the same extent investors are compensated for the risks they incur by investing in a similarly risky unregulated industry and (*b*) is high enough to attract the capital required by the regulated company. For further details see *Bluefield Water Works and Improvement Co.* v. *Public Service Commission* 263U.S.679(1923), *Federal Power Commission* v. *Hope Natural Gas* 320U.S.591(1944), and *Permian Basin Area Rate Cases* 390U.S.747(1968).

12. We obtained data from the *Argus Research Reports* published by the Argus Research Company. The companies are listed in Table 7.9 in the Appendix.

13. The term *test year* refers to the format of public utility hearings. In assembling the information needed to implement the formula given by equation (7.1), a commission uses data from a particular period, generally a calendar year, which is then called the test year.

14. The equation was estimated with data for test years 1974 through 1979. Argus reported only a few cases before 1974 and the 1980 data were incomplete at the time we conducted our analysis. We had a total of 196 observations.

15. That is, there is at least 1 chance out of 20 of observing a difference this large even if the true difference is actually zero.

16. For a nontechnical discussion of regression analysis see Franklin M. Fisher, "The Use of Multiple Regression Analysis in Legal Proceedings," *Columbia Law Review,* May 1980, pp. 702–736.

17. Calculated from data reported in FCC, *Statistics of Communications Common Carriers,* 1971–1979. The figure for the independents is based on independent companies which reported to the FCC during each year between 1971 and 1979. These companies are generally large independents which engage in interstate service. There were 31 such companies. Unless otherwise noted, the ratio of debt to total assets is referred to as the debt ratio.

18. A simple numerical example illustrates this phenomenon. Firm A is financed 40% by equity and 60% by debt. Its cost of equity capital is 12% and its cost of debt capital is 7%. Thus, its total cost of capital is 9% (= .40 × 12.0 + 60 × 7.0). Firm B is financed 50% by equity and 50% by debt. Its cost of equity capital is 11.5% and its cost of debt capital is 6.6% (= .50 × 11.5 + .50 × 6.6). Thus, its total cost of capital is 9.05%. Therefore, although firm B can obtain both debt and equity capital more cheaply than firm A can, firm B's cost of capital exceeds firm A's cost of capital.

19. The data in Table 7.2 were calculated as asset-weighted averages of data from commission decisions reported by Argus Research Corporation. Each commission decision pertains to one company (or part of a company when a company operates in more than one state) in one test year. To get an average figure for a test year the decisions for that test year must be combined. Because we are interested in rates of return on capital it makes sense to weight each decision by the total amount of capital (value of assets) to which it applies.

20. It is impossible to determine from this evidence whether the independents were more skillful or luckier than the Bell operating companies.

21. This correction should *not* be made if the independents were shrewder than the Bell operating companies. Moreover, as discussed in the Appendix, commissions that use the *comparable-earnings approach* purge these random fluctuations from their cost of capital estimates indirectly. The number 0.51 is based on the average of the yearly adjustments.

22. The exception was Pacific Telephone and Telegraph whose bond rating dropped to A in 1979. Southern New England Telephone Company, in which AT&T has minority ownership, had an Aa rating.

23. These independents include 22 of the 31 independents for which the data needed to compute the figures reported in Table 7.3 were available in the FCC's *Statistics of Communications Common Carriers,* 1971–1979. Credit ratings were not available for 9 of the 31 independents. The independents are listed in Table 7.8 in the appendix.

24. Under the terms of the settlement, as approved by the Court on August 11, 1982, the divested operating companies will have a debt ratio of approximately 45%, which is approximately equal to the debt ratio of the Bell System as a whole. The exception to this requirement is Pacific Telephone and Telegraph Company which, because its debt ratio prior to divestiture was 57%, must have a debt ratio of only 50% after divestiture. (These debt ratios refer to the value of outstanding debt divided by the value of outstanding debt and equity and therefore differ from the debt ratios reported in Table 7.2, which refer to the value of outstanding debt divided by the book value of net assets.) Under the terms of the settlement, the operating companies will be left with debt issues that reflect the terms and conditions comparable to the debt held by the Bell System as a whole.

25. See Lawrence Fisher, "Determinants of Risk Premiums in Corporate Bonds," *Journal of Political Economy,* June 1959, pp. 217–237. This classic study used variability in earnings, an index of firm reliability in meeting obligations to creditors, the debt ratio, and the total value of the firm's outstanding debt to estimate interfirm differences in risk premiums. (The last variable measures the ease with which a buyer can expect to find a seller for a bond if he chooses to sell it.) He found these variables explained much of the differences in interest rates. We argue that the Bell operating company bonds will continue to carry high ratings because the successors to the operating companies will be large companies with low debt ratios. The same reasoning shows that the divested operating companies will command low interest rates relative to the independents because they will be stable companies with low debt ratios, long histories of meeting their obligations to creditors, and large volumes of bonds outstanding. We chose to analyze bond ratings rather than interest rates themselves because, as Kaplan and Urwitz have

noted, "Bonds are complicated instruments which differ from one another not only because of risk differences among the issuing firms but also because of differences in important features of bonds. These features include coupon rate, maturity, call protection, sinking fund provisions and covenants. . . . " Unfortunately, little is known about the quantitative impact of these features. See Robert S. Kaplan and Gabriel Urwitz, "Statistical Models of Bond Ratings: A Methodological Inquiry," *Journal of Business,* February 1974, pp. 255–256.

26. *Ibid.,* pp. 231–291.

27. Kaplan and Urwitz did not attempt to predict the ratings of utility bonds because these ratings are heavily influenced by the actions of public utility commissions. Pacific Telephone and Telegraph (PT&T) illustrates how important this point is. In the face of an unfriendly regulatory environment, ratings of PT&T bonds tumbled from Aaa in 1976 to A in 1981. While the regulatory climate changed the values of some of the variables of the Kaplan and Urwitz model, it probably did not change enough variables quickly enough to make the model predict these rating changes.

28. See Shyan Bhandar, Robert M. Soldofsky, and Warren J. Boc, "Bond Quality Changes: A Multivariate Analysis," *Financial Management,* Spring 1979, pp. 74–81. This study examined bond rating changes for electric utilities. We prefer the Kaplan and Urwitz model because it is statistically more sophisticated and economically more sensible than the Bhandari *et al.* model.

29. Bond raters must predict the actions of public utility commissions when evaluating utility bonds. But unless the commissions systematically treat Bell operating companies differently from independents, this fact does not affect our conclusion that the bond rating differences between Bell and non-Bell companies are due to the different financial characteristics of the companies.

30. They also included a measure of the bond's subordination status. Subordinated bonds, those which give their owners a claim on the firm's assets that, in the event of a default, can only be satisfied after the holders of the unsubordinated bonds have been paid, have lower ratings than unsubordinated bonds. Because subordination explains differences in the ratings of bonds of the same firm more than differences in the ratings of bonds of different firms and because the long-term debt obligations of telephone companies are generally not subordinated, we have ignored subordination status.

31. The reported averages are calculated by averaging across years by company and then averaging by company across Bell and non-Bell companies, respectively.

32. The coefficient of variation equals the standard deviation divided by the mean.

33. We performed a *t*-test of whether the differences in column 3 of Table 7.3 are statistically significant. That is, we assume the observations on each variable are drawn from a normal distribution with mean μ_B and variance σ_B^2 (for the Bell Companies) and from a normal distribution with mean μ_I and variance σ_I^2 (for the independent companies.) If we assume $\sigma_B^2 = \sigma_I^2$, then we can use the following procedure to test the hypothesis that $\mu_I = \mu_B$ against the alternative that $\mu_I \neq \mu_B$. Let $t = (\bar{x}_B - \bar{x}_I)/S$ where $\bar{x}_B$ is the sample mean for the Bell companies, where $\bar{x}_I$ is the sample mean for the non-Bell companies, n_B is the number of Bell companies in our sample, n_I is the number of independent companies in our sample, s_B^2 is the sampling variance of the Bell companies in our sample, and s_I^2 is the sampling variance of the non-Bell companies in our sample, and

$$S^2 = \left[\frac{n_B + n_I}{n_B n_I} \right] \left[\frac{(n_B - 1)s_B^2 + (n_I - 1)s_I^2}{n_B + n_I - 2} \right]$$

Under the null hypothesis that $\mu_B = \mu_I$, t has a Student's t-distribution with $n_B + n_I - 2 = 42$ degrees of freedom. The probability of getting a t greater than 1.96 is 5%. The t values are CFBIT/INT (7.41), LTD/TA (8.73), NI/TA (1.33), TA (4.91), CVTA (1.06), and CVNI (.807). This CFBIT/INT, LTD/TA, and TA are statistically significant and NI/TA, CVTA, and CVNI are statistically insignificant.

If we drop the assumption that Bell and non-Bell companies have the same variance then we must use a different procedure to perform the same test. The most conservative procedure is, for each variable, to form the statistic

$$z = \frac{|x_B - x_I|}{\left[\dfrac{t_B^2 s_B^2}{n_B} + \dfrac{t_I^2 s_I^2}{n_I}\right]^{\frac{1}{2}}}$$

where t_B and t_I are the upper .025 points of the Student's distribution with n_B^{-1} and n_I^{-1} degrees of freedom respectively. Then under the null hypothesis that $\mu_B = \mu_I$, the probability that $z >$ 1 is less than .05. The values of z are: CFBIT/INT (3.78), LTD/TA (4.45), TA (2.50), CVTA (0.54), CVNI (0.42), NI/TA (0.66), so we get the same conclusion as before. For a discussion of these testing procedures, see C. Radhakrishna Rao, *Linear Statistical Inference and Its Applications,* 2nd ed. (New York: John Wiley and Sons, 1973), pp. 463–464. Note that both procedures identify the same differences as significant.

34. The Kaplan–Urwitz model is of the form

$$A_f = b_0 + b_1 x_{1f} + b_2 x_{2f} + \ldots + b_n x_{nf}$$

where A_f is firm f's score (bond rating on a numerical scale), x_{if} is the value of variable i for firm f and the b's are coefficients. Roughly speaking, about 1.6 points of score are equivalent to one step in the bond rating scale. We could simply compute the differences between the score of the average Bell and average non-Bell companies, as estimated by the Kaplan–Urwitz model. If we do this we find that the difference in score attributable to total asset size alone is about 13, which is too large to be believable. On the other hand, the contributions of the debt ratio to the score is about .2 while the contributions of the different stability measures are about 0.5. The two other variables, CFBIT/INT and NI/TA each contribute less than .01 to A_f.

35. One of the exceptions to the pattern of both positives or both negatives in column 5 and 6 is the result for NI/TA. There are two reasons for disregarding this exception. First, as discussed above, the difference in column 6 is too small to be of either practical or statistical significance. Second, this variable measures the rate of return. Public utility commissions set tariffs so as to determine the rates of return earned by utilities. A substantial difference in this variable between the Bell and non-Bell companies would imply a difference in the rates of return earned by Bell and non-Bell companies. Because, from another point of view, the rate of return on capital equals the cost of capital, the finding that Bell and non-Bell companies have similar NI/TA ratios implies that these two kinds of companies have the same costs of capital. The other two exceptions to the pattern of positives and negatives in columns 5 and 6 are the two stability variables, CVTA and CVNI. The differences for these variables in column 6 are small and statistically insignificant.

36. Other models of bond ratings might lead to different results. We have used the Kaplan–Urwitz model because it is more sophisticated and believable than other models. Kaplan and Urwitz praise James O. Horrigan's model, "The Determination of Long-Term Credit Standing with Financial Ratios," *Journal of Accounting Research* 4 (suppl.), pp. 44–62, which would give different results. Horrigan's model uses total asset size, the debt–equity ratio, and three financial ratios involving sales to predict bond ratings. While the size and debt–equity variables support the conclusions reached here, differences in two of the three ratios involving sales imply that non-Bell companies should have lower ratings than Bell companies. We believe that Horrigan model is inappropriate for use with public utilities. Sales ratios should not explain bond ratings for public utilities. Horrigan explains their significance as capturing profit margins and returns. However, profit margins and returns are regulated and probably not of as much significance for public utilities as size, stability, and the debt–equity variables which are not directly regulated.

37. Franco Modigliani and Merton Miller, "The Cost of Capital, Corporation Finance, and the Theory of Investment," *American Economic Review,* June 1958, pp. 261–297.

38. The form of the argument given here is a simplification of that originally given by Modigliani and Miller. See Joseph E. Stiglitz, "A Re-Examination of the Modigliani–Miller Theorem," *American Economic Review,* December 1969, pp. 784–793.

39. For simplicity, we shall assume that before the new investment takes place, the firm has no outstanding debt. Abandoning this assumption complicates our argument but does not change our conclusion.

40. This argument is due to James Tobin and is sometimes referred to as Tobin's q theory of investment. See James Tobin, "A General Equilibrium Approach to Monetary Theory," *Journal of Money Credit and Banking*, February 1969, pp. 15–29. For an application of this approach to the questions considered here as well as references to empirical research using Tobin's q see Roger H. Gordon and Burton G. Malkiel, "Corporation Finance" in Henry J. Aaron and Joseph A. Pechman, eds. *The Effects of Taxation on Economic Behavior* (Washington, DC: Brookings, 1981).

41. This method of analyzing the effects of divestiture on the value of the firm is well established in the finance literature. Our analysis largely follows that of James H. Scott, Jr., "On the Theory of Conglomerate Merger," *Journal of Finance*, September 1977, pp. 1235–1250 which summarizes and extends a large literature. For a standard textbook treatment of this topic see, for example, James C. VanHorne, *Financial Management and Policy*, 5th ed. (Englewood Cliffs, NJ: Prentice-Hall, 1980), pp. 243–248. On p. 243, VanHorne writes that if capital markets are perfect, "pure diversification by the firm is not a source of value. The whole will simple equal the sum of the parts." According to VanHorne, the only significant (for this argument) capital market imperfections are taxes and bankruptcy. We consider both below.

42. Bonds are riskless because, so long as AB Company issues five or fewer bonds, bondholders are guaranteed to receive $1000. If AB Company issued more than five bonds, bonds would be subject to default if the Picturephone failed. In this case, bonds consist of unequal combinations of the first and second basic risky commodity.

43. Otherwise, the assortments are identical (e.g., they come in the same quality packaging).

44. To see this, let p_1 be the price of the first basic risky commodity and let p_2 be the price of the second. Then the prices of stocks and bonds we have specified imply that the following two equations hold:

$$\text{Bonds } 1000\, p_1 + 1000\, p_2 = 900$$

$$\text{Stocks } 220\, p_1 + 70\, p_2 = 153$$

This system of equations has the unique solution $p_1 = .60$ and $p_2 = .30$.

45. An individual sells a stock or bond by going short—by exchanging a promise to give back one bond or one unit of stock next period in exchange for a payment now. Organized security markets permit many opportunities for some traders to go short. While some individuals may find it difficult to go short, large financial institutions experience no such problems. For our argument to hold, it is necessary only that some investors have opportunities to engage in short sales.

46. Let x be the percent of equity and y the number of bonds required to produce this pattern of return. The first equation specifies that the investor will get $1 if the Picturephone is a success; the second equation specifies that he will get nothing if it is a failure.

$$220x + 1000y = 1$$

$$70x + 1000y = 0$$

The unique solution to these equations is $x = .00667$ and $y = .00467$.

47. See the discussion in VanHorne, *op. cit.*, pp. 243–248 which concludes that mergers can change the value of the merged firms only if synergism (VanHorne's term for what we have called monopoly power or economies of scale or scope) is present or through the effect of bankruptcy and taxes. Similarly Gordon and Malkiel, *op. cit.*, analyze the effect of the federal tax structure on corporate financial policies including the cost of capital. They start with a simple model like that set out above and find it inadequate. When they account for the effects of taxes and bankruptcy they find it fits the available facts.

48. AT&T, *Defendant's Contention 396*, in *US* v. *AT&T*, Episode 62.

49. Calculated from data reported in Moody's *Public Utilities Manual*, 1979 and in Stephen M. Goldfeld and Lester V. Chandler, *The Economics of Money and Banking*, 8th ed. (New York: Harper & Row, 1981), p. 46.

50. See AT&T, *Defendant's Third Statement of Contentions and Proof*, in *US* v. *AT&T*, p. 337.

51. See *Wall Street Journal,* June 25, 1979, p. 29.

52. This is the Moody's rating. During the same period, Standard and Poor lowered its rating of Pacific Telephone bonds from AA in 1977 to A− in 1980.

53. The many backtrackings, retractions, and reversals in the history of AT&T's relations with its troubled subsidiary, Pacific Telephone and Telegraph, is a good illustration of the uncertainty with which the financial umbrella operates. See Donald P. Holt, "The Tax Break that Turned into a Nightmare," *Fortune,* September 10, 1981, pp. 110–115, *The New York Times,* February 9, 1980, p. 28, *The Wall Street Journal,* April 27, 1981, p. 26, August 25, 1981, p. 37, and September 2, 1981, p. 24.

54. AT&T, *Defendant's Contention* 396, Episode 62, in *US* v. *AT&T.*

55. These simplifications do not change the thrust of our argument.

56. Since, by assumption, there is no uncertainty, the firm's profits are known with certainty. Thus it is easy to calculate exactly what share of ownership will provide a payment of exactly $p(1 + r)$.

57. In "Debt and Taxes," *Journal of Finance,* May 1977, pp. 261–275, Merton H. Miller has tried to show that this conclusion is modified when due account is taken of the special tax status of capital gains. Miller's argument is roughly that investors will prefer equity to debt because they can take returns on equity in the form of capital gains which the personal income tax taxes more lightly than interest income. This preference leads investors to accept a lower pre-tax rate-of-return on equity than debt. Gordon and Malkiel, in "Corporation Finance," *op. cit.* argue that the effect of this adjustment is not large enough to nullify the tax advantage of equity financing.

58. By *real costs* we mean expenses like lawyer's and receiver's fees which the firm would not incur if it did not go bankrupt.

59. It would be impossible for the merged firm to go bankrupt in any situation in which one of its now independent divisions did not go bankrupt.

60. Scott, *op. cit.,* shows how a merger can lower the optimal debt–equity ratio of the firm.

61. It is by no means a settled question whether the Modigliani–Miller theorem must be modified to take account of taxes and bankruptcy. In "Debt and Taxes," *op. cit.,* Miller argues that bankruptcy costs were small and avoidable while the supposed tax advantages of debt were illusory. In "Corporation Finance," Gordon and Malkiel argue that bankruptcy costs are large and that unavoidable debt does have tax advantages even when Miller's adjustments are made. We believe that Gordon and Malkiel got the better of the argument but the evidence while convincing is not conclusive.

Chapter 8
Pricing, Predation, and Entry Barriers in Regulated Industries

William A. Brock

This chapter analyzes several aspects of pricing and predation in regulated industries and applies the results to some of the issues raised in the recent litigation against AT&T. It has four sections. The first examines Ramsey pricing by firms that face a competitive fringe, can realize learning by doing, and can exploit consumer habit formation. It shows that it is extremely difficult to infer from estimated Ramsey numbers whether a firm acts like a benevolent or a nefarious monopoly. The second examines whether regulated firms are more likely to engage in certain predatory practices than unregulated firms. It shows that regulated firms are no more likely to engage in predatory R & D programs and predatory cross subsidization than unregulated firms. The third analyzes the incentives of a regulated firm with increasing returns to scale to invest in barriers to entry. The fourth formalizes *pricing without regard to cost,* a novel theory relied upon by the Government in its antitrust act against AT&T. It shows that randomizing prices may deter socially beneficial entry.

Ramsey Pricing

In static models with independent demands, Ramsey numbers, as defined below, are zero at the social optimum, unity at the monopoly optimum, and between zero and unity at the Ramsey optimum.[1] In view of this standard result, it is tempting to conclude from empirical evidence that estimated Ramsey numbers are much less than unity that the firm under scrutiny is acting more like a socially meritorious Ramsey optimizer than a socially nefarious monopoly.[2] This section shows that profit-maximizing monopolies may generate small, even negative, Ramsey numbers in a dynamic model or when a competitive fringe is present. This point is made here by developing a sequence of models.

The standard, n-goods Ramsey problem is to

$$\text{maximize } B(x) - C(x) \tag{8.1}$$

$$\text{s.t. } R(x) - C(x) \geq \pi_0$$

where B denotes social benefit from producing $x \equiv (x_1, \ldots, x_n)$, C denotes cost, R denotes revenue, and π_0 denotes a target level of profit, usually taken to be zero.

In order to make relevant economic points rapidly with a minimum of technical tangentialities we assume *integrability*: For all $x \in R^n_+$ we have

$$\frac{\partial B}{\partial x} = D(x) \tag{8.2}$$

where $D:R^n_+ \to R^n_+$ is a multimarket system of inverse demand functions. Here R^n_+ denotes the nonnegative orthant of n-dimensional Euclidean space.

We have, for the case of independent demands,

$$\sum_{i=1}^{n} B_i(x_i) = B(x), \quad \partial B/\partial x_i = D_i(x_i), \quad R(x) \equiv \sum_{i=1}^{n} D_i(x_i)x_i \tag{8.3}$$

Let us quickly derive the Ramsey rule

$$\frac{(p_i - C_i)}{p_i}\varepsilon_i = \frac{\lambda}{1+\lambda}, \quad i = 1, 2, \ldots, n \tag{8.4}$$

where $\lambda \geq 0$ is the Lagrange multiplier for the constraint in (8.1), ε_i is the absolute value of the elasticity of demand for good i defined by

$$\varepsilon_i \equiv \frac{\partial \ln x_i}{\partial \ln p_i} = \frac{D_i}{x_i|D_i'|} \tag{8.5}$$

C_i denotes $\partial C/\partial x_i$, and $p_i \equiv D_i(x_i)$. The entity on the left-hand side (LHS) of (8.4) is called a Ramsey number. In order to see where (8.4) comes from, write the Lagrangian for (8.1)

$$L(x,\lambda) = B(x) - C(x) + \lambda[-\pi_0 + R - C(x)] \tag{8.6}$$

The first-order necessary conditions (FONC) for the critical point of L with respect to (wrt) x are given by

$$\frac{\partial L}{\partial x_i} = D_i - C_i + \lambda D_i\left(1 - \frac{1}{\varepsilon_i}\right) - \lambda C_i$$

$$= (1 + \lambda)(D_i - C_i) - \frac{\lambda D_i}{\varepsilon_i} = 0 \tag{8.7}$$

Rewriting (8.7) gives (8.4).

It is easy to see that $\lambda \geq 0$. Write $V(-\pi_0)$ for the maximum value of (8.1) given π. Then under sufficient regularity of B,C we have

$$\lambda = \lim_{h \to 0} \frac{V(-\pi_0 + h) - V(-\pi_0)}{h} = V'(-\pi_0) \qquad (8.8)$$

Hence, λ measures the value of relaxing the constraint by \$1.

Some remarks are necessary before proceeding. First the maximum of (8.1) with no target profit constraint is

$$\frac{\partial B}{\partial x_i} = D_i = C_i, \quad i = 1,2, \ldots ,n \qquad (8.9)$$

(i.e., "price equals marginal cost"). Let x^* solve the unconstrained problem. Revenue, when price equals marginal cost, is given by

$$R^*(x) \equiv \sum_{i=1}^{n} C_i x_i \qquad (8.10)$$

If

$$R^*(x) - C(x) \geq \pi_0 \qquad (8.11)$$

at the optimum x^*, then there is no "Ramsey problem." But if economies of scale are present, $R^*(x) < C(x)$ is typical so that we forced to "second best," that is, use Ramsey pricing as given by (8.1). Other methods of covering deficits from marginal cost pricing such as subsidies from general revenues will not be discussed here.

Second, for the case of nonindependent demands, the term $\partial R / \partial x_i$ is sometimes written, by definition, in the form

$$\frac{\partial R}{\partial x_i} = D_i \left(1 - \frac{1}{\varepsilon_i^*} \right) \qquad (8.12)$$

and ε_i^* is called a *superelasticity*.[3] By this device, the formula (8.4) is generalized to nonindependent demands by replacing ε_i by ε_i^*.

Third, a profit-maximizing monopoly solves

$$\text{maximize} \quad R(x) - C(x) \qquad (8.13)$$

which generates FONC

$$p_i \left(1 - \frac{1}{\varepsilon_i} \right) = C_i \qquad (8.14)$$

and Ramsey numbers

$$[(p_i - C_i)/p_i]\varepsilon_i = 1, \quad i = 1,2, \ldots ,n \qquad (8.15)$$

Fourth, notice that both the Ramsey optimum and the monopoly optimum mark up price over marginal cost relatively more for inelastic demands than for elastic demands (assuming demands are independent). This makes it difficult, at first blush, for a regulatory body or a court to distinguish a socially desirable Ramsey monopolist from a socially undesirable profit-maximizing monopolist. But there is a relatively simple test. Calculate Ramsey numbers and test the hypothesis that they are significantly less than unity. Indeed, a calculation of this type was performed by Willig and Bailey and discussed by Littlechild in his section on Ramsey pricing.[4]

Unfortunately, this neat test for Ramsey-like behavior is not valid when (a) there is a competitive fringe present, (b) learning by doing exists, (c) good will exists, (d) uncertainty is present, (e) asymmetric information exists. The remainder of this section outlines how some of these factors may be taken into account without undue stress on the data.

Competitive Fringe

We assume that a competitive fringe exists in each market i.[5] In order to keep things simple, model this fringe by a function $q_i(x_i)$ that solves

$$D_i[x_i + q_i(x_i)] = C'_{ei}[q(x_i)] \tag{8.16}$$

where $C_{ei}(q_i)$ is the minimum cost of having entrants produce q_i. Equation (8.16) treats the monopolist as a Stackelberg leader who realizes that, if he restricts output in an attempt to keep price high, he will attract entrants according to the supply relation[6]

$$p_i = C'_{ei}(q_i) \tag{8.17}$$

Facing such a fringe, a profit-maximizing dominant firm is assumed to solve

$$\text{maximize} \quad \sum_i D_i[x_i + q_i(x_i)]x_i - C(x) \tag{8.18}$$

The FONC are given by using suggestive notation for different marginal revenue concepts,

$$MR^*_i \equiv x_i D'_i(1 + q'_i) + D_i = C_i \equiv MR_i + x_i D'_i q'_i \tag{8.19}$$

$$MR_i \equiv x_i D'_i + D_i = D_i \left\{ 1 - \frac{x_i}{(x_i + q_i)\,(1/\varepsilon_i)} \right\} = p_i \left(1 - \frac{m_i}{\varepsilon_i} \right) \tag{8.20a}$$

$$\frac{(p_i - C_i)}{p_i}\varepsilon_i = m_i\,[1 + q'_i] \tag{8.20b}$$

where

$$\varepsilon_i(z_i) \equiv \frac{D_i(z_i)}{z_i|D_i'(z_i)|} \equiv \text{absolute value of the elasticity of}$$

$$\text{demand for demand curve } D_i$$

$$m_i \equiv \frac{x_i}{x_i + q_i} \equiv \text{market shares of dominant firm.}$$

We may use (8.19) and (8.20) to obtain

$$[(p_i - C_i)/p_i]\varepsilon_i = m_i[1 + \varepsilon_{q_ix_i}q_i/x_i] \tag{8.21}$$

where

$$\varepsilon_{q_ix_i} \equiv \frac{q_i'x_i}{q_i} = \frac{\partial \ln q_i}{\partial \ln x_i} \tag{8.22}$$

Notice that $\varepsilon_{q_ix_i}$ is the elasticity of supply of entrant output with respect to dominant firm output. This quantity is negative in the "normal" case $D_i' < 0$, $C_{ei}'' > 0$.

We are now ready to draw the main conclusion of this section. The right-hand side (RHS) of (8.21) is less than unity in the normal case. Furthermore, if the entrants' supply is highly elastic (i.e., $|\varepsilon_{q_ix_i}|$ is large), then the RHS (8.21) may be substantially less than unity for a profit-maximizing monopoly. But the LHS is the standard Ramsey number. Thus, it will not do to calculate Ramsey numbers for a dominant firm facing a competitive fringe, find that they are all substantially less than unity, and then attempt to conclude that the firm under scrutiny is not a profit-maximizing monopoly. What, then, is the rule that a Ramsey monopoly with a competitive fringe would satisfy?

In the spirit of Breautigam's *partially regulated second best* (PRSB) we posit that the Ramsey dominant firm is required to solve[7]

$$\text{maximize} \quad \Sigma B_i[x_i + q_i(x_i)] - C(x) \tag{8.23}$$

$$\text{s.t.} \quad \Sigma D_i[x_i + q_i(x_i)]x_i - C(x) \geq \pi_0 \tag{8.24}$$

The FONC are given by

$$D_i(1 + q_i') - C_i + \lambda[x_iD_i'(1 + q_i') + D_i - C_i] = 0 \tag{8.25}$$

But (8.25) reduces to

$$\frac{(p_i - C_i)}{p_i} \varepsilon_i = m_i \frac{\lambda}{(1 + \lambda)} \frac{1 + \varepsilon_{q_ix_i}q_i}{x_i} - \frac{\varepsilon_i q_i'}{(1 + \lambda)} \tag{8.26a}$$

[the RHS of (21) is multiplied by $\lambda/(1 + \lambda)$ and is augmented by a positive term], where

$$\varepsilon_i q_i' = \varepsilon_i - \frac{\varepsilon_{q_i x_i} x_i}{q_i} \tag{8.26b}$$

Notice that the modified Ramsey formula (8.26) and the monopoly formula (8.21) require only one more elasticity, $\varepsilon_{q_i x_i}$, as well as the market share of the dominant firm m_i in order to test for the presence of Ramsey behavior in the presence of a competitive fringe.

It is worthwhile to make one final comment concerning (8.26a) and (8.21) before moving on. Formula (8.26a) reduces to the *first-best social optimum*

$$\frac{(p_i - C_i)}{p_i} \varepsilon_i = - \varepsilon_i q_i' \tag{8.27}$$

when $\lambda = 0$ and reduces to the monopoly optimum (8.21) when $\lambda = +\infty$. Furthermore, the size of Ramsey numbers must be compared on a case-by-case basis in order to determine whether the dominant firm under scrutiny is behaving as a Ramsey-monopolist facing a competitive fringe or as a profit-maximizing monopolist facing a competitive fringe. A calculation such as that performed by Willig and Bailey may simply be estimating the entry elasticity $\varepsilon_{q_i x_i}$ and market share m_i of (8.21). Notice that the market share of a dominant firm would have to be quite small and the entry elasticity would have to be quite large in absolute value in order for (8.21) to generate Ramsey numbers near $\frac{1}{2}$, as obtained by Willig and Bailey, for the case of AT&T's intercity service market. We turn to other factors to see if a profit-maximizing monopolist could generate small Ramsey numbers.

Learning by Doing and Good Will

We model *good will* (due perhaps to consumer habit formation) as

$$\frac{dG_i}{dt} \equiv \dot{G}_i = - \alpha_i G_i + x_i, \quad G_i(0) = G_{i0} \text{ given } \alpha_i > 0 \tag{8.28}$$

and *experience* by

$$\dot{E}_i = - \beta_i E_i + x_i, \quad E_i(0) = E_{i0} \text{ given } \beta_i > 0 \tag{8.29}$$

Demand is assumed to depend upon good will as follows

$$D_i \equiv D_i(x_i, G_i), \quad \frac{\partial D_i}{\partial G_i} > 0 \tag{8.30}$$

and cost is assumed to depend upon experience according to

$$c_i = C_i(x_i, E_i), \quad \frac{\partial c_i}{\partial E_i} < 0 \tag{8.31}$$

These specifications, although crude, correspond to the common practice of measuring shifts in demand and cost as functions of distributed lags of past output. One can think of (8.28) and (8.30) as a learning curve for consumers and (8.29) and (8.31) as a learning curve for producers. In any event, this specification is rich enough to make basic economic points quickly, but simple enough to be analytically tractable. We leave it to the reader to decide if we have optimized the tradeoff between precision, generality, realism, empirical implementation in the above for the purposes of this chapter.

We shall assume that the dominant firm faces a competitive fringe modeled by

$$D_i[x_i + q_i, G_i] = C_{ei}[q_i, E_{ei}] \tag{8.32}$$

$$\dot{E}_{ei} = -\gamma_i E_{ei} + q_i, \quad E_{ei}(0) = E_{eio}, \quad \text{given} \quad \gamma_i > 0 \tag{8.33}$$

$$\frac{\partial C_{ei}}{\partial E_{ei}} < 0 \tag{8.34}$$

The competitive fringe has its own learning curve which is modeled by (8.33). In order to emphasize the differences between static Ramsey pricing and dynamic Ramsey pricing, we suppress the competitive fringe at first.

We suppose that first best is described by the optimal control problem

$$V^1(G_0, E_0) \equiv \underset{x(\cdot)}{\text{maximum}} \int_0^\infty e^{-rt}[B(x, G) - C(x, E)] \, dt \tag{8.35}$$

where V^1 denotes the current value function, the maximum is taken over the class of piecewise-continuous control functions, B satisfies

$$\frac{\partial B}{\partial x_i} = D_i(x_i, G_i) \tag{8.36}$$

G denotes $(G_1, \ldots, G_n)$, E denotes $(E_1, \ldots, E_n)$, and r is the rate of interest which is assumed constant.

Elementary control theory yields[8]

$$r V^1(G, E) = H^o(G, E, \mu, \xi) \tag{8.37}$$

$$H^o \equiv \underset{x}{\max} H(G, E, \mu, \xi, x) \tag{8.38}$$

$$H \equiv B(x, G) - C(x, E) + \mu \cdot \dot{G} + \xi \cdot \dot{E} \tag{8.39}$$

Here $a \cdot b$ is the dot product of two vectors a, b.

$$0 = \frac{\partial H}{\partial x_i} = D_i(x_i, G_i) - C_i(x_i, E_i) + \mu_i + \xi_i \tag{8.40}$$

$$\dot{\mu}_i = r\mu_i - \frac{\partial H^o}{\partial G_i} = (r + \alpha_i)\mu_i - \frac{\partial B_i}{\partial G_i} \tag{8.41}$$

$$\dot{\xi}_i = r\xi_i - \frac{\partial H^o}{\partial E_i} = (r + \beta_i)\xi_i + \frac{\partial C_i}{\partial E_i} \tag{8.42}$$

$$\mu_i = \frac{\partial V^1}{\partial G_i,} \xi_i = \frac{\partial V^1}{\partial E_i} \tag{8.43}$$

$$\mu_i(t) = \int_t^\infty e^{-(r + \alpha_i)(s - t)} \frac{\partial B_i}{\partial G_i} ds, \quad t \geqslant 0 \tag{8.44}$$

$$\xi_i(t) = \int_t^\infty e^{-(r + \beta_i)(s - t)} \frac{-\partial C_i}{\partial E_i} ds, \quad t \geqslant 0 \tag{8.45}$$

The main conclusion to be drawn is that, at each date t, the marginal cost pricing rule

$$D_i = C_i \tag{8.46}$$

must be amended to read

$$D_i = C_i - \mu_i - \xi_i < C_i \tag{8.47}$$

But the quantities μ_i, ξ_i, which are positive, are given by the prospective (that is, forward looking) calculation (8.44), (8.45). The complexity of operationalizing (8.44), (8.45), and (8.47) over operationalizing the simple rule (8.46) underlies the torturous proceedings of the FCC in the Docket 18128.[9] It also underlies the skepticism that many economists bear toward the Areeda–Turner marginal cost pricing test for predatory pricing.[10] The problem is this. In the Docket 18128 proceedings the regulated firm's (AT&T's in this case) data must be used to calculate the analogues of μ_i, ξ_i. In the Areeda–Turner test, defendants' data must be used to calculate μ_i, ξ_i. Since the firm under scrutiny knows its data best, it has an enormous advantage to juggle its data in its own favor. Turn now to the problem of second-best Ramsey pricing in a dynamic world.

We model the Ramsey problem as

$$V^2(G_0, E_0) = \int_0^\infty e^{-rt} [B(x,G) - C(x,E)] dt \tag{8.48}$$

$$\text{s.t} \quad R(x,G) - C(x,E) \geqslant \pi_0 \tag{8.49}$$

Elementary control theory yields (using the same notation as above for convenience)

$$rV^2 (G,E) = H^o(G,E,\mu,\xi) \tag{8.50}$$

$$H^o \equiv \max_x H(G,E,\mu,\xi,x) \quad \text{s.t. (8.49)} \tag{8.51}$$

$$H = B(x,G) - C(x,E) + \mu \cdot \dot{G} + \xi \cdot \dot{E} \tag{8.52}$$

The Lagrangian for (8.51) is

$$L = H + \lambda[- \pi_o - C + R], \quad \lambda \geqslant 0 \tag{8.53}$$

The FONC for a critical point of L w.r.t. x are

$$D_i - C_i + \mu_i + \xi_i - \lambda C_i - \lambda D_i \left(\frac{\varepsilon_i - 1}{\varepsilon_i} \right) = 0 \qquad (8.54)$$

which can be rewritten

$$\left(\frac{p_i - C_i}{p_i} \right) \varepsilon_i = \frac{\lambda}{1 + \lambda} - \frac{(\mu_i + \xi_i)\varepsilon_i}{p_i(1 + \lambda)} < \frac{\lambda}{1 + \lambda} \qquad (8.55)$$

The last inequality follows because (8.41)–(8.45) remain valid.

It is important to realize that Ramsey numbers are *not* identical in the dynamic case even when demands are independent. Furthermore, a correct calculation of Ramsey numbers by (8.55) requires knowledge of the prospective quantities μ_i, ξ_i. Let us compare (8.55) with the Ramsey numbers generated by a profit-maximizing monopoly.

Monopoly solves

$$V^3(G,E) = \max \int_0^\infty e^{-rt}[R(x,G) - C(x,E)]dt \qquad (8.56)$$

We obtain

$$0 = \frac{\partial R_i}{\partial x_i} - C_i + \mu_i + \xi_i = 0 \qquad (8.57)$$

$$\mu_i(t) = \int_t^\infty e^{-(r + \alpha_i)(s - t)} \frac{\partial R_i}{\partial G_i} ds, \quad t \geq 0 \qquad (8.58)$$

The rest is the same as before. Equation 8.57 may be rewritten thus

$$\frac{(p_i - C_i)}{p_i} \varepsilon_i = 1 - \frac{(\mu_i + \xi_i)\varepsilon_i}{p_i} < 1 \qquad (8.59)$$

Hence, we see that Ramsey numbers in the dynamic case are always less than one and may even be negative. This makes economic sense. C_i is no longer the opportunity cost. Each extra unit of production generates a capitalized stream of demand augmentation and cost reductions. Call

$$C_i - \mu_i - \xi_i \equiv IC_i$$

where IC_i denotes *true incremental cost*. Then (8.59) may be rewritten

$$\frac{(p_i - IC_i)}{p_i} \varepsilon_i = 1 \qquad (8.60)$$

Notice, however, that it is not true that (8.55) may be rewritten as

$$\frac{(p_i - IC_i)}{p_i} \varepsilon_i = \frac{\lambda}{(1 + \lambda)} \qquad (8.61)$$

because the measure of C in (8.49) does not satisfy

$$\frac{\partial C}{\partial x_i} = IC_i \tag{8.62}$$

This brings us naturally to the present-value Ramsey problem. Suppose the monopoly is regulated so that it is expected to generate a present value of profit that achieves a given target level of date 0, i.e., the regulator solves

$$\text{maximize} \int_0^\infty e^{-rt} [B(x,G) - C(x,E)] \, dt \tag{8.63}$$

$$\text{s.t.} \int_0^\infty e^{-rt} [R(x,G) - C(x,E)] \geq \left(\frac{\pi_0}{r}\right) : \Lambda \tag{8.64}$$

Here Λ is the Lagrange multiplier for the *isoperimetric constraint* (8.64). Write the Lagrangian

$$L = \int_0^\infty e^{-rt} [B + \Lambda R - (1 + \Lambda)C] \, dt - \Lambda \frac{\pi_0}{r} \tag{8.65}$$

We solve this problem by locating a Λ such that (8.64) holds with equality at the maximum of (8.65) subject to

$$\dot{G}_i = -\alpha_i G_i + x_i, \quad \dot{E}_i = -\beta_i E_i + x_i, \ G_i(0) = G_{i0}, \quad E_i(0) = E_{i0} \tag{8.66}$$

The FONC are given by

$$D_i + \Lambda p_i (1 - 1/\varepsilon_i) - (1 + \Lambda)C_i + \mu_i + \xi_i = 0 \tag{8.67}$$

Equation (8.67) may be written in the more illuminating form

$$\frac{(p_i - C_i)}{p_i} \varepsilon_i = \Lambda/(1 + \Lambda) - \frac{(\mu_i + \xi_i)\varepsilon_i}{p_i(1 + \Lambda)} \tag{8.68}$$

Superficially, there appears to be no difference between (8.68) and (8.55). The quantity λ, however, measures the value of relaxing the target current profit constraint (8.49) at *each* point in time. Hence, λ varies in possibly complicated ways over time. In contrast, Λ is a *constant*. It measures the value of relaxing the target *present value* profit constraint (8.64).

It may be argued that (8.63) is a more sensible formulation of the dynamic Ramsey problem than (8.48). It seems silly not to recognize the existence of capital markets for borrowing and lending and thus require that the utility break even at each and every point in time. But there is a problem. The solution of (8.63) may be *time inconsistent*. That is, if we solve (8.63) at date t starting at G_t^o, E_t^o instead of G_0, E_0, where G_t^o, E_t^o are the optimum values of G, E starting at G_0, E_0 at date 0, it will typically be the case that for dates $s > t$, the optimum values G_s^t, E_s^t, for dates s starting from $t < s$, G_t^o, E_t^o will differ from the optimum values, G_s^o, G_s^o for date s starting at date 0 and G_0, E_0.

Time inconsistency is a potentially serious problem with a solution concept. There is little point in laying out an optimal plan over time if it is not optimal for the planner to follow the rest of the plan starting at subsequent dates. There is no scope in this chapter to do anything with the problem of time inconsistency except refer the reader to some literature on analytically similar problems.[11] A measure of the seriousness of the problem may be generated by calculating G'_s, E'_s and checking sensitivity to changes in initialization date $t < s$. Ramsey problems of this type are studied by Brock and Dechert.[12]

Let us now probe more deeply into the relative magnitude of Ramsey numbers in different environments by examining steady states. In what follows we shall want to be able to study the impact of changes in the speed of learning and consumer habit formation and the rate of decay of old knowledge and old habits. To do this we replace (8.28), (8.29) by the parameterizations

$$\dot{G}_i = -\alpha_i G_i + a_i x_i \tag{8.28'}$$

$$\dot{E}_i = -\beta_i E_i + b_i x_i \tag{8.29'}$$

Since we are mainly interested in comparing monopoly with Ramsey pricing, let us compute Ramsey numbers for the steady states in these two cases. In order to keep things simple we shall neglect the competitive fringe. In order to do steady-state analysis set all the time derivatives equal to zero.

For monopoly, use (8.56)–(8.62) modified by (8.28'), (8.29') to obtain

$$\frac{(p_i - C_i)}{p_i}\varepsilon_i = 1 - \frac{(a_i\mu_i + b_i\xi_i)\varepsilon_i}{p_i} \tag{8.69}$$

$$G_i = \frac{a_i x_i}{\alpha_i}, \quad E_i = \frac{b_i x_i}{\beta_i}, \quad \mu_i = \frac{\dfrac{\partial R_i}{\partial G_i}}{r + \alpha_i}, \quad \xi_i = \frac{\dfrac{\partial c_i}{\partial E_i}}{r + \beta_i} \tag{8.70}$$

In order to say something concrete, examine the cost specification

$$C(x, E) = F + \sum_{i=1}^{n} c_i(E_i)x_i, \quad c'_i < 0, \ c''_i > 0 \tag{8.71}$$

and the demand specification

$$D_i(x, G) = \overline{D}_i(x_i)h_i(G_i), \quad h'_i > 0, \ h''_i < 0, \quad h_i(0) = 1 \tag{8.72}$$

Then (8.69) becomes

$$\Delta_i(x_i) \equiv \frac{a_i x_i \overline{D}_i(x_i)h'_i(a_i x_i/\beta_i)}{r + \alpha_i} + \frac{-b_i x_i c'_i(b_i x_i/\beta_i)}{r + \beta_i} \tag{8.73a}$$

$$\frac{p_i - c_i(b_i x_i/\beta_i)}{p_i}\varepsilon_i = 1 - \frac{\Delta_i(x_i)\varepsilon_i}{p_i} \tag{8.73b}$$

Formula (8.73) looks complicated but contains a lot of economics. It says that Ramsey numbers are always less than unity for a profit-maximizing monopoly when habit formation of consumers (good will) and learning by doing (endogenous technical change) are present. All other things being equal, the Ramsey number for service i is small when (a) output x_i is large so that market i is a *large market*, (b) the effectiveness, a_i, of sales x_i in building the stock of consumer habits (e.g., brand loyalty or product familiarity) is large, (c) underlying demand, $\overline{D}_i(x_i)$, is large, (d) the effectiveness of consumer habits in strengthening demand h_i' is large, (e) the real interest rate r plus the rate of decay α_i of consumer habits is small, (f) the marginal drop $-c_i'$ in unit cost is large, (g) the effectiveness b_i of production x_i in augmenting the *stock of technical knowledge* E_i is large, (h) the rate of decay β_i of learning-by-doing knowledge is small.

A strong message emerges from equation (8.73). In an industry like long-distance telecommunications where consumer demonstration effects are probably large (large a_i), and opportunities for technical change possess high potential for unit cost reduction for accumulated experience (large $|b_i c_i'|$), a profit-maximizing monopoly may generate small, possibly negative Ramsey numbers.

This conclusion follows from the economics. Let us explain. Each unit of output causes measured incremental cost c_i. But each unit of output allows the incumbent to entrench itself with users as well as uncover new opportunities for cost reduction. The net, which may even be negative in a rapidly growing market that is being quickly infused with new technology, is the true opportunity cost to the monopoly of producing the extra unit.

The policy relevance of Formula (8.73) and its analogues in more complex models is clear. After measuring the long-run incremental costs (LRIC) of maintaining an extra unit of production (a tough task to say the least), the estimate must be corrected for the more subjective but possibly more important effects of habit formation and endogenous technical change. Such a correction of LRIC estimates cuts both ways for public utilities. On the one hand, it will make it difficult for a public utility to make a convincing argument that it is a Ramsey monopolist rather than a profit-maximizing monopolist masquerading in Ramsey clothing by presenting Ramsey number estimates à la Willig and Bailey that are significantly less than unity. On the other hand, submission of corrected LRIC estimates that are smaller will allow a multiproduct public utility to defend itself effectively against charges of predatory cross subsidization. This defense will ring true because it is socially useful for society to take advantage of intertemporal economies of scope in technical progress. Furthermore, if it can be shown that the entrenchment effects of prior sales on the demand side are due to users' learning curves in product usage rather than irrational brand loyalty effects caused by manipulative advertising, then it is in society's interest to capture the economies of consumption-learning-by-doing as well.

These qualitative results are likely to remain in more elaborate models that recognize uncertainty, peak loads, adjustment costs, and the like. The point remains that each unit of output generates valuable knowledge on the production and demand side that can be captured by the firm. It is also worth remarking that the firm that is most efficient at exploiting these dynamic effects should get the business. To put it another way, true marginal cost of i, MC_i, is given by

$$C_i - a_i\mu_i - b_i\xi \equiv MC_i \tag{8.74}$$

not C_i. Turn now to developing the analogue of (8.73) for the Ramsey case.

We calculate the analogue of (8.73) for the formulation (8.49). That is, the utility is required to meet its profit target each period. From (8.55), making use of (8.28′), (8.29′), we obtain in steady state

$$\frac{p_i - C_i}{p_i}\varepsilon_i = \frac{\lambda}{1 + \lambda} - \frac{(a_i\mu_i + b_i\xi_i)\varepsilon_i}{p_i(1 + \lambda)} \tag{8.75}$$

$$= \frac{1}{1 + \lambda}\left[\lambda - \frac{\Delta_i(x_i)\varepsilon_i}{p_i(1 + \lambda)}\right]$$

where the last follows from (8.73a). Notice that Ramsey numbers are not equated across product lines. Furthermore, (8.75) shows that differences in Ramsey numbers between stagnant markets and growing markets may be quite large. For lack of space, we leave it to the reader to extract more implications from (8.75).

Predation by a Rate-of-Return Regulated Firm

It has been alleged that multiproduct dominant firms may indulge in crash R & D programs for services threatened by competition and finance this predatory R & D with monopoly profits from noncompetitive services. It has also been alleged that when the courts or regulators use a long-run incremental cost pricing standard, dominant firms have an incentive to tilt R & D funding towards developing technologies with low LRIC for threatened services but high fixed costs overall. The presence of fixed common costs gives the dominant firm greater pricing freedom and places overburdened courts and regulators at a disadvantage in ascertaining the dominant firm's true LRIC.

Some observers have argued that a dominant rate-of-return regulated firm has a stronger incentive to invest in costly predatory strategies than a dominant unregulated firm because the regulated firm can partly write off losses from predation against the rate payers.[13] But the profits from predation are partly passed on to the rate payers. The net impact of regulation on the cost–benefit ratio for predation by dominant firms is ambiguous. We shall investigate this and several related issues by using a sequence of models developed below.

Consider the monopoly problem first:

$$\text{maximize} \int_0^\infty e^{-rt}\left(\sum D_i[x_i + q_i(x_i)]x_i - \sum_{i=1}^n c_i(F,y_i)x_i - \sum_{i=1}^n y_i - f_0I\right) dt \tag{8.76}$$

$$\text{s.t.} \quad D_i[x_i + q_i(x_i)] = C'_{ei}[q_i(x_i)], \quad i = 1, 2, \ldots, n \qquad (8.77)$$

$$F = I - \eta F, \quad F(0) = F_0 \qquad (8.78)$$

Here symbols that have not been previously defined are (1) c_i which denotes the incremental cost of i, (2) y_i which denotes incremental cost reducing R & D, (3) F which denotes fixed (common) costs, (4) I which denotes gross investment in *capacity* or fixed cost, (5) f_0 which denotes cost of a unit of capacity, and, (6) η which is the depreciation rate of capacity.

We assume

$$\frac{\partial c_i}{\partial F} < 0, \quad \frac{\partial c_i}{\partial y_i} < 0 \qquad (8.79)$$

Insert (8.78) into the objective (8.76), integrate by parts, and observe that (8.76) is equivalent to the following static maximization problem at each point of time

$$\text{maximize} \quad \sum D_i x_i - \sum c_i x_i - \sum y_i - f_0 (r + \eta)F \qquad (8.80)$$

Whatever the choice of $x \equiv (x_1, \ldots, x_n)$, it is obvious from (8.80) that the monopoly will produce x at minimum cost. Hence, define C by

$$C(x) = \underset{F, y}{\text{mimimum}} \sum c_i(F, y_i) x_i + \sum y_i + f_0 (r + \eta)F \qquad (8.81)$$

where $y \equiv (y_1, \ldots, y_n)$.

We compare four regimes: (*a*) social optimum, (*b*) Ramsey optimum, (*c*) rate-of-return regulation, (*d*) monopoly. The social optimum solves

$$\text{maximize} \quad \sum B_i[x_i + q_i(x_i)] - C(x) - \sum C_{ei}[q_i(x_i)] \qquad (8.82)$$

Note that it is assumed that social control of entry is only indirect through x, that is, through the free entry condition (8.77). This modeling respects the costliness of detailed regulation of a multitude of small entrants. See Breautigam's *partially regulated second best* (PRSB) for a similar idea.[14]

The Ramsey optimum solves

$$\text{maximize} \quad \sum B_i[x_i + q_i(x_i)] - C(x)$$
$$\text{s.t.} \quad \sum D_i x_i - C(x) \geq \pi_0 \qquad (8.83)$$

Parenthetically, we remark that we are entitled to insert the minimum cost function $C(x)$ into (8.82) and (8.83) because cost will be minimized for each level of output at the optimum for the social and Ramsey problem.[15]

Under rate-of-return regulation, the firm solves

$$\text{maximize} \quad \sum D_i x_i - \sum c_i x_i - \sum y_i - f_0(r + \eta)F$$
$$\text{s.t.} \quad \sum D_i x_i - \sum c_i x_i - \sum y_i - f_0(r + \eta)F < v f_0 F : \lambda \qquad (8.84)$$

Here v is the premium above the cost of capital r that is allowed on rate base F. We are assuming, for simplicity only, that all rate base is common cost.

The Lagrangian, L, for (8.84) may be written in the form

$$L = (1 - \lambda) \{\sum D_i x_i - \sum c_i x_i - \sum y_i - f_0[(r + \eta) - v^*]F\} \quad (8.85)$$

where

$$C^*(x) \equiv \underset{x,F}{\text{minimum}} (\sum c_i x_i + \sum y_i + f_0[(r + \eta) - v^*]F) \quad (8.86)$$

and

$$v^* \equiv \lambda v/(1 - \lambda) \quad (8.87)$$

We are now ready to compare the solutions to the four regimes. Take up (8.85) first. It is a standard result that $0 \leq \lambda \leq 1$ in (8.85) above.[16] This result is intuitively obvious because a relaxation of a profit constraint by \$1 can never be worth more than \$1 of profit. The conclusion is that the rate-base-regulated firm acts as if it solves

$$\text{maximize} \quad \sum D_i x_i - C^*(x) \quad (8.88)$$

But, $v > 0$ implies that the "as if" cost of capital is less than the true cost of capital

$$r + \eta - v^* < r + \eta \quad (8.89)$$

Hence

$$C^*(x) \leq C(x), \quad \text{for all } x \quad (8.90)$$

Equation (8.85) brings out dramatically the point that "predatory" R & D is not obviously more advantageous for the rate-of-return regulated firm. The only input that is "as if" cheaper is *capital* not R & D. Furthermore, the "as if" revenue function is not changed by rate-of-return regulation. The conclusion is that the rate-of-return constraint itself has little to do with predatory R & D. A profit-maximizing monopolist is as likely to indulge in as much predatory R & D as a rate-of-return constrained firm even though the monopolist is paying "full expenses" of the predation.

The same conclusion holds for the case of predatory cross subsidy. The rate-of-return regulated firm has little incentive to use profits in one product line that is not threatened by competition in order to subsidize production in another product line which is threatened by competition unless a substantial augmentation to the rate base is made by such activity. This observation is made transparent in (8.85) because $1 - \lambda$ multiplies both revenue and cost. There is a tendency in discussions of predation involving rate-of-return regulated firms to forget that $1 - \lambda$ deflates *revenue* as well as cost leaving the cost–benefit ratio to predation unchanged except for a possible augmentation of the rate base F. Let us investigate output and rate-base effects in more detail.

In order to compare the four regimes diagrammatically, we restrict ourselves to the one-dimensional case at first. The FONC are given below

social optimum $D = C'$ (8.91)

Ramsey optimum $D = C/x$ and $\pi_0 = 0$ (8.92)

monopoly $D + xD'(1 + q') = C'$ (8.93)

rate-of-return regulation $D + D'(1 + q') = C^{*'}$ (8.94)

Figure 8.1 displays these results for small $|q'|$. Figure 8.1 is drawn to reflect the assumption

$$C^{*'}(x) < C'(x), \quad \text{for all } x$$ (8.95)

Assumption (8.95) is a type of regularity assumption. It is tantamount to assuming that marginal cost falls when factor prices fall—a modest request.

Differentiate the relation

$$D[x + q(x)] = C_e'[q(x)]$$ (8.96)

totally wrt x to obtain

$$D'(1 + q') = C_e''q'$$ (8.97)

Figure 8.1 Comparison of four pricing regimes.

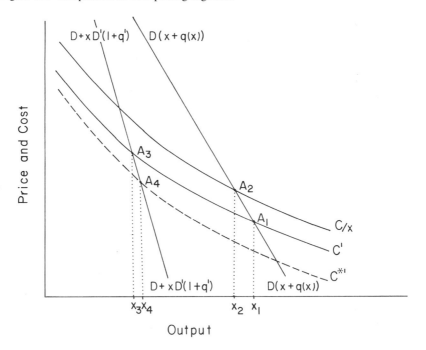

from which it follows that rising marginal entrant cost

$$C''_e > 0 \tag{8.98}$$

and $D' < 0$ implies

$$q' < 0 \tag{8.99}$$

Here we assume $1 + q' > 0$, which is a reasonable type of *stability* assumption.

We are now in a position to offer a more precise conception of predatory production by the rate-of-return regulated firm. The quantity we want to measure is the excess of output of service i produced by the rate-of-return regulated firm over the socially optimum amount of production by that firm. This brings up the issue of the relevant benchmark. In the one dimensional case, a possible measure is given by

$$x_p = x_4 - x_1 \tag{8.100}$$

It is clear from Figure 8.1 that $x_4 - x_1 < 0$ because x_4 is less than the Ramsey point x_2 which is less than x_1. Production level x_4 is less than Ramsey level x_2 because average cost of the rate-of-return regulated firm satisfies

$$AC = \frac{C}{x} + \frac{f_0 v F}{x} > \frac{C}{x} \tag{8.101}$$

because the premium v on the capital must be covered. Notice that no charge of output predation should arise in the one-dimensional case since output of the rate-of-return regulated firm is less than both of the reasonable benchmarks x_1, x_2. Turn now to the evaluation of welfare.

We measure welfare in the partially regulated world that we analyze here by

$$W(x) \equiv B[x + q(x)] - C(x) - C_e[q(x)] \tag{8.102}$$

where $q(x)$ solves

$$D[x + q(x)] = C'_e[q(x)] \tag{8.103}$$

Hence the welfare difference between x and $\bar{x}$ is given by

$$W(x) - W(\bar{x}) = \int_{\bar{x}}^{x} W'(y) dy \tag{8.104}$$

where

$$W'(y) = D(1 + q') - C' - C'_e q' = D[q + q(y)] - C'(y) \tag{8.105}$$

The last equality follows from (8.103).

Look at Figure 8.2 (drawn for constant marginal costs). The partially regulated second-best Ramsey point is given by x_2. If we measure the performance of the

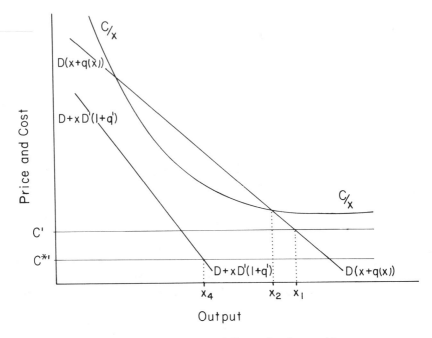

Figure 8.2 Rate-of-return regulation vs. partially regulated second best.

rate-of-return regulated firm relative to partially regulated second-best social optimum given by (8.91), we obtain

$$W(x_4) - W(x_1) = \int_{x_1}^{x_4} W'(y)dy = \int_{x_1}^{x_4} (D - C')dy < 0 \quad (8.106)$$

If we measure the performance relative to x_2, we obtain

$$W(x_4) - W(x_2) = \int_{x_2}^{x_4} W'(y)dy < 0 \qquad (8.107)$$

The conclusion is that, what is on the surface a very poorly performing rate-of return regulated firm which, allegedly, aggressively displaces entrants in order to retain rate base, performs worse than the apparently reasonable regime x_1 and performs worse than the apparently reasonable regime x_2, as Figure 8.2 is drawn. But the firm is producing too *little*, not too much. We must go to a multiproduct analysis to get anything out of the idea of predatory cross subsidy in this model.

The above analysis may be readily extended to multiproducts. Here we measure welfare by (assuming specialized entrants and independent demands)

$$B[x_1 + q_1(x_1), \ldots, x_n + q_n(x_n)] - C(x) - \sum C_{ei}(q_i(x_i)) \quad (8.108)$$

where $q_1, \ldots, q_n$ solve

$$D_i[x_i + q_i(x_i)] = C'_{ei}[q_i(x_i)], \quad i = 1, 2, \ldots, n \qquad (8.109)$$

Welfare change between two vectors x, $\tilde{x}$ is given by the line integral

$$W(x) - W(\tilde{x}) = \int_{t=0}^{t=1} \sum \frac{\partial w}{\partial x_i} dx_i = \int_{t=0}^{t=1} (D_i - C_i) \, dx_i \quad (8.110)$$

where $x(\cdot): [0,1] \rightarrow R_t^n$ is an arc in n-dimensional goods space which satisfies $x(0) = x$, $x(1) = \tilde{x}$. The last equality follows from differentiating (8.108) and using (8.109).

Two remarks are in order. First, a formula like (8.110) may be derived for the case of interdependent demands. But it is likely to be hard to use because cross-elasticity estimation on the demand side is even more difficult than it is on the cost side. Second, the line integral is independent of path so the path may be chosen to maximize convenience of calculation.

It is beyond the scope of this chapter to do much more than show that the problem of predatory cross subsidy may not be all that serious in the multidimensional case for much the same reasons as in the one-dimensional case. This is so for two reasons. First, because a rate-of-return regulated firm uses an inefficient input mix, it has a higher actual cost than an unregulated firm for the same output mix. As a result, the rate-of-return regulated firm tends to produce at least one product at a lower level than any reasonable social benchmark.

Second, a benchmark must be chosen from which to measure the distortion of the output mix. Welfare loss must be measured relative to such a benchmark. This procedure, together with its attendant discipline upon loose allegations about social injury due to predatory cross subsidy, will be illustrated in a two-product case below.

Let C be given by, putting $\alpha \equiv f_0 (r + \eta)$, $\alpha^* \equiv f_0(r + \eta) - f_0 v^*$, $v^* \equiv v\lambda/(1 - \lambda)$

$$C(x_1, x_2) = \alpha(F_0 + F_1 + F_2) + c_1(F_1)x_1 + c_2(F_2)x_2 \qquad (8.111)$$

Here F_0 is the common cost and F_i is i-specific fixed cost. All F cost goes into the rate base.

Assume demands are independent as in the framework of (8.108)–(8.109). The FONC for profit-maximizing monopoly, Ramsey monopoly, and rate-of-return regulated monopoly are given below

$$MR_i(x_i) \equiv D_i + x_i D'_i(1 + q'_i) = C_i = c_i(F_i) \quad \text{(profit maximizing)} \quad (8.112a)$$

$$\alpha = -c'_i(F_i)x_i \qquad (8.112b)$$

$$MR_i^\lambda(x_i) \equiv D_i + \frac{\lambda}{(1 - \lambda)} x_i D'_i(1 + q'_i) = C_i = c_i(F_i) \quad \text{(Ramsey)} \quad (8.113a)$$

$$\alpha = -c'_i(F_i)x_i \qquad (8.113b)$$

$$MR_i(x_i) = C_i^* = c_i(F_i) \quad \text{(rate-of-return regulated)} \qquad (8.114a)$$

$$\alpha^* = - c_i'(F_i)x_i \qquad (8.114b)$$

The differences in the three regimes are displayed in Figure 8.3. Notice that the profit-maximizing-monopoly quantity $x_i^M < x_i^*$ for all i. This is the purpose of rate-of-return regulation—to get the dominant firm to produce more. Service one has MR_1^λ close to demand D_1 which can occur if $1 + q_1'$ is close to zero (i.e., the service is heavily contested by entrants). Service two is not so heavily contested by entrants. Thus the spread between MR_2^λ and D_2 is large. It is clear that the larger the spread between MR_i^λ and D_i the smaller is x_i^R relative to x_i^*.

It is not likely, however, that $x_i^R < x_i^*$ for both $i = 1,2$ because actual costs are higher for the rate-of-return regulated firm which must cover v. Coverage of higher costs restrains demand. Furthermore the very nature of Ramsey optimization is to extract as much surplus as possible subject to break-even constraint. We have drawn a likely case in Figure 8.3.

Let us use x^R as the welfare benchmark. Then

$$W(x^*) - W(x^R) = \int_0^1 \sum \frac{\partial w}{\partial x_i}\, dx_i = \int_0^{\frac{1}{2}} (D_2 - C_2)\,(- 2)\,(x_2^R - x_2^*)\, dt$$

$$+ \int_{\frac{1}{2}}^1 (D_1 - C_1)\,(2)\,(x_1^* - x_2^*)\, dt \qquad (8.115)$$

Figure 8.3 Comparison of output under different regimes.

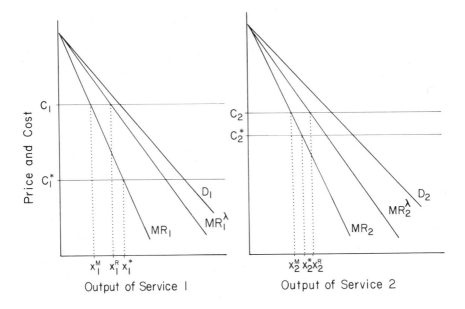

where we evaluated the line integral along the path

$$x_1(t) = x_1^R, \quad 0 \leq t \leq \tfrac{1}{2}, \quad x_1(t) = x_1^R + (2t - 1)(x_1^* - x_1^R), \quad \tfrac{1}{2} < t < 1$$

$$(8.116)$$

$$x_2(t) = x_2^R - 2t(x_2^R - x_2^*), \quad 0 \leq t < \tfrac{1}{2}, \, x_2(t) = x_2^*, \quad \tfrac{1}{2} \leq t \leq 1 \qquad (8.117)$$

Obviously, the RHS of (8.115) is just the sum of the areas between D_2 and C_2, D_1, and C_1 weighted by $2(x_2^R - x_2^*)$, $2(x_1^* - x_1^R)$ respectively. Notice that, although the first term is negative and the second term is positive, the sum is negative. The sum is negative because x^* breaks even and x^R, by definition, is the best break-even point.

The ambiguous consequences for social welfare of predatory activity that takes the form of a dominant firm aggressively displacing entrant output with its own are disappointing results for the policy maker. No clear policy directions are recommended—only careful calculation of welfare relative to a precisely delineated and carefully rationalized standard on a case-by-case basis. But such a conclusion is not surprising by hindsight. Abstracting from income distribution questions, society as a whole is indifferent as to which firm produces output so long as society gets plenty of output at a low cost. However, there is another type of predatory behavior that dominant firm regulation may stimulate, namely investment in barrier-to-entry capital.

Gerald Brock investigates the incentive to build barrier-to-entry capital by a profit-maximizing monopoly.[17] We shall address the same question here but with a somewhat different model that is designed to compare the impact of rate-of-return regulation upon the formation of barrier-to-entry capital. Barriers to entry come in many forms: (a) learning curves, (b) imperfect capital markets, (c) fixed costs, (d) certificate-of-need requirements imposed by regulators upon entrants, and (e) lack of access to "bottleneck" facilities are a few examples. The term barriers to entry in the older tradition in industrial organization usually denotes costs that a new entrant must bear that an incumbent does not have to bear. But we are interested here in a welfare analysis of barriers to entry. Recently C.C. von Weizsäcker proposed the following definition: A barrier to entry is a cost of producing that must be borne by a firm which seeks to enter an industry but is not borne by firms already in the industry and that implies a distortion in the allocation of resources from the social point of view.[18]

In the spirit of von Weizsäcker's notion we assume that a profit-maximizing monopoly solves (sticking to the one-product case)

$$\text{maximize} \quad D[x + q(x,k)] \, x - c(F,y)x - y - f_0(r + y)F - p_k \, (r + \delta)k$$

$$(8.118)$$

$$\text{s.t.} \quad D[x + q(x,k)] = C_e'[q(x,k),k] \qquad (8.119)$$

where k denotes barrier-to-entry capital, p_k is price per unit of k, δ is the rate of depreciation of k, and $\partial C_e'/\partial k > 0$.

The social optimum amount of k that corresponds to (8.82) and (8.91) is zero. This is because k is costly but produces nothing but a rise in cost for the entrant and for society as a whole. But the Ramsey optimum amount of k may be positive if the break-even constraint on the monopoly is tight enough, k is cheap enough, and entrants' marginal costs are close enough to the monopolists'. Consider

$$\text{maximize} \quad B[x + q(x,k)] - c(F,y)x - y$$
$$- f_0(r + \eta)F - p_k (r + \delta)k - C_e \qquad (8.120)$$

$$\text{s.t.} \quad D[x + q(x,k)]x - c(F,y)x - y - f_0(r + \eta)F - p_k(r + \delta)k \geq \pi_0$$

The Lagrangian is given by

$$L = B - \tilde{C} - C_e + \lambda(- \pi_0 + xD - \tilde{C})$$
$$= B + \lambda xD - (1 + \lambda)\tilde{C} - C_e \qquad (8.121)$$

where the total cost $\tilde{C}$ is obviously defined by (8.120) and (8.121). The FONC are given by (using subscripts to denote partial derivatives)

$$0 = \frac{\partial L}{\partial x} = D(1 + q_x) - C_{eq} q_x + \lambda[D + xD' (1 + q_x)] - (1 + \lambda) \tilde{C}_x$$

$$= D + \lambda [D + xD' (1 + q_x)] - (1 + \lambda)\tilde{C}_x \qquad (8.122)$$

$$0 = \frac{\partial L}{\partial k} = \lambda xD'q_k - (1 + \lambda)\tilde{C}_k - C_{ek} \leq 0 \, (= 0 \text{ if } k > 0) \qquad (8.123)$$

Condition (8.123) exposes the tradeoff that society as a whole must make in allowing a Ramsey monopoly to erect entry barriers in order to protect its revenue base. Let us explain. The higher λ is, the more important it is to society to protect the revenue base of the incumbent in order to exploit increasing returns to scale but yet break even. The quantity λ is the value of relaxing the break-even constraint by \$1.

Now it costs society $\tilde{C}_k$ to erect one unit of k. The quantity $\tilde{C}_k$ must be paid for out of the regulated firm's budget. Thus, its *social* cost is $(1 + \lambda)\tilde{C}_k$. The extra unit of k causes the surplus loss $(D - C_k)q_k$ for entrant output foregone. But, this is zero via the free entry condition. The extra unit of k causes the surplus loss $-C_{ek}$ because entrants' costs increase. Benefit is $\lambda xD'q_k$. This is the revenue protection term. Erection of the extra unit of k prevents $xD'q_k$ of monopoly revenue from being captured by entrants and each unit of revenue so saved relaxes the break-even constraint, thereby allowing more scale economies to be captured. Hence, the total value of such revenue saving is $\lambda xD'q_k$—the benefit term.

Of course, if λ is small, costs are large, and $xD'q_k$ is small, (8.123) will hold with strict inequality and $k = 0$. But, if λ is large, which is the case when large scale economies, heavy fixed costs and small marginal costs are present, and if revenue loss $xD'q_k$ from entry is large, then k may well be positive.[19] The next section treats barrier-to-entry capital in more detail.

Barrier-to-Entry Capital: Dynamic Analysis Under Increasing Returns

It requires theoretical dexterity to controvert the *Chicago School* argument that predatory price cutting is likely to be a poor investment by incumbent firms.[20] Witness the toil spent in a recent book on predation to erect convincing arguments that price predation is a problem with which policy makers should concern themselves.[21] The argument that rational entrants and even the customers of the entrants will "see through" the predator's attempt to chase them from the market by price cutting is a powerful one.[22] Indeed, far-sighted entrants and their customers can simply invest for the big day when the incumbent jacks up his prices to recoup his losses. Such "rational expectations" on the parts of entrants and customers tilt the cost–benefit ratio of the incumbent away from price predation. In other words, $1 of expenditure by an incumbent in price cutting typically does less than $1 of damage to the entrant. One can show that price predation may be profitable when asymmetric information, uncertainty, risk aversion, and reputation are important.[23]

The power of the Chicago School argument extends to rate-of-return regulated firms generally and to predatory cross subsidy by regulated firms specifically. That is, predatory cross subsidy is likely to be a poor investment on the part of a rate-of-return regulated firm. This is so because the same Lagrange multiplier that measures tightness of regulation multiplies both costs and benefits in the case of predatory cross subsidy at a point in time (we call this cross-sectional predatory cross subsidy). Hence, the net gain to predation can only work through the indirect effect of rate-base augmentation.

The argument is more complicated for the case of price predation or longitudinal predatory cross subsidy. This is so because costs incurred at date t are multiplied by $1 - \lambda_t$ where the Lagrange multiplier λ_t measures tightness of regulation at date t. Revenues received at date $t + s$ when attempted recoupment takes place are multiplied by $1 - \lambda_{t+s}$. If regulation is expected to be looser at $t + s$ (i.e., $\lambda_{t+s} < \lambda_t$), then price predation is more likely to be profitable, other things being equal, than in the case of unregulated monopoly.

But, the record seems to show that regulation has grown tighter over the last decade. Indeed, Joskow argues convincingly that regulation was "loose" with utilities "lying low" until the recent inflation took off.[24] Then rate cases bloomed. Hence, especially for AT&T, the argument that longitudinal predation would be a foolish investment on the part of a rational utility expecting inflation seems quite strong. But predation in the form of investment in barrier-to-entry capital may be a good strategy for a regulated utility to follow. Unlike predatory pricing, a dollar of expenditure by an incumbent is likely to do more than a dollar of injury to entrants.

Any firm may invest in political activity or productive activity. If it is rational, it will equate margins in both activities. We want to look first at barrier-to-entry capital of a political form. This includes all the techniques of obstruction of competitors that operate through the political process. Even conservative commentators such as Bork are concerned about this type of predation which is particularly insidious because it can be cloaked in the First Amendment and rights to due process.[25] A regulated firm necessarily accumulates political-specific capital in its legitimate coping with the regulatory process. Unfortunately, such expertise is complementary with the low-cost production of barrier-to-entry capital. The same lawyers and economists that walk a reasonable tariff through the FCC can also be used to show that a certificate-of-need application by an entrant is inadequate. For some rather colorful allegations of such tactics and rebuttal of same, see the trial record in *US* v. *AT&T*. One can feel sympathetic with a frustrated economist who, after being dragged through tortuous FCC proceedings where an applicant for entry argues that existing service is inadequate and the incumbent argues that the applicant's facilities are not needed, jeopardizes his livelihood by exclaiming, "This is all a giant expensive waste of the taxpayers' and rate payers' money. Why not let the consumer decide?"

In a competitive industry or a loose oligopoly, the incentives to build barrier-to-entry capital that injures other competitors are weak. This is so because free-riding problems abound. If I spend resources to injure prospective entrants, my fellow oligopolists benefit without paying the cost. Hence, from the point of view of the oligopoly as a whole, I will underinvest in barrier-to-entry capital. The more firms there are in the oligopoly the greater the problem of free riding and smaller the amount of barrier-to-entry capital that is built. The analytics of the incentives of an oligopoly to build barrier-to-entry capital are similar to the tariff-building problem of Brock and Magee.[26] The story is different for a dominant firm, however.

This section shows that, under plausible assumptions on barrier-to-entry production technology, there is a cutoff level $\hat{k}$ of such captial such that if initial capital $k_0 < \hat{k}$ then the firm will allow its stock of barriers to entry to evaporate. If $k_0 > \hat{k}$ it will build them up to a higher steady-state level. The importance of this finding for policy purposes is that it demonstrates that "shock treatment" may be necessary to squash the incentive to build barrier-to-entry capital. If the firm is close to its cutoff level, however, a "small" remedy such as an injunctive remedy may produce a dramatic reduction in barrier to entry capital. Let us get into the model.

The dominant firm is assumed to solve

$$\text{maximize} \quad \int_0^\infty e^{-rt} \left[(y - q)(p - c) - p_I I \right] dt \tag{8.124}$$

$$\text{s.t.} \quad \dot{k} = I - \delta k, \, k(0) = K_0$$

$$\text{given} \quad 0 \leq I \leq \overline{M} \tag{8.125}$$

where q denotes output of the competitive fringe, y is inelastic demand, p_I is the price of a unit of barrier to entry capital, I is gross investment in barrier to entry capital, k, and δ is the depreciation rate. Constraint (8.125) says that investment cannot proceed faster than $\overline{M}$ and k cannot be sold.

The fringe faces no adjustment costs. Hence, it solves the static problem

$$\max_{q} pq - C_e(q,k) \tag{8.126}$$

We specialize (8.126) as follows

$$C_e(q,k) = \frac{q^2}{2L(k)} \tag{8.127}$$

where

$$L(\infty) > 0, \quad L(0) > 0, \quad L_k(k) < 0, \quad L_{kk}(k) < 0, \quad k < \bar{k},$$
$$L_{kk}(k) > 0, \quad k \geq \bar{k} \tag{8.128}$$

The restrictions on L given by (8.128) ensure that L looks like Figure 8.4. Use (8.126), (8.127) to obtain

$$q = pL(k) \tag{8.129}$$

Figure 8.4 Shape of the supply function of entrants.

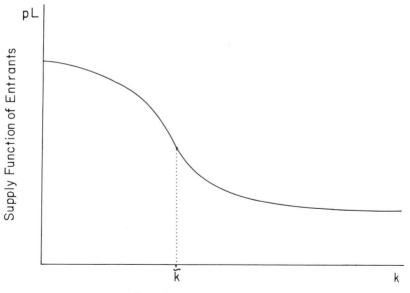

Barrier to Entry Capital

Notice from (8.129) that the supply function has the same general shape of a falling function of k that is first concave then convex. The point $\hat{k}$ separates the zone of concavity from the zone of convexity.

Optimal control theory yields the following FONC for the dominant firm's problem

$$H = (y - pL)(p - c) - p_I I + \mu(I - \delta k) \tag{8.130}$$

$$H_p = 0 = -2pL + cL + y, \quad p = \frac{y + cL}{2L} \tag{8.131}$$

$$\dot{k} = \overline{M} - \delta k, \quad p_I < \mu, \quad \dot{k} = -\delta k, \quad p_I > \mu \tag{8.132}$$

$$\dot{\mu} = (r + \delta)\mu + L_k p(p - c) = (r + \delta) + L_k \left[\left(\frac{Y}{2L}\right)^2 - \left(\frac{c}{2}\right)^2 \right] \tag{8.133}$$

The phase diagram of the system is depicted in Figure 8.5. The curve $\dot{k} = 0$ is the horizontal line $\mu = p_I$. The curve $\dot{\mu} = 0$ is drawn so that it intersects the curve $\dot{k} = 0$. The fact that the curve $\dot{\mu} = 0$ starts at 0 and falls back down to zero as k tends to infinity follows from the hypotheses on L. Hence, except for hairline cases, there are an even number of intersections of $\dot{\mu} = 0$ and $\dot{k} = 0$. We restrict outselves to the case where there are only two. We also assume that $\overline{M}/\delta > \overline{k}$ in order to have a nondegenerate problem.

Figure 8.5 Phase diagram for investment in barrier to entry capital.

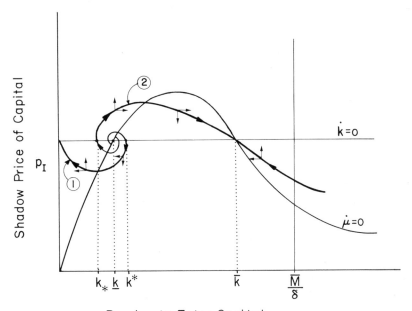

Barrier to Entry Capital

The intersections of the curves $\dot{k} = 0$ and $\dot{\mu} = 0$ are steady states. The phase diagram is drawn in Figure 8.5 with two particular paths drawn in black. The following theorem can be established: There is a critical level of k, call it $\hat{k}$ such that if $k(0) < \hat{k}$ the optimal path is path 1 and if $k(0) > \hat{k}$ the optimal path is 2.[27] Furthermore,

$$k_* < \hat{k} < k^* \tag{8.134}$$

While the mathematics needed to establish this theorem are rather complex, the logic behind it is easy. Common sense suggests that there should be a cutoff level $\hat{k}$ below which it is not profitable to build and maintain barrier-to-entry capital when there are increasing returns present. The shape of L depicted in Figure 8.4 captures the idea that one gets very little return from small levels of k. At midsized levels near $\bar{k}$, a small increment in k yields a large return in deterring entry. This property corresponds to the notion of critical level of knowledge or critical mass of staff if k corresponds to a staff of lawyers and economists steeped in regulatory expertise. Figure 8.5 and the theorem show that if the dominant firm's initial level of k is small then it will simply take too long and it will cost too much money to build k up to the level where it will do much good in retarding entry. In this case it is optimal to accomodate entry.

Any factor that raises p_I or lowers the curve $\dot{\mu} = 0$ lowers $\bar{k}$ and is likely to raise $\hat{k}$. Indeed, if the curve $\dot{\mu} = 0$ falls below p_I for all k, then decumulation of k is always optimal. By (8.124), the steady state $\bar{k}$ is given by the largest solution to

$$\mu = p_I = -L_k \frac{\left(\dfrac{y}{2L}\right)^2 - \left(\dfrac{c}{2}\right)^2}{r + \delta} \tag{8.135}$$

It is obvious from (8.135) that increases in market size Y, decreases in LRIC c, decreases in $(r + \delta)$, and decreases in p_I all act to increase $\bar{k}$.

Policy action may be taken to shift L and to increase p_I. It is difficult for policy makers to get a handle on barrier-to-entry capital. The concept is subjective and hence difficult to measure. The analytical model makes the point that a "big push" is needed to dislodge the incentive to form k if $k_0 > \hat{k}$. But k is hard, if not impossible, to measure. Furthermore, First Amendment and due-process rights make it difficult to label elements of k barrier-to-entry capital. The main conclusions that can be drawn from an analytical exercise like that above is that the formation of k should be discouraged with as much vigor as possible. Two possible methods of doing this follow.

The first method operationalizes an idea of Vernon Smith: There should be a constitutionally guaranteed freedom of contract.[28] While a regulatory agency like the FCC cannot call a constitutional convention, it can change its posture concerning entry. Indeed, the natural-monopoly argument for entry restrictions should be used only to admit the possibility that it may be in the public interest to restrict freedom of contract. A two-stage procedure would be useful. Freedom of contract is not to be tampered with under the first stage. All entry is not restricted. Under the second stage, a case for entry restriction can get off the ground if natural monopoly can be shown. However, under the second stage, the burden of proof would be placed upon those who desire to restrict entry and hence restrict freedom of contract. The burden of proof should be extremely stringent.

Second, in view of the difficulty of detecting and proving the existence of barrier-to-entry capital, the economics of crime teaches us that, in situations where probability of detection and apprehension of an offense is low, the penalty upon conviction should be high, all other thing being equal.[29] Indeed, if triple damage awards provide optimal antitrust deterrent under normal circumstances, greater awards should be made in cases involving barrier-to-entry capital such as obstruction through the regulatory process.

A vivid example of the difficulty of establishing and securing conviction on barrier-to-entry grounds involving use of regulatory procedures to obstruct entry is contained in Judge Greene's *Opinion* on AT&T's motion to have the court dismiss the Government's case.[30] The Government trotted out episodes of AT&T's use of regulatory procedures to frustrate entrants in an attempt to establish the sham exception to the *Noerr–Pennington* doctrine.[31] After discussing the tension between First Amendment concerns and quoting Bork on the high potential for predation through regulatory processes, Judge Greene held that, with one exception, all of AT&T's petitioning activities were protected by *Noerr–Pennington*. He states, "to be a sham, the representation must go beyond the normal and legitimate exercises of the right to petition; it must amount to a subversion of the integrity of the process. And, absent special circumstance, this standard is not breached unless there is evidence of a series of misleading statements, of representations having the effect of actually barring access to an official body, or of an intent to mislead the body concerning central facts."[32]

The exception mentioned by Greene was AT&T's opposition to an FCC application by Datran where AT&T claimed that Datran had not demonstrated a need for its service and the economic and technical possibility of Datran's proposal was highly questionable.[33] He found, "Internal Bell documents introduced into evidence revealed, however, that defendants recognized at that very time that the Datran proposal was carefully planned and well financed, that they viewed it as a threat to AT&T's monopoly in network transmission; and that, in their opinion, a strategy of delay was necessary, to be implemented by a petition to the FCC requesting a general inquiry into the public interest aspects of the subjects raised by the Datran application. The court can reasonably infer from this evidence that AT&T's sole purpose in opposing the Datran application was to preserve its monopoly and that it well knew that the positions it took before the FCC were baseless."[34] This is an example of how tall an order it is to overturn a *Noerr–Pennington* defense—especially if the defendant's legal staff is alert in advising employees not to leave a trail of evidence.

It would probably enhance economic efficiency to give large multiple damages to any private litigant who overturns a *Noerr–Pennington* defense. This is so because the incentive to predation through regulatory processes is very strong but the likelihood of getting caught, if one is alert, is low. Furthermore, predation through regulatory processes is pure waste. Price predation, in contrast, makes goods available at lower cost. Regulatory-process predation, like war, is just pure waste. It should be dealt with severely.

Ironically, the exception mentioned by Greene to AT&T's successful *Noerr–Pennington* defense is the one where the economic case for entry restriction was the strongest. Let us explain. Datran wanted to build a digital network requiring a high fixed cost (i.e., a lumpy investment). It can be shown, under plausible assumptions, that a high fixed-cost investment tends to be built too soon in a competitive environment. There is a very strong *first-mover advantage* in building an entire digital network. Competition between two firms causes a race to build first. There is a tendency to build too soon. We show this using a simple model below.

Suppose that net surplus yielded by Datran's digital network is given by

$$\pi(t) = \pi_0 e^{\alpha t}$$

and let F be the fixed cost of building the network. Assume that, once built, the network is big enough to serve the entire market for the forseeable future. Assume that economies of scale are so large that it is optimal to build the whole network at once. This is just an idealization of scale economies and intertemporal economies of scope in construction.

Given these assumptions and given real interest rate r, the socially optimal time to build solves

$$\text{maximize} \int_t^\infty \pi_0 \, e^{(\alpha \, - \, r)t} \, dt \, - \, e^{-rt}F \tag{8.136}$$

The solution is given by

$$\pi_0 e^{\alpha t_2} = rF, \quad t_2 = \frac{1}{\alpha} \ln \frac{rF}{\pi_0} \tag{8.137}$$

The value to society of initializing the project at date τ_2 is given by

$$V(t_2) = \int_{t_2}^\infty \pi_0 \, e^{(\alpha-r)t} \, dt \, - \, e^{-rt_2} \, F \, = \, \pi_0 \frac{e^{-(r \, - \, \alpha)t_2}}{r-\alpha} \, = \, -e^{-rt_2}F \tag{8.138}$$

Now let competition ensue. Each firm will try to get the jump on the other. Suppose, in order to make our point quickly, that price discrimination is allowed so that the winner may sweep out the entire surplus each period. In this case competition to be first mover forces net profit $V(t_1) = 0$, i.e., $\pi_0 e^{\alpha t_1} = (r - \alpha)F$. In this case, competition costs society the entire surplus $V(\tau_2)$ given by (8.138). The welfare cost of competition, given by $V(\tau_2)$, explodes when r falls to the rate of growth of demand α. It falls in F and rises in π_0 for the economically meaningful case $r > \alpha$.

We abstracted from many aspects of reality in making the above argument. In reality the network depreciates, price discrimination may not be allowed, uncertainty may exist, etc. But the basic principle that competition to get in and get market share early in the game when large fixed installation costs are present or when large intertemporal economies of scope are present leads to social waste remains. This argument, which has a long history in the economics literature, is developed for a wide class of environments, including uncertainty and regulation, by Brock, Miller, and Scheinkman.[35]

To our knowledge AT&T never made the above public-interest argument in opposing Datran's application. Datran estimated that the network would cost $349 million, obviously a large F. The network was to be nationwide. The quantities π_0, α could have been estimated off of Datran's demand projections. Hence a rough, back-of-the-envelope calculation of welfare loss due to competition could have been made.

By hindsight, an argument like the public-interest argument laid out above might have had some chance in persuading the FCC that a potentially socially damaging race to build a network to serve the burgeoning demand for digital transmission should not take place. Once the FCC accepted the idea that development of digital transmission facilities should be regulated, AT&T could have argued that there was a substantial economy of scope between its own network and a digital data transmission network. Indeed, Data Under Voice seems to be an attempt to exploit such a scope economy. Once AT&T established that digital and voice transmission was a natural monopoly, due to networking economies perhaps, then AT&T could have opposed Datron on the grounds that competition would force AT&T to waste ratepayers' funds by installing its own network too early. This argument would follow that laid out above.

Pricing Without Regard to Cost

In *US* v. *AT&T*, the Government's pricing case relied on a novel theory called *pricing without regard to cost*. The allegation was that the only possible explanation for the practice of pricing without knowing one's costs was intent to drive out competition. The trial record reveals rather close questioning by Judge Greene of Government witness Bruce Owen on the argument that pricing without regard to cost (even though prices may not have systematically violated cost based standards such as Areeda–Turner) should be evidence of anticompetitive intent.[36] Judge Greene, in his opinion on dismissal, reserved judgment on the grounds that evidence of pricing without regard to cost was evidence of anticompetitive intent until defendants presented their side of the case.[37]

The precise meaning of *pricing without regard to cost* is not clear in the trial record. It seems to mean that AT&T priced in response to competitive threats without carefully calculating its own costs.[38] Consider the following dialogue between Government attorney J. Denvir and Government witness B. Owen.[39]

> DENVIR: Dr. Owen, assume that in fact AT&T has priced in the way we say that they have, and that after the fact someone comes along, for example AT&T, and can demonstrate that the prices were above cost, however measure[sic]; would that show, in your opinion, that there was no anticompetitive intent at the outset?"

> OWEN: No, that's just an accident. What matters is whether they set the prices with relationship to a knowledge of costs in the beginning, and not what turned out to be the case *ex post*.

In view of this background, one way to capture the spirit of the notion of pricing without regard to cost and to differentiate it from systematically pricing below cost is to represent it by a *price distribution*. Suppose that the dominant firm is risk neutral, potential entrants are risk averse, and all have access to the same cost function. To keep things simple, we shall assume constant marginal costs, zero fixed costs, and no capacity limitations. Suppose, for simplicity, that demand is linear

$$D(p) = a - \alpha p, \quad a > 0, \quad \alpha > 0 \tag{8.139}$$

and marginal cost is zero. Profit is given by

$$\pi = ap - \alpha p^2 \tag{8.140}$$

In the absence of entry, a concave profit function such as (8.140) generates a dominant firm incentive for stable prices. Let us explain. If p is a random variable with mean μ and variance σ^2, expected profit is given by

$$E\pi = a\mu - \alpha(r^2 + \mu^2) \tag{8.141}$$

Obviously, this is maximum when $\sigma^2 = 0$, $\mu = a/2\alpha$. Hence, stable prices are optimal.

Now let the dominant firm face a fringe of n identical, risk-averse, potential entrants having the same costs. In order to generate risk-averse behavior suppose that utility of profit of each entrant is given by

$$U(\pi) = \pi - \frac{1}{2b \, \pi^2}, \quad b > 0 \tag{8.142}$$

$$\pi \equiv px \tag{8.143}$$

Suppose that entrants must choose output before p is revealed. Then expected utility to an entrant producing x is given by

$$EU(\pi) = \mu x - \tfrac{1}{2} b \, x^2(\sigma^2 + \mu^2) \tag{8.144}$$

Optimum x is given by[40]

$$x^o = \frac{\mu}{b(\sigma^2 + \mu^2)} \qquad (8.145)$$

for $\mu \geq 0$. Hence, the dominant firm's expected profits are given by[41]

$$J(\mu, \sigma^2) \equiv E\{p(a - \alpha p - nx^o)\} \qquad (8.146)$$

$$= a\mu - \alpha(\sigma^2 + \mu^2) - \frac{n\mu^2}{b(\sigma^2 + \mu^2)}$$

Notice that, if $b = 0$ or $n = \infty$, then $\mu = 0$ and $J = 0$ so that the dominant firm is forced to price at cost ($p = c$) if entrants are risk neutral ($b = 0$) or there are infinitely many entrants ($n = \infty$).

Turn to the case of infinitely inelastic demand ($\alpha = 0$). In this case, the dominant firm can achieve arbitrarily large profit by choosing $\sigma^2 = +\infty$ and setting $p = \mu$ at any finite level achieving $J = ap$. It does not seem instructive to work out the optimum for general $\alpha > 0$. This is so because the quadratic utility function generates a silly supply function in the argument μ for entrants with no capacity constraints. We look at a more reasonable class of utility functions below. But, first, let us look at possible causes of risk aversion in the entrant's behavior relative to the dominant firm. If the dominant firm has many activities as well as the contested activity, if sunk costs of entry are present, if exit barriers exist, if capital markets are imperfect, and if entrants are small, then the entrants are likely to be more risk averse than the dominant firm. It is important to understand why the entrants may be risk averse.

If entrants can hit and run (i.e., the market is perfectly contestable), then they will not be risk averse.[42] If actuarially fair insurance can be written against price variation generated by the dominant firm, then entrants can purchase such insurance and thereby avoid financial stress should a loss emerge. Therefore, entrants would *not* be risk averse against dominant firm price variation if actuarially fair insurance could be written. The strategy of randomizing prices by a dominant firm would not deter entrants in the presence of such insurance. How likely is it that such insurance or a workable substitute for such insurance is present in actual practice?

Insurance is easy to write if the insured event is easily auditable, the probability of the insured event may be agreed upon by the contracting parties, the presence of insurance does not affect the probability of the insured event (moral hazard), and adverse selection effects are not present. It is also quite likely that there are at least initial economies of scale in the technology of insurance service. Hence, insurance against a very specialized event such as price randomization by a dominant firm is likely to be expensive. Other difficulties are likely to emerge. Costs must be audited if the excess of price over cost is to be measured and insured. Moral hazard and adverse selection will probably render such profit insurance too difficult to operationalize.

What about insurance on price alone? Once the dominant firm knows about the existence of insurance against one price distribution they could simply change the distribution. This would frustrate insurer and insuree alike. The power of the dominant firm to change the distribution will probably make contract agreement between insurer and insuree a costly process. Imperfections in contract compliance such as bankruptcy statutes may create problems. Are there substitutes for market insurance?

Entrants can self-insure against small variations by use of their own funds. They would be approximately risk neutral against such variations. But if the variation is large relative to entrant liquidity, risk aversion would appear unless they could borrow at reasonable terms. However, any lender could realize that the dominant firm's randomization could dramatically affect the ability of the entrant to pay off the loan. Realizing this, the lender would mark up the interest rate. The facing of upward-sloping interest-rate schedules in the event of financial embarrassment would lead to "as if" risk aversion behavior by entrants. Are there substitutes for lenders? Since we are dealing here with a form of predation, it is natural to search through Easterbrook's extensive list for counterstrategies.[43]

One strategy that entrants might try is to approach their customers for "stay alive" insurance. Customers have a strong incentive to preserve entrant viability. But free-rider problems abound. Free-rider problems may be attenuated by entrants writing long-term supply contracts contingent on business volume. Indeed, it may be possible to extend Easterbrook's and Grossman's ideas on competition in the space of contracts and rational expectations on predator's and predatee's part to show that predation via price randomization is fruitless.[44] The argument that long-term contracts and rational expectations render most predatory strategies incredible permeates Easterbrook's persuasive piece.

There are two issues neglected by Easterbrook and, to our knowledge, the predation literature, in general. That is (a) the role of "reforms" in the bankruptcy law such as the 1979 Bankruptcy Reform Acts; and (b) complementary effectiveness of a diversified program of predatory strategies taken by a multiproduct regulated firm against specialized competitors.

Let us look at bankruptcy law and contract enforceability. Loose bankruptcy law makes it difficult to write enforceable contracts with lenders and potential customers. If lenders and potential customers realize that the threat of potential declaration of bankruptcy looms large in the event of contract noncompliance, then the likelihood of this unhappy event will be factored into the contract terms. Indeed, if reform of bankruptcy law leaves creditors with little power to collect payment on unfulfilled obligations, the viability of Easterbrook's contract strategy against attempted predation may be threatened. Williamson's argument that integration is a substitute for inherently faulty contracts lays out other practical problems that must be solved in writing viable contracts.[45] Turn now to regulated dominant firms.

Regulated dominant firms can indulge in legal predation. They can also generate uncertainty in the terms of access to bottleneck facilities (e.g., network access in telecommunications). These are cases where the predator can spend x dollars to create $x + y$ dollars worth of injury to competitors. This injury-expenditure ratio is usually not available for price predation.[46] Hence, a regulated dominant firm can generate a lot of uncertainty in the operating environment of a potential competitor. Although the competitor can retaliate with his own injection of uncertainty (e.g., urging that an FDC floor be placed under AT&T's tariffs in the Docket 18128 proceedings), it can be argued that incumbent's accumulated experience (at least some of which was accumulated at rate-payer expense) would give it an advantage at playing the regulation game.[47]

In any event, the addition of price uncertainty to the cloud of uncertainty that already exists may be enough to induce risk-averse behavior by entrants even after all the countermeasures listed by Easterbrook adjusted for practical implementation have been exhausted. It may generate an injury–expenditure ratio greater than one for a predator. It is important to keep in mind, however, that the more concave the revenue function in price the more costly to the dominant firm is a given amount of variance. After this long digression, let us return to an examination of the impact of price uncertainty upon a more general class of utility functions.

It is worthwhile to consider more general utility functions than the quadratic because the quadratic has theoretical flaws.[48] We saw one flaw in the rather odd behavior of entrant supply as a function of μ. Consider $U(\pi)$, where $U(0) = 0$, $U'(\pi) > 0$, $U''(\pi) < 0$ (maintaining the assumption of zero marginal cost and fixed cost $F \geq 0$ for simplicity). Each entrant chooses x^o to solve

$$\underset{x \geq 0}{\text{maximize}}\ E\ U(px - F) \tag{8.147}$$

FONC are given by

$$E[U'(px^o - F)p] = 0, \quad \text{if } x^o > 0 \tag{8.148}$$

$$E[U(px^o - F)] > 0, \quad \text{if } x^o > 0 \tag{8.149}$$

It is clear that, if the utility function has a vertical asymptote at same level π, possibly negative, then the dominant firm may block entry by simply choosing a distribution of prices with arbitrary mean but placing mass below the survival threshold π.

In general, the impact of a *small* increase in riskiness upon the choice of x^o of an active entrant depends upon the shape of the function

$$\psi(p,x,F) = U'(px - F)p \tag{8.150}$$

as a function of p. If ψ is concave in p an increase in riskiness of p leads to a fall in ψ and, hence, a fall in x^o. The reverse is true if ψ is convex in p.

Although the impact upon x^o (of an ongoing firm) of a small increase in riskiness in p is ambiguous, in the sense that it depends on such abstruse notions as convexity or concavity of ψ, the impact of a large enough increase in riskiness is unambiguous. That is, the entrant will shut down if the variability of p is large enough and U has a vertical asymptote. This result extends to general concave utilities. Let us show why.

Let p denote the price random variable. Suppose by way of contradiction that optimum $x^o > 0$ exists for $\bar{p}$. Define

$$J(\bar{p}) \equiv \max_x EU(px) \equiv EU(\bar{p}x^o) \tag{8.151}$$

Pick $\bar{p}$ such that $E\bar{p} = 0$. Then $J(\bar{p}) < 0$. This is so because, if the variance of $\bar{p}$ is nonzero, $U'' < 0$ everywhere implies

$$J(\bar{p}) = EU(\bar{p}x^o) < U(E\bar{p}x^o) = 0 \tag{8.152}$$

by strict concavity of U. Since $J(\bar{p}) < 0$ the entrant will shut down if the dominant firm chooses $\bar{p}$. This is a contradiction to the presumed optimality of $x^o > 0$. It is a general proposition that a firm facing the distribution $\bar{p}$ will produce less than it would if it faced the mean $\mu \equiv E\bar{p}$.[49]

One may extend the above analysis to include nonzero marginal and variable cost as well as random fixed cost. The conclusion remains the same.[50] The dominant firm may find random-pricing strategies that assure it positive expected profit but will induce risk-averse entrants to stay out. Hence, a dominant firm who is less risk averse than its competitive fringe, may eliminate the fringe simply by randomizing price.

This conclusion requires comment. First equally efficient, indeed more efficient risk-averse entrants, may be eliminated by random pricing strategies. If profits are concave in p, the dominant firm injures itself by using randomization. If profits are not concave in p, the strategy is costless to the dominant firm. The tradeoff is subtle when dominant firm profits are strictly concave as a function of p. The viability of a random-pricing strategy depends upon the degree of concavity of profits as a function of p versus the degree of risk aversion of entrants. Let us explain further.

If profits are approximately linear in p (as they would be if demand were perfectly inelastic and variable costs were linear), then a random pricing strategy appears promising. If entrants have a threshold level of profits that they need to survive, then a random strategy placing enough mass below the threshold to sink the entrants yet concentrating the bulk of the mass at one point to avoid loss from the demand side appears promising.

Second, it may be difficult for courts and other agencies to detect and prosecute random pricing accurately. Costs change. They are not easy to calculate. Market conditions change. Pricing is a dynamic trial-and-error process in the real world. It is hard to distinguish innocent trial-and-error pricing from random pricing with intent to exclude competitors. Presumably one would look for undue price movement relative to movements in measures of costs in an attempt to deter random pricing.

We have shown that pricing without regard to cost (i.e., random pricing) is profitable to the dominant firm when (a) entrants must decide output in advance of price, (b) entrants are more risk averse than the dominant firm, and (c) demand is relatively inelastic. Randomizing price will fail to deter entrants if they can costlessly adjust output in response to price. Our model was primitive (i.e., we just assumed that entrants had to produce before price was revealed). We do not believe, however, that a more realistic model such as an adjustment-cost model would overturn the basic result that a dominant firm facing relatively more risk-averse entrants and a linear (in price) profit function may find it profitable to randomize price.

It is important to keep in mind, however, that if entrants can produce in response to price, the result may be reversed. Let us explain. If entrants are risk neutral, randomizing price benefits them. This is so because maximized profits is convex in p. Hence, an increase in riskiness increases value. Therefore, if entrants are not very risk averse, an increase in riskiness may actually increase value rather than decrease it. The economics is easy to understand. Risk-neutral entrants, which can easily adjust output in response to price increases, gain more in periods of high prices than they lose in periods of low prices.

We conclude that pricing without regard to cost is an idea that must be handled with caution. If entrants can adjust output rapidly in face of fluctuating prices, then the dominant firm creates bonuses for entrants in periods of high prices and does little injury to entrants who contract during periods of low prices. Random prices need sunk costs as an ally in order to be an effective weapon in the predatory arsenal. Brock and Evans documented the important role that sunk costs played in the MCI story.[51] Furthermore, it would seem to be relatively costly for MCI to adjust output in the short run. Hence, a strategy of randomizing price, if actually used by AT&T, may have helped to slow MCI's growth. However, it is difficult to actually measure the extent of price randomization, if actually used by AT&T.

NOTES

1. For a discussion of Ramsey pricing see F.P. Ramsey, "A Contribution to the Theory of Taxation," *Economic Journal*, March 1927, pp. 47–61; William Baumol and David Bradford, "Optimal Departures from Marginal Cost Pricing," *American Economic Review*, June 1970, pp. 265–283; and E.E. Zajac, *Fairness or Efficiency: An Introduction to Public Utility Pricing* (Cambridge: Ballinger, 1978).

2. Elizabeth Bailey and Robert D. Willig have made such an argument in "Ramsey Optimal Pricing of Long-Distance Telephone Service," in John Wenders, ed., *Pricing in Regulated Industries*, Vol. I (Keystone, CO: Mountain States Telephone and Telegraph Co., 1977.)

3. This *superelasticity* is a function of the own and cross-price elasticities of demand. See Baumol and Bradford, *op. cit.*, for formulas.

4. Bailey and Willig, *op. cit.*, and S.C. Littlechild, *Elements of Telecommunications Economics* (London: Institute of Electrical Engineers, 1979).

5. See William A. Brock and David S. Evans, *Federal Regulation of Small Business* (New York: Holmes & Meiers, forthcoming, 1984), who develop a model along these lines.

6. See J. Friedman, *Oligopoly and the Theory of Games* (Amsterdam: North Holland, 1977).

7. Ronald Breautigam, "Optimal Pricing with Intermodal Competition," *American Economic Review,* March 1979, pp. 219–240. Another possible formulation of PRSB would be to replace C by $C + C_e$ in (8.25).

8. See for example I. Gelfand and S. Fomin, *Calculus of Variations* (Englewood Cliffs, NJ: Prentice Hall, 1963).

9. For a discussion of Docket 18128 see Walter G. Bolter, "The FCC's Selection of a 'Proper' Costing Standard after Fifteen Years—What Can We Learn from Docket 18128," in Harry Trebing, ed., *Assessing New Pricing Concepts in Public Utilities* (East Lansing, MI: Graduate School of Business Administration, Michigan State University, 1978), pp. 333–372.

10. P. Areeda and D. Turner, "Predatory Pricing and Related Practices Under Section 2 of the Sherman Act," Harvard Law Review, February 1975, pp. 697–733. For critiques of this test see Frederic Scherer, "Predatory Pricing and the Sherman Act: A Comment," *Harvard Law Review,* March 1976, pp. 869–900; Oliver Williamson, "Predatory Pricing: A Strategic and Welfare Analysis," *Yale Law Journal,* December 1977, pp. 284–340. Scherer noted that the marginal cost test advocated by Areeda and Turner would be a "defendant's paradise" because of the difficulty of estimating marginal cost.

11. Stephen Turnovsky and William A. Brock, "Time Consistency and Optimal Government Policies in Perfect Foresight Equilibrium," *Journal of Public Economics,* April 1980, pp. 183–212; William A. Brock and W.D. Dechert, "Dynamic Ramsey Pricing," SSRI Working Paper No. 8253 (Madison, WI: Social Systems Research Institute, University of Wisconsin at Madison, June 1982).

12. Brock and Dechert, *ibid.*

13. This chapter focuses on regulations that limit the rate of return on the rate base. See Harvey Averch and Leland L. Johnson, "Behavior of the Firm under Regulatory Constraint," *American Economic Review,* December 1962, pp. 1052–1069.

14. Breautigam, *op. cit.*

15. As shown below, this insertion is incorrect under rate-of-return regulation because cost is not minimized under regulation. Also note that a different formulation of the Ramsey problem would be to replace C by $C + C_{ei}$ in (8.83).

16. William Baumol and Alvin Klevorick, "Input Choices and Rate-of-Return Regulation: An Overview of the Discussion," *Bell Journal of Economics and Management Science,* Autumn 1970, pp. 162–190.

17. Gerald Brock, *The Telecommunication Industry* (Cambridge: Harvard University Press, 1981), Chapter 2.

18. C.C. von Weizsäcker, "A Welfare Analysis of Barriers to Entry," *Bell Journal of Economics,* Autumn 1980, p. 400. Also see his *Barriers to Entry* (Berlin: Springer-Verlag, 1980).

19. This result suggests a possible line of defense that AT&T might have taken for its famous interconnection restrictions through customer premises requirements. This model, together with relevant empirical evidence, might have generated more convincing evidence than testimony about how the Bell System protects us from the Russians.

20. For statement of this argument see Robert Bork, *The Antitrust Paradox* (New York: Basic Books, 1978); Frank H. Easterbrook, "Predatory Strategies and Counterstrategies," *University of Chicago Law Review,* Spring 1981, pp. 263–311; and John McGee, "Predatory Price Cutting: The Standard Oil (N.J.) Case," *Journal of Law and Economics,* October 1958, pp. 133–167.

21. S. Salop, ed., *Strategic Predation and Antitrust Analysis* (Washington, DC: Federal Trade Commission, 1981).

22. See Easterbrook, *op cit.*

23. See the articles in Salop, *op. cit.* Even the recent articles by Milgrom and Roberts and Kreps and Wilson rely on asymmetric information to reach the conclusion that price predation may pay in Selten's chain store game. See P. Milgrom and J. Roberts. "Predation, Reputation, and Entry Deterrence," Research Paper No. 600 (Evanston, IL: Graduate School of Business, Northwestern University, 1981) and D. Kreps and R. Wilson, "Reputation and Imperfect Information," *Journal of Economic Theory,* August 1982, pp. 280–312.

24. Paul Joskow, "Inflation and Environmental Concern: Structural Change in the Process of Public Utility Regulation," *Journal of Law and Economics,* October 1974, pp. 291–328.

25. See Bork, *op. cit.,* for a discussion of predation through the political process.

26. William A. Brock and Stephen P. Magee, "The Economics of Special Interest Politics: The Case of a Tariff," *American Economic Review,* May 1978, pp. 246–250.

27. For proof, see Brock and Dechert, *op. cit.*

28. Vernon Smith, "Comments," in Salop, *op. cit.,* pp. 579–605.

29. Gary Becker, "The Economics of Crime and Punishment," *Journal of Political Economy,* March/April 1968, pp. 169–237.

30. Judge Harold Greene, *Opinion on Defendants' Motion for Involuntary Dismissal Under Rule (41b),* in *US* v. *AT&T,* September 11, 1981.

31. *Eastern Railroad Presidents' Conference* v. *Noerr Motor Freight, Inc.,* 305U.S.127 (1961) and *United Mine Workers* v. *Pennington,* 318U.S.657 (1965). Also see, US Department of Justice, *Plaintiff's Memorandum in Opposition to Defendants' Motion for Involuntary Dismissal Under Rule 41(b),* in *US* v. *AT&T,* August 16, 1981.

32. Greene, *op. cit.,* p. 41.

33. Greene, *op. cit.,* p. 42.

34. Greene, *op. cit.,* p. 42.

35. William A. Brock, Robert Miller, and José A. Scheinkman, "Natural Monopoly and Regulation," unpublished paper.

36. Bruce Owen, testifying in *US* v. *AT&T,* Tr. 10960–10969.

37. Greene, *op. cit.*

38. See, for example, Owen Tr. 10962.

39. Owen Tr. 10968-10969.

40. A well-known problem involved in trading off the analytic tractability of quadratic utility functions against sensible economic results appears in the supply function (8.145). Put $r^2 = 0$. Then

$$x^o = 1/b\mu$$

(i.e., supply is falling in price μ). This is clearly silly.

41. Notice that the formulation of profit function (8.146) assumes that the dominant firm can choose its price first and then use commonly available technology to produce the demand, $a - \alpha p - nx^o$, that is generated. Yet, the entrants are assumed to have to produce their output in advance, although they are assumed to have rational expectations in the sense that they know the distribution of p. Obviously, if entrants can produce after p is announced, then random pricing will not deter them. The reader might well complain that an artificial advantage to the dominant firm above and beyond price leadership is built into the very formulation of our problem. Such an objection is irrelevant for the case of infinitely inelastic demand. The objection is more important the more elastic is demand. It is beyond the scope of this article to discuss a more elaborate model which would be needed to justify our type of formulation of the problem.

42. On contestability, see William Baumol, John Panzar, and Robert Willig, *Contestable Market and the Theory of Industry Structure* (San Diego: Harcourt, Brace, Jovanovich, 1982). Obstacles to contestable market are discussed in Chapters 4 and 9 of the present volume.

43. Easterbrook, *op. cit.*

44. Easterbrook, *op. cit.,* and S.J. Grossman, "Nash Equilibrium and the Industrial Organization of Markets with Larged Fixed Costs," *Econometrica,* September 1981, pp. 1149–1172.

45. Oliver Williamson, "The Vertical Integration of Production: Market Failure Considerations," *American Economic Review,* May 1971, pp. 114–115.

46. See our discussion in the previous section.

47. Bruce Owen and Ronald Breautigam, *The Regulation Game* (Cambridge: Ballinger, 1978).

48. See M. Rothschild and J. Stiglitz, "Increasing Risk I: A Definition," *Journal of Economic Theory*, September 1970, pp. 225–243 and "Increasing Risk II: Its Economic Consequences", *Journal of Economic Theory*, March 1971, pp. 66–84 and P. Diamond and M. Rothschild, *Uncertainty in Economics: Readings and Exercises* (New York: Academic Press, 1978), p. 141.

49. Rothschild and Stiglitz, *op. cit.*, p. 82.

50. For example, using (8.144), let variable cost be 0 and let fixed cost be F. Then

$$EU(\pi) = EU(px - F) = \frac{1}{2b} \left\{ \frac{\mu^2}{(\mu^2 + \sigma^2)} [(1 + bF)^2 - 2bF - b^2F^2] \right\}$$

at optimum x^o of

$$x^o = \frac{\mu(HbF)}{b(\mu^2 + \sigma^2)}$$

But

$$EU(\pi) < 0 \quad \text{if and only if} \quad \mu^2 < \sigma^2 bF(2 + bF)$$

Hence the dominant firm can keep entrants out by choosing σ^2 large enough relative to μ^2.

51. See Chapter 4 of this volume.

Chapter 9

Free Entry and the Sustainability
of Natural Monopoly:
Bertrand Revisited by Cournot*

William A. Brock and José A. Scheinkman

The sustainability literature has generated wide professional interest.[1] This literature examines whether a natural monopoly can ward off uninnovative entry and finds many situations where a natural monopoly cannot and is therefore nonsustainable.[2] Regulators who believe these situations are plausible face an uncomfortable tradeoff. On the one hand, they must prohibit entry in order to maintain the natural monopoly, which is the most economical organization of production. On the other hand, they must forego competition as a tool for disciplining the natural monopoly.

This literature suggests that policy makers should take the nonsustainability problem seriously. Papers by Baumol, Panzar, and Willig cast an even darker cloud of doubt over the desirability of free entry.[3] Willig and Baumol show that a natural monopoly that must build up capital in anticipation of growing demand is often nonsustainable. Panzar shows that a natural monopolist who faces a piggyback entrant has a serious nonsustainability problem. A piggyback entrant buys cheap inputs from the monopolist and sells output that competes with one of the monopolist's outputs. Panzar has argued that airlines may have had nonsustainable natural monopolies when airline markets were smaller and that deregulation could have been harmful earlier in this century.[4]

Before we outline the results offered in this chapter, let us define sustainability heuristically. An n-goods natural monopoly is said to be sustainable at $(\bar{p}, \bar{q})$, $\bar{p} \in R^n_+$, $\bar{q} \in R^n_+$ if any entrant contemplating production plan $q^e \in R^n_+$ anticipates negative profits at price $\bar{p}$. Here R^n_+ denotes the nonnegative orthant of R^n. This notion of sustainability is reminiscent of Bertrand's price equilibrium in oligopoly theory.[5] Bertrand assumes that entrants anticipate that the monopolist's prices

*We thank Avinash Dixit, R. J. Reynolds, and Paul Romer for helpful and perceptive comments on an earlier draft of this essay.

remain fixed. It is the only notion used in the sustainability literature we have seen.

Cournot's quantity equilibrium is another concept used in oligopoly theory. His notion is "dual" to that of Bertrand in that entrants anticipate that the monopolist's quantities remain fixed but that prices adjust to absorb the extra output of the entrant. This dual notion of equilibrium motivates a dual notion of sustainability. We say that $(\bar{p}, \bar{q})$ is quantity sustainable if any contemplated production plan q^e makes negative profits at the prices that absorb $\bar{q} + q^e$. Quantity sustainability is the natural multiproduct extension of the Sylos Postulate.[6] According to the Sylos Postulate, entrants conjecture that the incumbent will maintain output at the preentry level and prices will adjust to absorb the total of the entrant's and incumbent's output.[7]

This chapter compares the notions of price sustainability and quantity sustainability. We derive five major results. First, if demand satisfies a gross substitutes property, then price sustainability implies quantity sustainability. Second, a single-product natural monopoly is quantity sustainable at the Ramsey point provided that a mild assumption on average cost is met. Consequently, Panzar and Willig's price nonsustainability example is always quantity sustainable.[8] Third, in the one-good case, every natural oligopoly is quantity sustainable at the Ramsey allocation under the same mild condition on average cost. Fourth, we present partial results giving sufficient conditions for quantity sustainability when there are many goods. Fifth, if a price-quantity pair is not quantity sustainable then, unlike the price nonsustainable case, a welfare-improving reallocation can always be made.

Quantity sustainability is important to regulators for two reasons. First, a natural monopoly that is not price sustainable but is quantity sustainable presents a less serious policy problem than a monopoly that is not quantity sustainable when the technology requires large sunk costs.[9] Price sustainability assumes that the monopolist's prices remain fixed. But if entry forces losses upon the monopolist and if the monopolist has large sunk costs, an entrant has little reason to anticipate that monopoly prices will not change. Quantity sustainability assumes that the monopolist's quantities remain fixed. A monopolist who is not quantity sustainable cannot keep entrants out even when they anticipate that he will leave his entire output on the market and let prices fall.[10] Thus, quantity nonsustainability is a much more serious problem than price nonsustainability.

Second, price and quantity sustainability provide crude measures of the monopolist's ability to withstand entry threats without protection by the regulator. On the one hand, if he is price sustainable he is secure indeed. He is in such a strong position that entrants anticipate negative profits without taking into account the depressing effect of extra output on prices if the monopolist holds fast to his old production level. On the other hand, if he is not quantity sustainable he cannot even ward off entrants with the threat of crushing market prices by leaving his output at its present level even though he will make less if the entrant's threat is carried out. In order to ward off entrants, he must use more sophisticated strategies that we do not discuss in this chapter. He is a feeble monopolist. Hence, the dual concept of quantity sustainability helps isolate serious cases of price nonsustainability.

Before we get into the substance of this chapter, we wish to make a disclaimer. A proper treatment of the problem would endogenize the formation of entrants' conjectures. We are operating more in the spirit of early literature (e.g., Modigliani), by treating quantity maintenance as an incumbent's most unfavorable response toward the entrant.[11] Price maintenance can be viewed as the incumbent's most favorable response toward the entrant. It began to be realized that the incumbent could profitably threaten to make even more aggressive responses than output maintenance. Whether these threats are credible has been analyzed from the angles of capacity commitment[12] and reputation.[13]

Although price maintenance and quantity maintenance are naive conjectures on the part of entrants, economists have found it useful to study these simple strategies. Dixit has located conditions where the two conjectures make economic sense.[14] As he points out, the conjecture of price maintenance has economic relevance in the case of a perfectly contestable market. The theory of perfectly contestable markets finds conditions under which postentry oligopoly is irrelevant and strategic entry deterrence impossible.[15] A perfectly contestable market has the following properties: (a) all producers have access to the same technology; (b) this technology may have scale economies such as fixed costs, but must not involve any sunk costs; (c) incumbents can only change prices with a nonzero time lag; and (d) consumers respond to price differences with a shorter lag.[16] In this case, hit-and-run strategies by entrants can enforce a type of as-if competitive equilibrium with the attendant socially beneficial results of competitive equilibrium.

Dixit points out that the concept of quantity sustainability is relevant for the "polar" case where: (a) incumbent's costs are all sunk, in which case it is in the incumbent's self-interest to maintain output levels after entry; and (b) the market is an efficient auction market with the result that prices adjust quickly to absorb the additional output coming from the entrant. The resulting equilibrium does not have the strong socially desirable properties of the price-sustainable case above.[17]

This chapter has six sections. Section one presents basic definitions. Section two shows that price sustainability implies quantity sustainability under general conditions for a single-product monopoly. Extensions of this result are developed for the multiproduct monopoly. Section three demonstrates that nonquantity sustainability is a signal that society is better off with two producers rather than one. This result is always true but price discrimination may be necessary to ensure that both entrant and incumbent break even after entry has occurred. Postentry break-even profit levels for both entrant and incumbent may, under quite general conditions, be achieved without price discrimination for the single-product case. Section four presents partial results for the multiproduct case. Section five compares the necessary conditions for price sustainability with the necessary conditions for quantity sustainability. Section six discusses the relevance of sustainability to regulation of entry.

Definitions

In order to save space, we use the analytical framework and notation used by Panzar and Willig (PW).[18] We also follow Baumol, Bailey, and Willig (BBW) closely.[19] Recall that $N = \{1, 2, \ldots, n\}$ denotes the set of products and services provided by the monopolist.

Definition 1.[20] Production of the positive vector of outputs $y^m \equiv (y_1^m, \ldots, y_n^m)$ is characterized by natural monopoly if and only if

$$C(y^m) < C(y^1) + \cdots + C(y^k) \tag{9.1}$$

$$\forall\, y^1, \ldots, y^k \text{ s.t. } \sum_{i=1}^{k} y^i = y^m$$

and at least two $y^i \neq 0$.

Definition 1a.[21] The *industry* is a natural monopoly if condition (9.1) holds throughout the relevant range (i.e., for all y^m that are consistent with (at least) zero economic profits for the monopolist).

We repeat the PW definition for sustainability but we shall call it price sustainability.

Definition 2.[22] The price vector p^m is *price* sustainable if and only if $p_S^e y_S^e - C(y_S^e)$, ≤ 0, for all $y_S^e \geq 0$, for all $S \subseteq N$, $p_S^e \leq p_S^m$, $y_S^e \leq Q_S(p_S^e, p_{\bar{S}}^m)$, with $y_S^e \neq Q_S(p^m)$. It is also required that $\pi(p^m) \equiv Q(p^m) - C[Q(p^m)] \geq 0$. Here $y = Q(p)$ denotes a system of n continuously differentiable market demand functions, and y_S denotes the vector $\{y_j\}_{j \in S}$.

Definition 2a. The *natural monopoly* is price sustainable if and only if there exists at least one sustainable price vector.

In contrast to Panzar and Willig and the rest of the sustainability literature cited above we offer

Definition 2′. The price vector p^m (with associated quantity vector $y^m \equiv Q(p^m)$) is *quantity* sustainable if and only if $D_S(y^m + y_S^e)y_S^e - C(y_S^e) \leq 0$, for all $y_S^e \geq 0$, for all $S \subseteq N$. It is also required that $\pi^m \equiv p^m y^m - C(y^m) \geq 0$. Here D denotes the inverse function of $Q(\cdot)$ which is assumed to exist.

Definition 2′a. The natural monopoly is *quantity* sustainable if and only if there exists at least one quantity sustainable price vector.

Roughly speaking it is "easier" for a monopolist to be quantity sustainable than it is to be price sustainable if demand is *downward sloping*. The property of downward sloping that we want is embodied in

Assumption 1. Demand D is downward sloping (DSD) at y^m i.e., $[D(y^m + y_S) - D(y^m)]y_S < 0$ for all $y_S \geq 0$, $y_S \neq 0$, $S \subseteq N$.

Price Sustainability Implies Quantity Sustainability

We are now ready for

Theorem 1. *Consider the one-dimensional case. Suppose that demand is downward sloping at y^m. Suppose further that p^m, y^m is price sustainable. Then p^m, y^m is quantity sustainable.*

PROOF: Assume quantity nonsustainability but price sustainability at p^m, y^m. Consider an entry plan $y^e > 0$ such that $D(y^m + y^e)y^e \geq C(y^e)$. By downward-sloping demand we have

$$D(y^m + y^e)y^e \leq D(y^m)y^e \equiv p^m y^e \qquad (9.2)$$

But price sustainability implies

$$p^m y^e < C(y^e)$$

provided that

$$p^e \equiv D(y^m + y^e) \leq p^m, \; y^e \leq Q(p^e) \qquad (9.3)$$

The first part of inequality (9.3) holds by DSD. The latter part holds because $y^m + y^e = Q(p^e) > y^e$. Hence we have

$$C(y^e) \leq D(y^m + y^e)y^e \leq p^m y^e < C(y^e)$$

a contradiction.

Theorem 1 is contained in the more general Theorem 2 stated below. But the ideas are clearer in the single-product case. Some restrictions on demand are necessary to ensure that price sustainability implies quantity sustainability in the multigood case. One needs to control the impact that quantity increases of one set of goods can have on demands for the rest of the goods. We have

Theorem 2. *Assume that the n goods demand system satisfies, for all i,j*

$$\text{(a)} \; \frac{\partial D_i}{\partial y_j} \leq 0, \; i \neq j, \; \frac{\partial D_i}{\partial y_i} \leq 0; \qquad \text{(b)} \; \frac{\partial Q_i}{\partial p_j} \geq 0, \; \frac{\partial Q_i}{\partial p_i} \leq 0 \qquad \text{(GS)}$$

Then price sustainability implies quantity sustainability.

PROOF: We proceed by way of contradiction. Suppose that (p^m, q^m) is not quantity sustainable. Then there is $y_S^e \geq 0$, $y_S^e \neq 0$ such that

$$D_S(y_S^m + y_S^e, y_{(S)}^m)y_S^e > C(y_S^e) \tag{9.4}$$

Prices p_S^e, $p_{(S)}^e$ are determined by

$$p_S^e \equiv D_S(y_S^m, y_S^e, y_{(S)}^m), \quad p_{(S)}^e \equiv D_{(S)}(y_S^m + y_S^e, y_{(S)}^m) \tag{9.5}$$

If we can find $p_S \leq p_S^m$, $y_S \leq Q_S(p_S, p_{(S)}^m)$, $p_S y_S - C(y_S) > 0$ we shall have a contradiction to price sustainability of p^m, y^m. A good candidate is

$$p_S = p_S^e, \quad y_S = y_S^e \tag{9.6}$$

Gross substitutes (GS) implies (by its very definition)

$$p_S^e \equiv D_S(y_S^m + y_S^e, y_{(S)}^m) \leq D_{(S)}(y_S^m, y_{(S)}^m) \equiv p_S^m , \tag{9.7}$$

$$p_{(S)}^e \equiv D_{(S)}(y_S^m + y_S^e, y_{(S)}^m) \leq D_{(S)}(y_S^m, y_{(S)}^m) \equiv p_{(S)}^m \tag{9.8}$$

The gross-substitutes assumption and (9.8) imply

$$y_S^e \equiv Q_S(p_S^e, p_{(S)}^e) \leq Q_S(p_S^e, p_{(S)}^m) \tag{9.9}$$

Inequality (9.9) follows from the gross-substitutes assumption because $p_{(S)}^e \leq p_{(S)}^m$ causes a rise in demand for goods S in a gross-substitutes system.

The break-even constraint $p_S y_S - C(y_S) > 0$ is satisfied for p_S^e, y_S^e by definition of quantity nonsustainability. Hence, we have found a contradiction to price sustainability of p^m, q^m. This ends the proof.

REMARK 1: The above proof may easily be adapted to the case of independent demands.

REMARK 2: The assumption (GS) is obviously related to the assumption of gross substitutes well-known from general equilibrium theory. Under general conditions $\partial D_i / \partial y_j \leq 0$, $i \neq j$, $\partial D_i / \partial y_i \leq 0$ is implied by $\partial Q_i / \partial p_j \geq 0$, $\partial Q_i / \partial p_i \leq 0$.[23]

There is a strong presumption that price sustainability implies quantity sustainability even if demand satisfies only the weaker condition DSD. For example, consider BBW's *pseudo-revenue function* $R^e(y^e) = \sum_{i=1}^{n} D_i(y^m)y_i^e$.[24] Since entrants must charge less than the monopolist, entrant revenue from y^e is less than $R^e(y^e)$. BBW impose enough assumptions on demand and cost to ensure that the pseudo-revenue hyperplane H tangent to the *hyperbagel* of points where revenue equals cost lies below the cost function for all entry threats lying in the potentially profitable set $T \equiv \{y | R(y) \geq C(y)\}$.[25] They show that the point of tangency of H and the hyperbagel is the Ramsey point.

Now $H \equiv \{(R^m, y) | R^m \equiv R(y^m) \equiv \sum_{i=1}^{n} D_i(y^m)y_i\}$. We have

Theorem 3. *Assume that H lies below the cost function over T. Also suppose that demand satisfies DSD. Then the Ramsey point is price sustainable. The Ramsey point is also quantity sustainable against challenges $y \in T$.*[26]

PROOF: By BBW's discussion, price sustainability is just the property that H be below cost over T. That is, for all $y \in T$, $y \neq y^m$ we must have

$$D(y^m)y < C(y) \qquad \text{(PS)}$$

But, quantity sustainability concerns the y^m-*adjusted pseudo-revenue function* $R(y;y^m) \equiv D(y^m + y)y$. DSD implies that

$$R(y;y^m) \leqslant D(y^m)y$$

Hence (PS) implies quantity sustainability for challenges $y \in T$. This ends the proof.

We have shown that price sustainability implies quantity sustainability provided that "demand" is well behaved. It is well known, however, that demand functions with arbitrary derivatives can be constructed when there is a finite collection of consumers with arbitrary preferences and income.[27] Of course, in such a case, consumer surplus types of arguments are no longer valid but one may still study Ramsey points. Thus, it seems that, in general, a Ramsey point could be price sustainable but not quantity sustainable.[28]

Quantity Nonsustainability Implies the "Natural" Monopoly Is Illusory

In what follows we need to be able to measure *welfare*. All we want to do is compare the notions of price sustainability and quantity sustainability. Here we assume

Integrability. There is $B:R^n_+ \to R$ such that $\nabla B(y) \equiv (\partial B/\partial z)(y) = D(y)$ for all $y \in R^n_+$.

We are aware that integrability is a strong assumption. Ideally, one would like to conduct our exercise replacing B by a welfare function defined on a vector of indirect utility functions. But we doubt that this refinement will blunt the basic point that we want to make. That is, "true" natural monopolies are much easier to quantity sustain than to price sustain.

Strong results may be had for the single-product case even if *AC* is not falling but is U-shaped. As Panzar and Willig and Baumol, Bailey and Willig point out, subadditivity does not imply that average cost is falling even in the single-product case.[29,30] This possible U-shaped behavior of average cost is what allows these authors to construct examples where the Ramsey point is not price sustainable in the single-product case.

Recall Panzar–Willig's one-dimensional example. In the PW example, (p^m, y^m)

is the Ramsey point and entrant y^e takes a profit at price p^m because AC is U-shaped on $(0,y^m)$ even though cost is subadditive on $(0,y^m)$. The example is not price sustainable. Can the demand curve be changed on (y^m,∞) so that we get an example that is not quantity sustainable? If demand is downward sloping, the answer is "No!" If demand is assumed downward sloping the existence of y such that

$$D(y^m + y)y > C(y)$$

signals that a lot of consumer surplus may be had by expanding output to $y^m + y$. In the single-good case, this observation leads us to

Theorem 4. *Consider the single-good case. Assume average cost is quasi-convex. Let (p^m,y^m) be the one-plant Ramsey point and let demand be non-increasing everywhere. Then, if (p^m,y^m) is not quantity sustainable, there exist two plants y_1,y_2 such that both break even at p, $p \equiv D(y_1 + y_2)$ and net social surplus is greater than at the one-plant Ramsey point.*

PROOF: Suppose (p^m,y^m) is not quantity sustainable; then there exists $y > 0$ such that

$$D(y^m + y)y > C(y) \tag{9.10}$$

By concavity of the benefit function (it is the area under a declining demand curve)

$$B(y + y^m) - B(y^m) \geq D(y + y^m)y > C(y) \tag{9.11}$$

since the entrant makes a profit. Thus, if $D(y + y^m)y^m \geq C(y^m)$, so that the old plant still breaks even, the two-plant Ramsey point dominates. Thus we may assume

$$D(y + y^m)y^m < C(y^m) \tag{9.12}$$

and in particular

$$AC(y^m) > AC(y) \tag{9.13}$$

Now, the rest of the proof is divided into three cases—case 1: $y \geq y^m$; case 2: $y \leq y^m$, and $2y \geq y^m$; case 3: $y < y^m$ and $2y < y^m$. In case 1 consider a single producer producing y. Thus

$$B(y) - B(y^m) \geq D(y)(y - y^m) \tag{9.14}$$
$$> D(y + y^m)(y - y^m) \geq C(y) - C(y^m)$$

which is a contradiction to y^m, a Ramsey point, since

$$D(y)y \geq D(y + y^m)y > C(y)$$

In case 2, consider two plants each producing at y. Thus

$$B(2y)y - B(y^m) \geq D(2y)(2y - y^m)$$
$$\geq D(y + y^m)(2y - y^m) > 2C(y) - C(y^m)$$

Thus the two-plant Ramsey point dominates the single plant one since

$$D(2y)y \geqslant D(y + y^m)y > C(y)$$

In case 3, we have $y < y^m - y < y^m$. Consider two plants, one producing y, the other $y^m - y$. By quasi-convexity of average cost we have

$$AC(y^m - y) \leqslant \max\{AC(y), AC(y^m)\} = AC(y^m) \qquad (9.15)$$

Then

$$\text{Total cost} = (y^m - y)AC(y^m - y) + y\, AC(y) < y^m\, AC(y^m) = C(y^m)$$

Furthermore, both plants break even: Thus, again a two-plant Ramsey point dominates a single-plant one. This ends the proof.

Theorem 4 has a simple message. Suppose that a one-plant Ramsey point was not quantity sustainable in the single-product case. Then society is better off with two break-even plants rather than one. Furthermore, it is not necessary to introduce price discrimination for the two plants to break even. Notice that costs cannot be subadditive everywhere. Theorem 4 has content only when cost is not subadditive everywhere. Note, however, that cost may be subadditive on $\{0,y^m\}$.

REMARK: Existence of y such that $D(y^m + y)y > C(y)$ always signals that society is better off with two plants y^m, y than with only the one plant.[31] This is so because there is a net gain

$$B(y^m + y) - C(y^m) - C(y) - [B(y^m) - C(y^m)] \qquad (9.16)$$
$$\geqslant D(y^m + y)y - C(y) > 0$$

The trouble is that price discrimination may be necessary to allow both plants to break even.

Limited results like Theorem 4 may be had in the multidimensional case. In particular, let (p^m, y^m) be an n-goods one-plant Ramsey optimum. Consider entry along the ray through $0, y^m$. If there is an entry threat y^e on the ray through $0, y^m$ such that $D(y^m + y^e)y^e > C(y^e)$, then we can show that y^m is Ramsey dominated by two plants as in the one good case.

We need more notation. Let $\|y^m\|$ denote the Euclidean norm of $y^m \in R_+^n$. Put $b(q) \equiv B(q\, \bar{v})$, $\bar{v} \equiv y^m/\|y^m\|$, $c(q) \equiv C(q\bar{v})$, $d(q) \equiv D(q\bar{v})\bar{v}$, $q \in R^1$.

Theorem 5. *Assume DSD and quasi-convexity of $c(q)$. By definition the one-plant Ramsey optimal point y^m solves*

$$\text{maximize} \quad b(q) - c(q) \qquad (9.17)$$
$$\text{s.t.} \qquad d(q) \geqslant c(q)$$

If there is $y^e \in r_+^1$ such that

$$d(q + y^e)y > c(q)$$

then two plants may be found that Ramsey dominate y^m.

PROOF: Just mimic the proof of the one-dimensional case, recognizing that DSD implies concavity of B.

How quantity sustainable is a Ramsey optimum when one plant is Ramsey optimal in the multiproduct case? Entry threats must lie in the set $E \equiv \{y \in R^n_+ | D(y^m + y)y \geq C(y)\}$. Following BBW $\rightarrow$ let $T \equiv \{x | D(x) \geq C(x)\}$. We may prove

Lemma 1. *Consider the single-product case. If DSD holds then $E \subseteq T$.*

PROOF: Let $y \in E$. Then

$$C(y) \leq D(y^m + y)y \leq D(y)y \qquad (9.18)$$

by DSD in one dimension. This ends the proof.

Furthermore, in the scalar case we have

Lemma 2. *Consider the single-product case. Assume DSD and that y^m is the one-plant Ramsey optimum. Then if $y \in E$, $y \neq 0$ satisfies either*

$$\text{(a) } D(y + y^m)y^m \geq C(y^m)$$

or

$$\text{(b) } y^m \leq 2y$$

there exist plant outputs y_1 and y_2 such that both plants break even and which yield a net social surplus at least as great as the one-plant Ramsey optimum.

PROOF: The proof simply repeats the relevant parts of the proof of Theorem 4. By the concavity of B,

$$B(y + y^m) - B(y^m) \geq D(y + y^m)y > C(y)$$

If (a) holds, the original plant will break even at the new prices and the proposed outputs are y, y^m.

If (b) holds and (a) does not hold then it must be true that $y \leq y^m$. For if not,

$$B(y) - B(y^m) \geq D(y)(y - y^m) \geq D(y + y^m)(y - y^m) > C(y) - C(y^m)$$

Furthermore $D(y)y \geq D(y + y^m)y > C(y)$, so y^m could not have been the one-plant Ramsey optimum.

Thus $y \leq y^m \leq 2y$. Not let each plant produce y.

$$B(2y) - B(y^m) \geq D(2y)(2y - y^m)$$
$$\geq D(y + y^m)(2y - y^m) > 2C(y) - C(y^m)$$

This ends the proof.

The importance of Theorem 4, above, its relative, Theorem 5, and their demonstrations is that, in the one-dimensional case and special multidimensional cases, lack of quantity sustainability at the Ramsey point signals that society is better off with two plants. Furthermore, the two plants may be chosen so that they both break even. Hence the rationale for entry restrictions to protect the quantity nonsustainable one-plant Ramsey optimum disappears. What about M-plant Ramsey optima?

Generalizations to M-Plant Ramsey Optima

The notions of price sustainability and quantity sustainability may be extended to cover M-plant Ramsey optima as well as the case of one-plant Ramsey optima developed above. Some care must be taken with the M-plant case. The cost function cannot be subadditive everywhere, else the M plants could be replaced by one plant with a net gain to society. The domain of subadditivity of cost is even restricted by the presence of more than one optimal Ramsey plant.

For example, suppose two plants x_1, x_2 break even at $p = D(x_1 + x_2)$. Then $D(x_1 + x_2)(x_1 + x_2) \geq C(x_1) + C(x_2)$. If C was subadditive on $[0, x_1 + x_2]$, obviously both plants could be replaced with one plant at $x_1 + x_2$ with a cost savings to society. Multiplant Ramsey optima become interesting when the domain of subadditivity of cost is small relative to demand.

Definition 3. A multiplant, n-goods Ramsey optimum $(\bar{x}_1, \bar{x}_2, \ldots, \bar{x}_M)$ solves

$$\underset{x_i \in R_+^L}{\text{Maximize}} \quad B(x_1 + \cdots + x_M) - \sum_{i=1}^{M} C(x_i)$$

$$\text{s.t.} \quad D(x_1 + \cdots + x_M) x_i \geq C(x_i)$$

$$i = 1, 2, \cdots, M$$

Notice that M is endogenous in the definition.

As indicated before, it may be more appropriate to treat Ramsey optima from the point of view of social welfare functions, but our desire to make contact with the sustainability literature, especially BBW and PW, leads us to treat optimal taxation in a second-best world of uniform prices as in (9.2) above. Sustainability concepts may now be defined.

Definition 4. A multiplant Ramsey optimum $(\bar{x}_1, \ldots, \bar{x}_M)$ is price sustainable if and only if $p_S^e y_S^e - C(y_S^e) < 0$ for all $S \subseteq N$, $p_S^e \leq p_S^m$, $y_S^e \leq Q_S(p_S^e, p_{(S)}^m)$, with $y_S^e \neq Q_s(p^m)$. It is also required that the M plants break even at prices $p^m = D(y^m)$, $y^m \equiv \bar{x}_1 + \ldots + \bar{x}_M$. A multiplant Ramsey optimum $(\bar{x}, \ldots, \bar{x}_M)$ is quantity sustainable if and only if

$$D_s(y^m + y_S^e)y_S^e - C(y_S^e) \leq 0, \quad \text{for all } y_S^e \geq 0, \ y_S^e \neq 0, \quad \text{for all } S \subseteq N$$

It is also required that the $\bar{M}$ plants break even at prices $p^m = D(y^m)$, $y^m \equiv$

$\bar{x} + \ldots \bar{x_M}$. If demand is "well behaved," price sustainability implies quantity sustainability.

Theorem 6. *Assume that demand satisfies the gross-substitutes assumption of Theorem 2. Furthermore, suppose that the multiplant Ramsey optimum $(\bar{x}_1, \ldots, \bar{x_M})$ is price sustainable. Then it is quantity sustainable.*

PROOF: Exactly the same as that of Theorem 2.

In the one-dimensional case when demand is nonincreasing and average cost is quasi-convex, then all multiplant Ramsey optima are quantity sustainable. In particular we have

Theorem 7. *Consider the single-product case. Assume DSD and that average cost is quasi-convex. Given fixed M, let $(\bar{x}_1, \ldots, \bar{x}_M)$ be the M-plant output which maximizes net social surplus. If $(\bar{x}_1, \ldots, \bar{x}_M)$ is not quantity sustainable, then there is a set of $M + 1$ outputs such that all plants break even and which yields at least as much surplus as the original M outputs.*

PROOF: The proof again follows that of Theorem 4. If

$$y^m = \sum_{i=1}^{M} \bar{x}_i$$

is not quantity sustainable then there is a $y > 0$ such that

$$D(y^m + y)y > C(y)$$

Suppose first that, for all $i = 1, \ldots M$,

$$D(y^m + y)\bar{x}_i \geq C(\bar{x}_i)$$

By concavity of B,

$$B(y + y^m) - B(y^m) \geq D(y + y^m)y \geq C(y)$$

Then output $(\bar{x}_i, \ldots, \bar{x}_M, y)$ yields at least as much net social surplus as $(\bar{x}_1, \ldots, \bar{x}_M)$ and all $M + 1$ firms break even.

Thus, assume there is a plant $\bar{x}_j$ such that

$$D(y^m + y)\bar{x}_j < C(\bar{x}_j)$$

Let

$$\hat{q} = \sum_{\substack{i=1 \\ i \neq j}}^{M} \bar{x}_i$$

As before, distinguish three cases—case 1: $y > \bar{x}_j$; case 2: $y \leq \bar{x}_j$, $\bar{x}_j \leq 2y$; case 3: $2y \leq \bar{x}_j$. In case 1, replacing $\bar{x}_j$ by y contradicts the assumption that $(\bar{x}_1, \ldots, \bar{x}_M)$

maximizes social surplus given M.

$$B(y + \hat{q}) - B(y^m) \geqslant D(y + \hat{q}) (y - \bar{x}_j)$$
$$\geqslant D(y + y^m) (y - \bar{x}_j) > C(y) - C(\bar{x}_j)$$

In case 2, $\bar{x}_j$ is replaced by two plants each producing y

$$B(2y + \hat{q}) - B(y^m) \geqslant D(2y + \hat{q}) (2y - \bar{x}_j)$$
$$\geqslant D(y + y^m) (2y - \bar{x}_j) > 2C(y) - C(\bar{x}_j)$$

In case 3, the fact that $\bar{x}_j$ does not break even at the prices $D(y^m + y)$ implies $AC(y) < AC(\bar{x}_j)$. Because $y < \bar{x}_j - y < \bar{x}_j$, convexity of AC implies

$$AC(\bar{x}_j - y) < AC(\bar{x}_j)$$

Thus

$$\text{Total cost} = \sum_{i \neq j} C(\bar{x}_i) + y \, AC(y) + (\bar{x}_j - y)AC(\bar{x}_j - y)$$
$$< \sum_{i \neq j} C(\bar{x}_i) + \bar{x}_j \, AC(\bar{x}_j)$$

and replacing $\bar{x}_j$ by $\bar{x}_j - y$ and y gives the same output at less total cost. This ends the proof.

We have, by now, generated the impression that "all" Ramsey optimal $\bar{M}$-opolies are quantity sustainable. Hence, if policy makers can be persuaded that they should be concerned about quantity nonsustainability but not to worry about price nonsustainability there should be no cause for concern. The focus of concern about nonsustainability would shift to a debate about the policy relevance of price nonsustainability versus quantity nonsustainability.

It is, however, apparently not true that all Ramsey monopolies are quantity sustainable in the two-goods case. We sketch a counterexample.

Consider a pair of technologies defined by

    i. $C_1(x_1, x_2) = 1$,   for all $x_1 \geqslant 0$, $x_2 \geqslant 0$

    ii. $C_2(x_1, x_2) = \alpha x_1$,   for all $x_1 \geqslant 0$, $x_2 = 0$

Choose preferences so that demands are independent, continuous, and it is Ramsey optimal to produce both goods using (i). To find a counterexample to quantity sustainability just choose $\alpha < P_1$ where P_1 is the Ramsey price for good one. Clearly, by continuity of demand, an entrant using (ii) could enter with a small output ε and turn a profit. However, it is in society's interest to bar such entry. We leave the details of construction of such a counterexample to the reader. The economics is clear.

We may, however, pose the important question: What conditions on cost and demand ensure that quantity nonsustainability of the one-plant Ramsey optimum y^m against specialized (i.e., one-product) entrants implies that society is better off with two break-even plants?

Theorem 8. *Let (p^m, y^m) be a one-plant, n-goods Ramsey optimum and suppose that the cost function can be written*

$$C(y_1, \ldots, y_m) = C_i(y_i) + C_{(i)}(y_{(i)})$$

and that AC_i is quasi-convex. Assume that the firm makes nonnegative profit on the set (i), that is,

$$\pi_{(i)}(y^m) \equiv \sum_{\alpha \neq i} D_\alpha(y^m)y_\alpha - C_{(i)}(y_{(i)}) \geq 0$$

Recall that (i) denotes the set $\{1, 2, \ldots, n\} - \{i\}$. Assume further that i and (i) are complementary (or independent) and that demand for a good is non-increasing in its own price. If there exists a y_i such that

$$D_i(y^m + y_i)y_i > C_i(y_i)$$

then (p^m, y^m) is Ramsey dominated by two plants.

PROOF: Suppose first that

$$D_i(y^m + y^i)y_i^m \geq C_i(y_i^m)$$

By assumption of complement,

$$D_\alpha(y_\alpha^m + y_i)y_\alpha^m \geq D_\alpha(y^m)y_\alpha^m$$

so the original firm will still make a profit on the set (i); that is,

$$\pi_{(i)}(y^m + y_i) \geq \pi_{(i)}(y^m) \geq 0$$

By the concavity of B,

$$B(y^m + y_i) - B(y^m) \geq D(y^m + y_i)y_i > C_i(y_i)$$

Thus, two plants producing y^m and y_i both break even and produce more surplus.
    Thus, suppose that

$$D_i(y^m + y_i)y_i^m < C_i(y_i^m)$$

and consider three cases—1: $y_i > y_i^m$; 2: $y_i \leq y_i^m \leq 2y_i$; 3: $2y_i < y_i^m$.

Case 1: Again we show that $y_i > y_i^m$ contradicts one-plant Ramsey optimality.

$$B(y^m - y_i^m + y_i) - B(y^m) \geq \frac{\partial B}{\partial y_i}(y^m - y_i^m + y_i)(y_i - y_i^m)$$

$$= D_i(y^m - y_i^m + y_i)(y_i - y_i^m)$$

$$\geq D_i(y^m + y_i)(y_i - y_i^m)$$

$$> C_i(y_i) - C_i(y_i^m)$$

To show that the firm still breaks even,

$$\pi(y^m - y_i^m + y_i) = D_i(y^m - y_i^m + y_i)y_i - C_i(y_i) + \pi_{(i)} (y^m - y_i^m + y_i)$$

$$\geqslant D_i (y^m + y_i) - C_i(y_i) + \pi_{(i)} (y^m - y_i^m + y_i)$$

$$\geqslant D_i (y^m + y_i)y_i - C_i(y_i) + \pi_{(i)} (y^m)$$

$$\geqslant 0$$

Case 2: Now $y_i \leqslant y_i^m < 2y_i$; implies

$$B(y^m - y_i^m + 2y_i) - B(y^m) \geqslant \frac{\partial B}{\partial y_i} (y^m - y_i^m + 2y_i) (2y_i - y_i^m)$$

$$= D_i (y^m - y_i^m + 2y_i) (2y_i - y_i^m)$$

$$\geqslant D_i (y^m + y_i) (2y_i - y_i^m)$$

$$> 2C_i (y_i) - C_i (y_i^m)$$

We must check that the firm producing $y^m - y_i^m + y_i$ still breaks even (the entrant producing only y_i breaks even by assumption). We have

$$\pi(y^m - y_i^m + 2y_i) = D_i(y^m - y_i^m + 2y_i)y_i - C(y_i) + \pi_{(i)} (y^m - y_i^m + 2y_i)$$

$$\geqslant D_i (y^m + y_i) y_i - C_i(y_i) + \pi_{(i)} (y^m - y_i^m + 2y_i)$$

$$\geqslant D_i (y^m + y_i)y_i - C_i(y_i) + \pi_i(y^m)$$

$$\geqslant 0$$

Case 3: $2y_i < y_i^m$ For this case produce the same total output as $y^m - y_i$ and y_i. For the firm producing $y^m - y_i$

$$\pi(y^m) = D_i(y^m) (y_i^m - y_i) - AC_i (y_i^m - y_i) (y_i^m - y_i) + \pi_{(i)} (y^m)$$

We are still working under the hypotheses that

$$D_i(y^m + y_i)y_i^m < C_i (y_i^m)$$

and

$$D_i(y^m + y_i)y_i \geqslant C_i(y_i)$$

so

$$AC_i(y_i^m) > AC_i(y_i)$$

As $y_i \leqslant y_i^m - y_i < y_i^m$, quasi-convexity of AC_i implies as well that

$$AC_i (y_i^m - y_i) < AC_i(y_i^m)$$

Now the total cost of producing y^m is

$$TC = AC_i(y_i^m - y_i) (y_i^m - y_i) + AC_i(y_i)y_i + C_{(i)}(y_{(i)}^m)$$

$$< AC_i(y_i^m)y_i^m + C_{(i)}(y_{(i)}^m)$$

so the original output is produced at less total cost.

We show that the firm producing $y^m - y_i$ still breaks even. Write

$$\pi(y^m) = D_i(y^m) (y_i^m - y_i) - AC_i(y_i^m - y_i) (y_i^m - y_i) + \pi_{(i)} (y^m)$$

If the firm originally made profit on the sale of i so that

$$D_i(y^m) \geqslant AC_i(y_i^m) > AC(y_i^m - y_i)$$

it still does at $y^m - y_i$ and by assumption $\pi_{(i)} (y^m) \geqslant 0$. If the firm lost on the sale of i so that $D_i(y^m) < AC_i(y_i^m)$

$$[D_i(y^m) - AC_i(y_i^m)]y_i^m \leqslant [D_i(y^m) - AC_i(y_i^m)] (y_i^m - y_i)$$

$$< [D_i(y^m) - AC_i(y_i^m - y_i)] (y_i^m - y_i)$$

Finally, by the assumption that y^m was a one-plant Ramsey optimum.

$$[D_i(y^m) - AC_i(y^m)]y_i^m + \pi_{(i)} (y^m) \geqslant 0$$

This ends the proof.

Notice that the above argument may be applied to consider social gains from allowing entrants whose costs are dependent upon the presence of the monopolist. In other words, it may be optimal to allow "parasitic" entrants. It may even be Ramsey optimal to allow such entrants. We develop this below.

Notice that nothing was fixed about the division of $C(y)$ into a sum of functions $C_j(y_j)$. All that was assumed was that such a division could be found so that (a) $D_i(y^m)y_i^m \leqslant C_i(y_i^m)$, (b) $D_i(y^m + y_i)y_i > C_i(y_i)$.

This observation has relevance to the case of "parasitic" entrants. Suppose that an entrant can produce y_i at $C_i(y_i)$ given that the monopolist is producing $\hat{y}_i^m$. If (a), (b) are satisfied then the entrant should be allowed into the industry.

The point of the above discussion is that our type of analysis may be extended to consider conditional technologies (i.e., technologies whose operations depend upon other production taking place).

Comparison of Necessary Conditions for Price Sustainability and Quantity Sustainability

It is useful to compare the implications of quantity sustainability with the necessary conditions for price sustainability developed by Panzar and Willig. We repeat them here. Let p^m, y^m be price sustainable, then: (1) The monopoly produces y^m at least cost; (2) $\pi(p^m) = 0$; (3) p^m must be undominated; (4) $C(y^m)$ $< \sum_{i=1}^{k} C(y^i)$, if $\sum_{i=1}^{k} y^i > y^m$ with at least two $y^i \neq 0$; (5) least-ray average cost is $(1/\lambda)\, C(\lambda y^m) > C(y^m)$ for $0 < \lambda < 1$; (6) prices must not be below marginal costs; (7) for all $S \subset N$ (proper subset), $\sum_{i \in S} p_i^m y_i^m < C(y_S^m)$, where $y_i^m \equiv Q^i(p^m)$.

The seven necessary conditions for price sustainability show the strong discipline imposed upon the monopolist by the Bertrand price competition embedded in the definition of price sustainability of (p^m, y^m). Indeed, the discipline imposed by Bertrand price competition is so strong that a Bertrand monopoly producing perfect substitutes is driven to zero profits or instability for $n \geq 2$. This lack of correlation between profits and n (a market share proxy) probably led economists to end up siding with Cournot in the famous Bertrand–Cournot controversy. The positive association between market share and profit is predicted by Cournot's model but not Bertrand's.

What happens to the necessary conditions for sustainability of (p^m, y^m) if we replace the notion of price sustainability (Bertrand) with quantity sustainability (Cournot)? One can construct counterexamples to all of them except condition (3).

The monopolist is not forced to produce at least cost. For let him satisfy $D(y^m)$ $> AC(y^m)$. To find a counterexample just make $D(y^m + y)$ fall faster than $AC(y)$ so that $D(y^m + y) < AC(y)$ for all $y > 0$. The same principle of counterexample construction can be applied to (2), (4), (5), (6). A counterexample to (7) may be found in the two-goods case by letting a specialized production of good one make a profit at p_1^m but constraining independent demand $D_1(y_1^m + y_1) < C(y_1)/y_1$ for all $y_1 > 0$.

The bottom line is that, in the above case, Bertrand competition competes away all rent whereas Cournot competition does not.

Theorem 9. *Assume gross substitutes, that is,*

$$\frac{\partial D_\alpha}{\partial y_\beta} \leq 0, \ \forall \alpha, \beta$$

Suppose that the cost function $C(y)$ is subadditive and concave and suppose that the system (D,C) is regular at y^m in the sense that if $\exists \ y \geq 0$, such that $\pi(y) \equiv D(y^m + y)y - C(y) > 0$ then a $\bar{y} \gg 0$ can be found such that $\bar{y} \in$ Argmax $\pi(y)$. Then the Ramsey point y^m is quantity sustainable.

PROOF: The point $\bar{y}$ being an interior maximum satisfies

$$\frac{\partial \pi}{\partial y_i} = 0 \equiv D_i + \sum_{\alpha=1}^{n} \frac{\partial D_\alpha}{\partial y_i} y_\alpha - \frac{\partial C}{\partial y_i} = 0, \quad i = 1, 2, \ldots, n$$

We shall show

$$B(y^m + \bar{y}) - B(y^m) \geq C(y^m + \bar{y}) - C(y^m)$$

and

$$D(y^m + \bar{y})(y^m + \bar{y}) \geq C(y^m + \bar{y})$$

Since, by subadditivity

$$B(y^m + \bar{y}) - B(y^m) \geq D(y^m + \bar{y})\bar{y} > C(\bar{y}) > C(y^m + \bar{y}) - C(y^m)$$

it will suffice to show that

$$D(y^m + \bar{y})\bar{y} > C(\bar{y}) \quad \text{implies } D(y^m + \bar{y})(y^m + \bar{y}) \geq C(y^m + \bar{y})$$

The first-order conditions will be used to demonstrate the last inequality. Compute

$$\pi^* \equiv D(y^m + \bar{y})(y^m + \bar{y}) - C(y^m + \bar{y}) = \pi(\bar{y}) + D(y^m + \bar{y})y^m$$

$$- [C(y^m + \bar{y}) - C(\bar{y})] > D(y^m + \bar{y})y^m - [C(y^m + \bar{y}) - C(\bar{y})]$$

$$= \sum_{i=1}^{n} y_i^m (\frac{\partial C}{\partial y_i}(\bar{y}) - \sum_{\alpha=1}^{n} \frac{\partial D_\alpha}{\partial y_i}) - C(y^m + \bar{y}) - C(\bar{y}) \equiv -\sum_{\alpha,i} y_i^m \frac{\partial D_\alpha}{\partial y_i}\bar{y}_\alpha$$

$$+ \frac{\partial C(\bar{y})}{\partial y} y^m - [C(y^m + \bar{y}) - C(\bar{y})]$$

Now *GS* at $y^m + \bar{y}$ implies

$$- \sum_{\alpha,i} y_i^m \frac{\partial D_\alpha}{\partial y_i}\bar{y} \geq 0$$

Concavity of cost implies

$$C(\bar{y}) - C(y^m + \bar{y}) + \frac{\partial C(\bar{y})}{\partial y} y^m \geq \frac{\partial C}{\partial y}(\bar{y})(\bar{y} - y^m - \bar{y}) + \frac{\partial C}{\partial y}(\bar{y})y^m = 0$$

Hence, $\pi^* \geq 0$. Thus, the plant $y^m + \bar{y}$ yields more welfare than y^m and breaks even. Contradiction to Ramsey optimality of y^m.

REMARK: A critical step in showing $\bar{\pi} \geq 0$ implies $\pi^* \geq 0$ was

$$D(y^m + \bar{y})y^m \geq \frac{\partial C}{\partial y}(\bar{y})y^m$$

(i.e., revenue at prices $D(y^m + \bar{y})$ was greater than or equal to revenue from marginal cost pricing of y_j^m using $\bar{y}$ marginal costs).

REMARK: If demands are independent, then GS holds. Also, if $C(y) \equiv F + cy$, then the hypotheses of the theorem are satisfied for C. An alternative proof is to consider the modified Ramsey problem.

$$\text{Maximize} \quad S(y^m + y) - S(y^m)$$

$$\text{s.t.} \quad D(y^m + y)y \geq C(y)$$

Let $\bar{y}$ solve this. Assume $\bar{y} \gg 0$, and the budget constraint is effective. We have: there is $\tilde{\lambda} \geq 0$ such that

$$[D(y^m + \bar{y}) - \frac{\partial C}{\partial y}(\bar{y})] = - \frac{\tilde{\lambda}}{1 + \tilde{\lambda}} \frac{\partial D}{\partial y}(y^m + \bar{y})\bar{y}$$

Now follow the steps of the previous proof.

Notice that we showed that y^m was Ramsey dominated.

Relevance to Regulation

The findings reported above are obviously relevant to regulation policy. For example, consider a petition for denial of entry before a regulatory commission. The petition has successfully done the following: (*a*) Cost subadditivity has been demonstrated on $(0, y^m)$ (a difficult econometric task to say the least); (*b*) Ramsey numbers and rate-of-return calculations have been prepared to rebut dark claims that the monopolist is a profit-maximizing monopolist masquerading in Ramsey clothing; (*c*) the monopolist is not price sustainable at $p^m = D(y^m)$, y^m. Should the petition be granted? A promising rebuttal is the following. Demonstrate that sunk costs loom large in the technology in the case at hand for both incumbent and entrant. Also show that, for all practical purposes, a large amount of pricing flexibility exists on the part of the incumbent. Hence, if an entrant is willing to sink costs, $C(y)$, into the technology in order to enter at level y, then it is rational for entrant to expect prices to fall to approximately $D(y^m + y)$ since incumbent's costs are largely sunk. Because the entrant has strong incentives to form correct expectations, his willingness to commit $C(y)$ to the enterprise is a strong signal to the regulator that $D(y^m + y)y - C(y) > 0$. Hence, by our results, $B(y^m + y) - B(y^m) - C(y) > 0$. That is, society is better off with the presence of the entrant.

At this point, the burden of proof is likely to shift to the incumbent. The incumbent may rebut by showing that price discrimination may be necessary in order to allow him to break even at $y^m + y$. But the entrant may counter by using our results to show that there are y_1, y_2 levels of production available for both incumbent and entrant that socially dominate y^m and, at the same time, allow both entrant and incumbent to break even at $p = D(y_1 + y_2)$ without price discrimination. Hence, the regulator, being aware that entry denial dampens the monopolist's incentives to be efficient (especially in a world of rapid change), might well allow entry in the scenario above.

Panzar has commented that his and Willig's sustainability analysis is "best interpreted as an analysis of the potential problems with a policy of free entry given that the regulator does not allow a price response—the simplest such regulatory policy to analyze. Their analysis demonstrated that there may indeed be serious problems with a policy of open entry under that condition. Therefore, an important direction for additional research is to examine open entry and sustainability under meaningful alternative assumptions about the response to entry the regulator will allow the monopolist to make."[32]

This chapter conducted such an analysis under the polar assumption of output maintenance. The general conclusion is, as expected, that sustainability problems are much less severe. Not quite so obvious, however, is the finding that the emergence of entry signals that society is better off with more than one plant and that, in a significant class of cases, production levels y_1, y_2 may be structured so that society is better off at y_1, y_2 than at y^m, and both plants break even without price discrimination at the prices that absorb $y_1 + y_2$.

Consider a situation where the regulator permits substantial pricing flexibility and technology is characterized by substantial sunk costs but the regulator does not know whether incumbent's costs are subadditive. If an entrant shows up and is willing to lay down his own funds to enter, then a real-world application of our results would be that there is a strong presumption that entry is socially beneficial. Notice here that the regulator may be poorly informed. All he needs to know is that sunk costs loom large in the technology.

NOTES

1. The pioneering articles on sustainability are John Panzar and Robert Willig, "Free Entry and the Sustainability of Natural Monopoly," *Bell Journal of Economics,* Spring 1977, pp. 1–22; and William Baumol, Elizabeth Bailey, and Robert Willig, "Weak Invisible Hand Theorems on the Sustainability of Natural Monopoly," *American Economic Review,* June 1977, pp. 350–365. An extensive review of the sustainability of natural monopoly appears in William Baumol, John Panzar, and Robert Willig. *Contestable Markets and the Theory of Industry Structure* (San Diego: Harcourt Brace Jovanovich, 1982).

2. See Panzar and Willig, *op. cit.,* for example.

3. William Baumol and Robert Willig, "Fixed Costs, Sunk Costs, Public Goods, Entry Barriers, and the Sustainability of Natural Monopoly," undated manuscript; John Panzar, "Sustainability, Efficiency, and Vertical Integration," D.P. No. 165 (Holmdel, NJ: Bell Labs, August 1980); and John Panzar and Robert Willig, "Sustainability, Efficiency, and Vertical Integration," manuscript dated August 1980. Some of this material also appears in Baumol, Panzar, and Willig, *op. cit.*

4. John Panzar, "Regulation, Deregulation, and Economic Efficiency: The Case of the CAB," *American Economic Review,* May 1980, pp. 311–315.

5. James W. Friedman, *Oligopoly and the Theory of Games* (Amsterdam: North-Holland, 1977).

6. P. Sylos-Labini, *Oligopoly and Technical Progress* (Cambridge: Harvard University Press, 1962).

7. There is some overlap in spirit between our analysis and K. Baseman's analysis in "Open Entry and Cross-Subsidization in Regulated Markets," in G. Fromm, ed., *Studies in Public Regulation* (Cambridge: MIT Press, 1981). Baseman considers a case where the entrant holds the Cournot–Nash expectation on the monopolist's short-run reaction function rather than on the monopolist's price. His conclusion is that, although sustainability problems may still exist, they are less severe when such reaction by the incumbent is allowed. This is due to the entrant's

expectation that the optimal response of the monopolist is to drop price when faced with the *fait accompli* of entry. Unlike our paper, however, Baseman's paper does not draw any welfare conclusions from the presence of an entrant willing to commit sunk costs. His paper's emphasis is more on the defects in incremental-cost types of price-setting rules and attentuation of sustainability problems by allowing the monopolist to react. Our paper presents welfare theorems for monopoly as well as for *N*-opoly. In contrast to Baseman, we analyze the output-maintenance reaction rather than the reaction of short-run profit maximization. An interesting problem for future research would be to analyze the welfare consequences of an entrant who is willing to sink costs and enter under Baseman's reaction function for the monopolist. Indeed, Lemma 1 of Knieps and Vogelsang carries out a closely related exercise. Their paper contains ideas related to ours although we were not aware of their paper when ours was written. The basic result of our paper, namely "profitable entry under the Cournot conjecture on incumbent's output signals that a new plant can be constructed and output may be reallocated over all incumbent plants such that all plants break even with no price discrimination" appears to be new. See G. Knieps and I. Vogelsang, "The Sustainability Concept Under Alternative Behavioral Assumptions," *Bell Journal of Economics*, Spring 1982, pp. 234–241.

8. Panzar and Willig (1977), *op. cit.*, Figure 4, p. 7.

9. Recently, Weitzman has shown "strictly speaking there is no such thing as a 'pure' fixed cost. Unless there are sunk costs located somewhere in the relevant production technology, all costs are variable. As a matter of formal theory, you cannot have a range of decreasing average costs without sunk costs." He argues that the contestability literature and its close cousin, the price sustainability literature, deals with the problem of decreasing returns by assuming it away. Criticisms such as Weitzman's strengthen the case for looking at a formulation of the sustainability problem, such as ours, which respects sunk costs. See M. Weitzman, "Comment on 'Contestable Markets: An Uprising in the Theory of Industry Structure,' " Working Paper (Cambridge: Department of Economics, MIT, May 1982). Also see the discussion of Dixit below.

10. In these situations, of course, the monopolist may be able to use more sophisticated strategies involving credible threats in order to deter entry. In order to stay with the spirit of the sustainability literature and in order to focus on the differences between price sustainability and quantity sustainability, only the naive conjectures (and strategies) of price and quantity maintenance will be considered here.

11. F. Modigliani, "New Developments on the Oligopoly Front," *Journal of Political Economy*, June 1958, pp. 215–232.

12. A. Dixit, "The Role of Investment in Entry Deterrence," *Economic Journal*, March 1980, pp. 95–106; and B. Eaton and R. Lipsy, "Capital, Commitment, and Entry Equilibrium," *Bell Journal of Economics*, August 1981, pp. 593–609.

13. D. Kreps and R. Wilson, "On the Chain-Store Paradox and Predation: Reputation for Toughness," IMSSS Technical Report No. 319 (Palo Alto, CA: Stanford University, 1980); and P. Milgrom and J. Roberts, "Predation, Reputation, and Entry Deterrence," Research Paper No. 600 (Evanston, IL: Graduate School of Business, Northwestern University, 1981).

14. A. Dixit, "Recent Developments in Oligopoly Theory," *American Economic Review*, May 1982, pp. 12–17.

15. See Baumol, Panzar, and Willig, *op. cit.*

16. Dixit (1982), *op. cit.*, outlines these properties.

17. That is, the Baumol, Bailey, and Willig weak invisible-hand theorem does not apply to markets in which sunk costs loom large and in which quantity maintenance is therefore a plausible strategy.

18. Panzar and Willig (1977), *op. cit.*

19. Baumol, Bailey, and Willig, *op. cit.*

20. Panzar and Willig (1977), *op. cit.*

21. Panzar and Willig (1977), *op. cit.*, p. 3.

22. Panzar and Willig (1977), *op. cit.*, p. 5.

23. A. Takayama, *Mathematical Economics* (Hinsdale, IL: The Dryden Press, 1974), Theorem 4.D.3 on p. 393 and the discussion on p. 405. Also see Appendix B of Baumol, Bailey, and Willig, *op. cit.*, for a discussion of the relation between GS and other popular hypotheses in demand theory.

24. Baumol, Bailey, and Willig, *op. cit.*, p. 356.

25. Baumol, Bailey, and Willig, *op. cit.*, p. 357.

26. The qualification "against challenges $y \subseteq T$" must be added because DSD does not allow us to conclude: $D(y^m + y) y > C(y)$ implies $D(y)y > C(y)$. This is true in the one-dimensional case but complementarities in demand may falsify it in the multidimensional case.

27. See H. Sonnenschein, "Market Excess Demand Functions," *Econometrica*, April 1972, pp. 549–563; R. Mantel, "On the Characterization of Aggregate Excess Demand," *Journal of Economic Theory*, March 1974, pp. 348–353; G. Debreu, "Excess Demand Functions," *Journal of Mathematical Economics*, March 1974, pp. 15–21.

28. It is worth pointing out here that there is a relation between price and quantity sustainability and supportability of the cost function in the Sharkey–Telser sense. As they point out, supportability is necessary but not sufficient for price sustainability. Notice that supportability is not necessary for quantity sustainability because one-dimensional examples with U-shaped average cost may be found that are quantity sustainable at the Ramsey point but are not supportable there. All that needs to be done to create an example is to generate a one-dimensional, U-shaped average cost Panzar–Willig price nonsustainable example but make demand fall off rapidly enough beyond the Ramsey point so that the Ramsey point is quantity sustainable. See W. Sharkey and L. Telser, "Supportable Cost Functions for the Multiproduct Firm," *Journal of Economic Theory*, June 1978, pp. 23–37.

29. Panzar and Willig (1977), *op. cit.*, p. 7.

30. Baumol, Bailey, and Willig, *op. cit.*, pp. 357–358.

31. This is not so for the case of price nonsustainability. Suppose there is y such that $D(y^m) y > C(y)$ then the net gain is

$$B(y^m + y) - C(y^m) - C(y) - [B(y^m) - C(y^m)] \leqslant D(y^m) - C(y)$$

That is, although the upper bound to net gain $D(y^m)y - C(y)$ is positive it is easy to construct examples even in the one-dimensional case where net gain is negative. The point is that quantity nonsustainability implies a positive lower bound to net gain. Price nonsustainability only implies the existence of a positive upper bound.

32. Panzar, "Comment on Baseman," in G. Fromm, ed., *op. cit.*

Chapter 10
Multiproduct Cost Function Estimates and Natural Monopoly Tests for the Bell System*

David S. Evans and James J. Heckman

This chapter (*a*) reports multiproduct cost function estimates for the Bell System; (*b*) describes an empirically tractable test for subadditivity of the cost function; and (*c*) reports the results of this test for alternative Bell System cost function estimates. Subadditivity of the cost function is a necessary and sufficient condition for natural monopoly. AT&T has argued that the telephone industry is a natural monopoly:[1]

> The scientific principles applicable to telecommunications, the organization of the nationwide telecommunications network, and the engineering principles and practices by which telecommunications services are provided make a single interactive and interdependent network the most efficient means for providing all of the Nation's telecommunications services. . . . [S]uch a network can be planned, constructed, and managed most efficiently by an integrated enterprise that owns the major piece-parts of the facilities network. . . .

This assertion is true if and only if the cost function for the Bell System is subadditive.

Previous studies of the Bell System provide little information concerning whether the telephone industry is a natural monopoly.[2] These studies aggregate diverse telecommunications service outputs into a single measure of output, estimate single-product cost or production functions, and determine whether there are scale economies. There are two major problems with this approach. First, cost or production functions based on an aggregate measure of output are valid only under highly restrictive assumptions, which these studies do not test. We

*We would like to thank Rod Smith and Lester Telser for comments; George Yates for developing the software for testing natural monopoly; and Thomas Coleman, Alan Brazil, and John Bender for fine research assistance.

have found that these assumptions do not hold for the Bell System. Second, the presence of scale economies for an aggregate measure of output does not imply the presence of scale economies for any of the components that comprise the aggregate measure of output. Christensen, Cummings, and Schoech claimed their finding of scale economies for an aggregate measure of output "is consistent with the view that the proliferation of suppliers of telecommunications would result in a large sacrifice of efficiency due to foregone scale economies."[3] Since aggregate scale economies do not imply that there are scale economies in intercity message toll or private line service—the two markets opened to competition by the FCC during the 1970s—their finding is consistent with the opposite view as well. As Baumol, Panzar, and Willig have observed, "We can see why analysts have attempted to use the analytically and statistically tractable concept of scale economies as a surrogate test of natural monopoly. Unfortunately, . . . such traditional tests simply can not do the job."[4]

This chapter is divided into three sections and three appendices. The first section reports multiproduct cost function estimates for the Bell System. The second section reports tests of whether these estimated cost functions as subadditive. The third section summarizes our results. Appendix A discusses alternative formulations of the cost function. Appendix B reports estimates from a cost function that was not restricted to satisfy the homogeneity and symmetry restrictions required by producer theory. Appendix C discusses the data that were used in this study.

Multiproduct Cost Function Estimates

The Bell System uses capital, labor, and materials to produce local and long-distance telephone service. Its cost function is[5,6]

$$C = f(L,T,r,m,w,t) \tag{10.1}$$

where L is local service output, T is long-distance service output, r is the capital rental rate, w is the wage rate, m is the price of materials, and t is an index of technological change. We have estimated a cost function rather than a production function in order to make our approach consistent with previous studies of the production characteristics of the telecommunications industry and because the theoretical literature on natural monopoly relies on the cost function rather than the production function. We have disaggregated output into local and long-distance service because these are the major services provided by the Bell System.[7] We decided upon this particular cost function because, as described in Appendix A, the data were not available to estimate cost functions that describe the structure of the Bell System more realistically.

Christensen, Jorgenson, and Lau claim that the translog cost function provides a useful second-order approximation to any twice differentiable cost function.[8] The translog cost function imposes fewer restrictions on the characteristics of the production structure than other commonly used cost function specifications and is therefore more suitable for testing alternative hypotheses concerning the characteristics of the production structure. The translog approximation to (10.1) is

$$\ln(C) = \alpha_0 + \sum_i \alpha_i \ln(p_i) + \sum_k \beta_k \ln(q_k) + \mu \ln(t)$$

$$+ \frac{1}{2} \sum_i \sum_j \gamma_{ij} \ln(p_i) \ln(p_j) + \frac{1}{2} \sum_k \sum_l \delta_{kl} \ln(q_k) \ln(q_l)$$

$$+ \sum_i \sum_k \rho_{ik} \ln(p_i) \ln(q_k)$$

$$+ \sum_i \lambda_i \ln(p_i) \ln(t) + \sum_k \theta_k \ln(q_k) \ln(t) + \tau [\ln(t)]^2$$

(10.2)

where p denotes the vector of input prices (r, m, w) and q denotes the vector of outputs (L, t). We apply Shephard's Lemma

$$x_i = \frac{\partial C}{\partial p_i}$$

(10.3)

where x_i is the quantity demanded of the ith input, in order to obtain the input cost share equations

$$S_i = \frac{p_i x_i}{C} = \alpha_i + \sum_i \gamma_{ij} \ln p_j + \sum_k \rho_{ik} \ln q_k + \lambda_i \ln t$$

(10.4)

Assuming AT&T operates efficiently, the cost function and the associated input cost share equations are consistent with production theory if (1) the cost function is linear homogeneous in input prices; (2) the Hessian of the cost function with respect to input prices is symmetric; and (3) the α's, ρ's, γ's, and λ's are identical across equations.[9] Homogeneity requires

$$\sum_j \alpha_j = 1, \quad \sum_j \gamma_{ij} = 0, \quad \sum_j \rho_{jk} = 0, \quad \sum_j \lambda_j = 0$$

(10.5)

Symmetry requires

$$\gamma_{ij} = \gamma_{ji}$$

(10.6)

This general specification of the cost function subsumes many special production structures. Table 10.1 lists alternative hypotheses concerning the structure of production and technological change and parameter restrictions these hypotheses imply.[10] Using standard statistical procedures, we can readily test between alternative hypotheses concerning the structure of production and technological change.

The separability and nonjointness hypotheses are of particular concern to this study. Separability implies that there exists an aggregate measure of output $Q = A(L,T)$ such that

$$C(L,T,r,w,m,t) = C[A(L,T),r,w,m,t] = C(Q,r,w,m,t)$$

(10.7)

Table 10.1 Alternative Hypotheses Concerning the Structure of Production and Technological Change

Hypotheses[a]	Coat function characteristics[b]	Translog parameter[c] restrictions	
Separability of inputs and outputs[d,e]	$C(q,p,t) = C[A(q),p,t]$	$\rho_{il}\beta_k = \rho_{ik}\beta_l$	$i = 1,2,3$ $k \neq 1$
Nonjointness[e]	$C(q,p,t) = \Sigma C_i(q_i,p,t)$	$\delta_{kl} = -\beta_k\beta_l$	$k \neq 1$
Homotheticity	$C(q,p,t) = A(q)g(p,t)$	$\rho_{il} = 0$	$i = 1,2,3$ $l = 1,2$
Homogeneity in outputs	$C(\kappa^r q,p,t) = \kappa^r C(q,p,t)$	$\rho_{il} = 0$ $\delta_{kl} = 0$	$i,l = 1,2,3$ $k,l = 1,2$
Unitary elasticities of substitution	$\dfrac{\partial^2 C}{\partial p_i \partial p_j} = 0$	$\gamma_{ij} = 0$	$i,j = 1,2,3$
Neutral technological change	$C(q,p,t) = h(t)C(q,p)$	$\theta_k = 0$ $\lambda_j = 0$	$i = 1,2$ $j = 1,2,3$
Nonfactor augmenting technological change	$C(q,p,t) = h(t)C[f(q,t),p]$	$\lambda_j = 0$	$j = 1,2,3$
Nonoutput augmenting technological change	$C(q,p,t) = h(t)C[q,g(p,t)]$	$\theta_k = 0$	$k = 1,2$

[a]The maintained hypothesis is that the homogeneity and symmetry restrictions implied by producer theory hold.
[b]p is the vector of input prices, q is the vector of outputs, t is technological change, and C is cost.
[c]See equation (10.2) in the text for a definition of the parameters.
[d]A sufficient condition for separability of inputs and outputs is the homotheticity of the cost function.
[e]Restriction holds at the point of expansion for the translog approximation to the cost function.

Denny and Pinto have shown that, at the point of expansion for the translog approximation to the cost function, the cost function is separable if[11]

$$\rho_{il}\beta_k = \rho_{ik}\beta_l, \quad i = 1,2,3, k \neq 1 \tag{10.8}$$

If these restrictions are accepted, it may be possible to form an aggregate output measure, estimate a single-product cost function, and use scale-economy estimates to test whether there is a natural monopoly.[12] Nonjointness implies that the cost of producing several outputs equals the sum of the costs of producing the outputs separately

$$C(L,T,r,w,m,t) = C_L(L,r,w,m,t) + C_T(T,r,w,m,t) \tag{10.9}$$

Denny and Pinto have shown that, at the point of expansion of the translog approximation to the cost function, the cost function is nonjoint if[13]

$$\delta_{kl} = -\beta_k\beta_l \tag{10.10}$$

If this restriction is accepted, the cost function exhibits neither economies nor diseconomies of scope.[14]

Several researchers have criticized the translog cost function specification. Burgess found that different results were obtained depending upon whether a translog cost function or a translog production function was estimated.[15] Gallant argued that the translog specification yields potentially biased parameter estimates.[16] Guilkey and Lovell found that "there is a pervasive, but not pronounced, upward bias to translog estimates of . . . returns to scale."[17] Fuss and Waverman found that a slight modification to the translog cost function reduced the aggregate scale elasticity estimate from around 1.4 to 1.0.[18]

In order to test the sensitivity of our results to the translog cost function specification, we estimated the modified translog cost function and the Box–Tidwell cost function.[19] The modified translog cost function performs a Box–Cox transformation on the output variables. The Box–Tidwell cost function performs a Box–Cox transformation on the right-hand side variables. The Box–Cox transformation of a variable y is given by

$$y^* = \frac{y^\eta - 1}{\eta} \tag{10.11}$$

The ordinary translog cost function is, following (10.2),

$$\ln(C_{OTL}) = g[\ln (L), \ln(T), \ln(r), \ln(w), \ln(m), \ln(t)] \tag{10.12}$$

The modified translog cost function is

$$\ln(C_{MTL}) = g[L^*, T^*, \ln(r), \ln(w), \ln(m), \ln (t)] \tag{10.13}$$

The Box–Tidwell cost function is

$$\ln(C_{BT}) = g(L^*, T^*, r^*, w^*, m^*, t^*) \tag{10.14}$$

With substitution of the transformed variables for the logarithmic variables in (10.2), the cost share equations (10.4) and cross equation restrictions (10.5) and (10.6) apply to the modified translog cost function. For convenience in estimation, we impose linear homogeneity in prices on the Box–Tidwell cost function by normalizing cost and input prices by the price of materials. Linear homogeneity in prices requires

$$C(\xi w, \xi r, \xi m, L, T, t) = \xi C(w, r, m, L, T, t) \tag{10.15}$$

Letting $\xi = \dfrac{1}{m}$ we have

$$C\left(\frac{w}{m}, \frac{r}{m}, 1, L, T, t\right) = \frac{1}{m} C\left(\frac{w}{m}, \frac{r}{m}, L, T, t\right) \tag{10.16}$$

Define

$$W = \frac{w}{m}, \quad R = \frac{r}{m}, \quad P = (W, R) \tag{10.17}$$

Then

$$\ln\left(\frac{C}{m}\right) = \alpha_0 + \sum_i \beta_i P_i^* + \sum_k \beta_k q_k^*$$

$$+ \frac{1}{2}\sum_i \sum_j \gamma_{ij} P_i^* P_j^* + \frac{1}{2}\sum_k \sum_l \delta q_k^* q_l^* + \sum\sum \rho_{ik} p_i^* q_k^* \quad (10.18)$$

$$+ \sum_i \lambda P_i^* t^* + \sum_i \theta_i q_i^* t^* + \tau(t^*)^2$$

Applying Shephard's Lemma (10.3)

$$S_i = P_i^\eta[\alpha_i + \sum_j \gamma_{ij} P_j^* + \sum_k \rho_{ik} q_k^* + \lambda_i t_i^*] \quad (10.19)$$

We specify additive disturbance terms $\varepsilon_C, \varepsilon_K, \varepsilon_L,$ and ε_M for the cost, capital share, labor share, and materials share equations. We assume that these disturbances are temporally uncorrelated, contemporaneously correlated, and multinormally distributed with

$$E\,\varepsilon_{it} = 0 \qquad i = C,K,L,M$$

$$E\,\varepsilon_{it}\varepsilon_{jt} = \sigma_{ij} \quad i,j = C,K,L,M \qquad (10.20)$$

$$E\,\varepsilon_{it}\varepsilon_{jt'} = 0 \quad i,j = C,K,L,M$$

$$t \neq t'$$

where t denotes time. Second, the disturbances are generated by the following first-order autoregressive process

$$\varepsilon_{it} = u_{it} + v_i \varepsilon_{i,t-1}, \quad i = C,K,L,M \qquad (10.21)$$

where the u_{it} are multinormally distributed with

$$Eu_{it} = 0 \qquad i = C,K,L,M$$

$$Eu_{it}u_{jt} = \sigma_{ij} \quad i,j = C,K,L,M \qquad (10.22)$$

$$Eu_{it}u_{jt'} = 0 \quad i,j = C,K,L,M$$

$$t \neq t'$$

Under (10.21), the disturbances are contemporaneously correlated across equations, temporally correlated within equations, and temporally uncorrelated across equations.[20]

For both error processes, the equations were estimated by iterating Zellner's two-step procedure for estimating seemingly unrelated regressions.[21] This iterated method is computationally equivalent to maximum likelihood estimation.[22] Because the input cost shares sum to one, the covariance matrix is singular. In order to provide a nonsingular covariance matrix, we deleted the materials share equation. Barten has shown that maximum likelihood estimates are invariant to the equation deleted when the error process is given by (10.20).[23] Berndt and

Savin have shown that maximum likelihood estimates are invariant to the equation deleted if $v_K = v_L = v_M$ in the error process given by (10.21).[24] The hypothesis $v_K = v_L = v_M$ was, unfortunately, always rejected at the 99% level or better. Nonetheless, rather than report the results for estimations based on the deletion of different share equations, we report estimates based on $v_K = v_L = v_M$ and note where results would differ when this restriction is not imposed.[25]

The equations were estimated with yearly data on the Bell System from 1947–1977. These data were obtained from Christensen who calculated Tornqvist indices of cost, outputs, and prices from detailed Bell System data. In order to represent the level of technology, this study used an index based on a distributed lag of research and development expenditures proposed by Vinod.[26] Several alternative measures of technological change were tested but yielded substantially less satisfactory results.[27] The data, their construction, and their limitations are described more fully in Appendix C.

Table 10.2 presents estimates for three cost function specifications under the

Table 10.2 Parameter Estimates for Alternative Cost Function Specifications with No Serial Correlation

Parameter[a]	Translog[b]		Modified[c] translog		Box–Tidwell[c]	
Constant	9.057	(.196)	9.054	(.004)	9.054	(.004)
Capital	.536	(.004)	.537	(.004)	.537	(.004)
Labor	.354	(.004)	.354	(.004)	.353	(.003)
Local	.294	(.261)	.260	(.350)	.542	(.204)
Toll	.420	(.197)	.462	(.299)	.110	(.140)
Technology	− .161	(.070)	− .193	(.108)	− .008	(.073)
Capital^2	.197	(.024)	.190	(.027)	− .145	(.085)
Labor^2	.176	(.025)	.171	(.027)	− .028	(.037)
Capital · Labor	− .163	(.021)	− .158	(.023)	− .246	(.027)
Toll^2	− 5.276	(1.700)	− 6.531	(4.905)	− 2.999	(1.432)
Local^2	− 2.640	(1.132)	− 3.951	(4.118)	.491	(.567)
Local · Toll	7.764	(2.700)	10.233	(8.828)	− .287	(1.185)
Technology^2	.412	(.799)	− .126	(1.547)	− .260	(.424)
Capital · Toll	.354	(.097)	.399	(.131)	.264	(.045)
Capital · Local	− .352	(.089)	− .390	(.114)	− .374	(.034)
Labor · Toll	− .221	(.087)	− .263	(.116)	− .038	(.028)
Labor · Local	.209	(.080)	.244	(.103)	.104	(.016)
Capital · Tech.	.106	(.037)	.119	(.044)	− .006	(.008)
Labor · Tech.	− .108	(.034)	− .120	(.039)	.020	(.074)
Tech. · Toll	− .967	(1.204)	− 1.924	(2.990)	2.440	(1.062)
Tech. · Local	.358	(1.202)	1.513	(3.130)	− .678	(.498)
v_C	—	—	—	—	—	—
$v_L = v_K$	—	—	—	—	—	—
η	—	—	− .031	(.114)	.725	(.110)

[a]Standard errors in parentheses.
[b]Maximum likelihood estimates obtained by an iterative Zellner method.
[c]Nonlinear seemingly unrelated regression estimates.

assumption that the disturbance terms are serially uncorrelated. The Durbin–Watson test for positive autocorrelation was inconclusive for the cost equation but indicated positive autocorrelation for the capital and labor share equations. Table 10.3 presents estimates for the three cost function specifications under the assumption that the disturbance terms are generated by a first-order autoregressive process, as given in (10.21). Table 10.4 reports summary statistics for each of the six estimated equations.

The translog cost function is a special case of the modified translog cost function and the Box–Tidwell cost function. To see this, observe that

$$\lim_{\eta \to 0} \frac{y^\eta - 1}{\eta} = \ln y$$

and substitute ln (y) for y^* in equations (10.13) and (10.14). Because of the numerical procedures we used, our likelihood estimates never converge to $\eta =$

Table 10.3 Parameter Estimates for Alternative Cost Function Specifications with First Order Serial Correlation

Parameter[a]	Translog[b]		Modified[c] translog		Box–Tidwell[c]	
Constant	9.054	(.005)	9.053	(.005)	9.053	(.006)
Capital	.535	(.008)	.535	(.009)	.538	(.009)
Labor	.355	(.007)	.354	(.007)	.352	(.007)
Local	.260	(.309)	.282	(.394)	.209	(.358)
Toll	.462	(.226)	.401	(.326)	.426	(.280)
Technology	− .193	(.086)	− .146	(.121)	− .121	(.108)
Capital^2	.219	(.024)	.216	(.027)	.211	(.062)
Labor^2	.174	(.027)	.162	(.030)	.154	(.067)
Capital · Labor	− .180	(.019)	− .179	(.019)	− .185	(.020)
Toll^2	− 8.018	(2.170)	− 6.837	(4.892)	− 14.545	(7.853)
Local^2	− 4.241	(1.314)	− 3.249	(3.788)	− 6.848	(5.983)
Local · Toll	11.663	(3.144)	9.411	(8.233)	9.969	(6.469)
Technology^2	− .176	(1.033)	− .007	(.057)	− .006	(.827)
Capital · Toll	.337	(.138)	.335	(.141)	.337	(.158)
Capital · Local	− .359	(.122)	− .355	(.118)	− .388	(.126)
Labor · Toll	− .179	(.083)	− .170	(.087)	− .132	(.091)
Labor · Local	.164	(.071)	.161	(.068)	.133	(.069)
Capital · Tech.	.083	(.053)	.074	(.057)	.002	(.015)
Labor · Tech.	− .057	(.047)	− .054	(.048)	.061	(.124)
Toll · Tech.	− 1.404	(1.497)	− .640	(2.464)	− .693	(2.158)
Local · Tech.	1.207	(1.431)	.591	(2.821)	.599	(2.443)
v_C	.187	(.105)	.219	(.056)	.212	(.112)
$v_L = v_K$	.712	(.094)	.713	(.050)	.724	(.074)
η	—	—	.038	(.175)	.032	(.134)

[a]Standard errors in parentheses.
[b]Maximum likelihood estimates obtained by an iterative Zellner method.
[c]Nonlinear seemingly unrelated regression estimates.

Table 10.4 Diagnostic Statistics on Alternative Cost Function Specifications

	R^2	Durbin–Watson[a] (Durbin-H)	Degrees of freedom Durbin–Watson[a] (Durbin-H)	Generalized variance[b] $\times 10^{-5}$
Translog − AR(0)				22.338
Cost function	.9998	1.17	15	
Capital share	.9463	.51	27	
Labor share	.9570	.49	27	
Translog − AR(1)				10.568
Cost function	.9999	(.65)	14	
Capital share	.9756	(1.50)	26	
Labor share	.9835	(1.37)	26	
Modified translog − AR(0)				23.962
Cost function	.9999	1.05	14	
Capital share	.9447	.53	26	
Labor share	.9612	.47	26	
Modified translog − AR(1)				10.790
Cost function	.9999	(1.24)	13	
Capital share	.9755	(1.77)	25	
Labor share	.9839	(1.77)	25	
Box–Tidwell − AR(0)				36.647
Cost function	.9997	1.78	14	
Capital share	.9443	.58	26	
Labor share	.9618	.73	26	
Box–Tidwell − AR(1)				11.560
Cost function	.9997	(1.09)	13	
Capital share	.9729	(1.34)	25	
Labor share	.9823	(1.52)	25	

[a]The upper and lower bounds for the Durbin–Watson statistic are .90 and 1.60 for the share equations and .45 and 3.38 for the cost equation. We accept the hypothesis of no positive autocorrelation when the Durbin–Watson statistic exceeds the upper bound. The Durbin–Watson statistic is biased for the equations estimated under the assumption that the error term follows an AR(1) process. For these equations, we calculated the Durbin-H statistic, which is asymptotically distributed as a standard normal deviate. We reject the hypothesis of no first-order autocorrelation when this statistic exceeds 1.96.

[b]Maximum likelihood estimation which was used for the translog specification minimizes the generalized variance. Seemingly unrelated nonlinear regression methods which were used for the modified translog and Box–Tidwell specifications provide an asymptotically unbiased estimate of the generalized variance.

0 even when this value provides a global maximum of the likelihood function. Direct comparison of the generalized variances for the specifications listed in Table 10.2 indicates that the translog specification provides the lowest generalized variance.[28] In order to check that $\eta = 0$ provides a global maximum for the likelihood function, we maximized the likelihood function conditional on several alternative values of η. Table 10.5 reports the generalized variances from

Table 10.5 Generalized Variance for Alternative Functional Forms

	−.03	−.01	0.0	.01	.03	.50	.75	1.00
Modified translog AR(0)	24.547	24.974	22.338	31.775	28.868	44.391	30.916	28.059
Modified translog[a] AR(1)	28.311	23.069	10.568	28.233	27.730	54.422	37.180	[b]
Box–Tidwell AR(0)	39.885	38.318	22.338	37.215	36.207	32.402	36.764	79.043
Box–Tidwell[a] AR(1)	393.279	385.807	10.568	378.942	1016.051	155.617	43.922	306.456

[a]Maximum likelihood estimates conditional on $v_C = .187$ $v_L = v_K = .712$
[b]Singular covariance matrix.

Table 10.6 Tests of Alternative Hypotheses Concerning the Structure of Production and Technological Change

	Likelihood ratio statistic for		Number of restrictions	Critical values of χ^2	
Hypothesis[a]	translog-AR(0)	translog-AR(1)[c]		.05	.01
Separability	11.73	8.95	2	5.99	9.21
Nonjointness	12.00	19.68	1	3.84	6.63
Homotheticity	33.96	11.95	4	9.49	13.30
Homogeneity in outputs	54.93	30.20	7	14.10	18.50
Unitary elasticities of substitution	51.98	50.20	3	7.81	11.30
Neutral technological change	14.61	4.70[b]	4	9.49	13.30
Nonoutput augmenting technological change	2.66[b]	3.36[b]	2	5.99	9.21
Nonfactor augmenting technological change	11.59	1.63[b]	2	5.99	9.21
Homogeneity and symmetry	210.97	188.71	21	32.70	38.90

[a]Except for the test of homogeneity and symmetry, the maintained hypothesis is that the general model with homogeneity and symmetry imposed is the true model.
[b]Indicates hypothesis acceptable at 5% level or better.
[c]Entries are equal to the likelihood ratio statistic $-2 \ln [|\Sigma_r|/|\Sigma_u|]^{-T/2} = T \ln[|\Sigma_r|/|\Sigma_u|]$ where $|\Sigma_r|$ denotes the generalized variance of the restricted system, $|\Sigma_u|$ denotes the generalized variance of the unrestricted system, and T denotes the number of observations which is always equal to 31. This statistic is distributed asymptotically as χ_r^2 where r is the number of restrictions.

this estimation. These statistics confirm that $\eta = 0$ minimizes the generalized variance and thereby maximizes the likelihood function.

For the translog specification, Table 10.6 reports tests of several hypotheses concerning the structure of production and technological change in the Bell System. We resoundingly reject the homogeneity and symmetry restrictions implied by producer theory.[29] We also reject these restrictions for the single-output cost function estimated by Christensen, Cummings, and Schoech. Rejection of homogeneity and symmetry may indicate that the translog cost function is a poor approximation to the true cost function, that the cost function is misspecified in some other basic way, or that firms do not behave as assumed by producer theory. We cannot resolve these issues in this chapter. Like other researchers, we restrict our cost function estimates to satisfy homogeneity and symmetry.

Given homogeneity and symmetry, we reject separability at the 1% level when the disturbances are assumed to be temporally uncorrelated and at the 5% level

Table 10.7 Parameter Estimates for Cost Functions with Alternative Technological Change Specifications

Parameter[a]	Translog-AR(0)[b] nonoutput augmenting technological change		Translog-AR(1)[b] neutral technological change	
Constant	9.057	(.005)	9.053	(.005)
Capital	.537	(.004)	.538	(.008)
Labor	.354	(.004)	.353	(.007)
Toll	.345	(.181)	.422	(.205)
Local	.368	(.247)	.237	(.282)
Technology	−.120	(.052)	−.120	(.055)
Capital^2	−.193	(.023)	.220	(.024)
Labor^2	.173	(.025)	.174	(.026)
Capital · Labor	−.161	(.021)	−.181	(.019)
Toll^2	−.231	(.084)	−.142	(.080)
Local^2	.216	(.078)	.139	(.067)
Local · Toll	.367	(.094)	.390	(.139)
Technology^2	−.361	(.086)	−.397	(.117)
Capital · Toll	−1.932	(.590)	−3.243	(.773)
Capital · Local	−4.178	(1.886)	−7.543	(1.947)
Labor · Toll	5.756	(1.843)	9.984	(2.483)
Labor · Local	.177	(.129)	.218	(.141)
Capital · Tech.	.107	(.037)	—	—
Labor · Tech.	−.109	(.033)	—	—
Toll · Tech.	—	—	—	—
Local · Tech.	—	—	—	—
ν_C	—	—	.191	(.081)
$\nu_L = \nu_K$	—	—	.712	(.070)

[a]Standard errors in parentheses.
[b]Maximum likelihood estimates obtained by an iterative Zellner method.

when the disturbances are assumed to be temporally correlated.[30] These results indicate that it is not possible to form a valid aggregate $A(L,T)$ of local and long-distance telephone service outputs and that scale-economy estimates based on such an aggregate provide no reliable evidence concerning whether the cost function exhibits subadditivity. We also reject nonjointness, homotheticity, and homogeneity in outputs for both the autoregressive and nonautoregressive specifications. The nonautoregressive specification exhibits no output-augmenting technological change while the autoregressive specification exhibits neither factor-augmenting nor output-augmenting technological change. Table 10.7 reports parameter estimates for the nonautoregressive specification with the output-technology variables excluded and for the autoregressive specification with the output-technology and price-technology variables excluded. Excluding these variables has only a modest impact on the parameter estimates for the remaining variables. Consequently, we focus our attention on the general translog cost function estimates reported in Table 10.2 and 10.3.

The own-price elasticities of demand for capital, labor, and materials were negative in all years as required by producer theory. It is useful to contrast these results with Christensen, Cummings, and Schoech's most general single-product specification.[31] They found that the own-price elasticities of demand for capital and labor were positive, contrary to producer theory. By relaxing their assumption that there exists a consistent translog aggregate of local and long-distance service output, we find that the translog cost function specification provides more reasonable estimates.[32] Table 10.8 reports factor demand elasticities calculated from our estimated general translog cost function specifications for the middle year of our sample.

Table 10.8 Factor Demand Elasticities[a] Evaluated at 1961 Observation[b]

| With respect to the price of | Elasticity of the demand for | | |
	Capital	Labor	Materials
Capital	− .097 / −.056	.076 / .28	.227 / .181
Labor	0.50 / .019	− .149 / − .151	.236 / .410
Materials	.047 / .183	.073 / .127	− .463 / − 590

[a]Top triangle is for the translog specification with no serial correlation (first column of Table 10.2); bottom triangle is for the translog specification with serial correlation (first column of Table 10.3).
[b]Own-price factor demand elasticities were negative for every year between 1947 and 1977

Natural Monopoly Tests

When all firms have access to the same technology, certain properties of the cost function reveal whether one firm can produce given levels of output more cheaply than several firms can. The cost function $C(Q_1, Q_2)$ is locally subadditive if and only if

$$\sum_{ij} C(a_i Q_1, b_j Q_2) > C(Q_1, Q_2), \quad i, j = 1, \ldots, n$$

$$\sum_i a_i = 1, \quad \sum_j b_j = 1$$

(10.23)

at least two a_i or b_j not equal to zero for given levels of Q_1 and Q_2. It is locally superadditive if the inequality is reversed ($<$) and locally additive if the inequality is replaced by equality ($=$). A firm with a locally additive cost function is not a natural monopoly. The cost function is globally subadditive if and only if the cost function is locally subadditive for all feasible levels of Q_1 and Q_2.[33] Local subadditivity is a necessary and sufficient condition for local natural monopoly. Global subadditivity is a necessary and sufficient condition for global natural monopoly.

Assuming that a firm acts efficiently, it will never produce an output configuration at which the cost function is superadditive, unless there are firm-specific fixed factors. Accordingly, the interesting statistical question centers on testing between local additivity and local subadditivity.

In order to determine whether the necessary and sufficient conditions for local natural monopoly are satisfied, we require global information concerning the cost function. The single-output average cost schedule drawn in Figure 10.1 illustrates this proposition. At point A, the cost function exhibits diseconomies of scale. Yet, as is readily verified, a single firm can produce Q^* more cheaply than two or more firms can. In order to determine whether the necessary and sufficient conditions for global natural monopoly are satisfied, we require global information concerning both the cost function and the demand function. If the demand curve in Figure 10.1 shifted rightward, perhaps because of increasing social wealth, this industry would be able to support several firms. At point B, two firms, each producing Q_m, could together produce $2Q_m$ more cheaply than a single firm could. Unfortunately, global information about cost and demand is seldom available.

Baumol, Panzar, and Willig have derived necessary conditions for subadditivity and sufficient conditions for subadditivity which require somewhat less information. Economies of scope is a necessary condition for subadditivity because, without economies of scope, two single-product monopolies could produce more cheaply than a multiproduct monopoly.[34] Economies of scope and product-specific scale economies are sufficient conditions for subadditivity; splitting production between several firms would reduce synergies from joint production and scale economies from volume production.[35] Baumol, Panzar, and Willig suggest testing the necessary and sufficient conditions separately. If the necessary con-

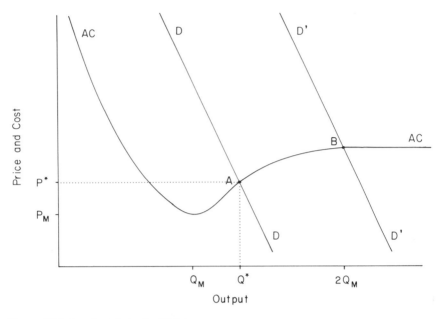

Figure 10.1 Local global subadditivity.

dition is rejected, subadditivity is decisively rejected. If the sufficient conditions
are accepted, subadditivity is decisively accepted. If the necessary condition is
accepted and the sufficient conditions are rejected, the test for subadditivity is
inconclusive.

The Baumol, Panzar, and Willig test of natural monopoly suffers from two
problems. First, estimates of economies of scope and product-specific scale
economies require information about the costs of separate production $C(0,Q_2)$
and $C(Q_1,0)$. Reliable data on separate production are seldom available for cases
of interest.[36] Second, when data are available we may frequently find that the
test is inconclusive.

We have developed a more direct and empirically less demanding test of
within sample subadditivity. The test is based on the definition of subadditivity
given in equation (10.23). Within the range of admissible sample variation
defined below, we evaluate equation (10.23) using the estimated cost function.
Rejection of (10.23) within a region leads to a rejection of global subadditivity.
Acceptance of (10.23) within a region obviously does not prove the existence
of global subadditivity. Our test is thus less demanding of the data than the one
proposed by Baumol, Panzar, and Willig and, for that reason, is more likely to
provide useful information.

We describe the test for the simple case of two-firm production versus one-
firm production of two outputs. It is straightforward to extend the test to the

multifirm and multiproduct case. Let $Q_t^* = (Q_{1t}^*, Q_{2t}^*)$ denote the output vector realized in year t. Let $Q_M = (Q_{1M}, Q_{2M}) = (\min_t Q_{1t}^*, \min_t Q_{2t}^*)$ be the smallest output vector observed in the sample. Firm A produces

$$Q_{At} = (\phi Q_{1t} + Q_{1M}, \omega Q_{2t} + Q_{2M}) \qquad (10.24)$$

Firm B produces

$$Q_{BT} = [(1 - \phi)Q_{1t} + Q_{1M}, (1 - \omega)Q_{2t} + Q_{2M}] \qquad (10.25)$$

The parameters (ϕ, ω) satisfy $0 \leq \phi \leq 1$ and $0 \leq \omega \leq 1$. In order to avoid extrapolating the cost function to unobserved output configurations, we require both firms A and B to produce Q_1 and Q_2 in a ratio within the range of the data. Thus, we require

$$R_L \leq \frac{\phi Q_{1t} + Q_{1M}}{\omega Q_{2t} + Q_{2M}} \leq R_U$$

$$R_L \leq \frac{(1 - \phi)Q_{1t} + Q_{1M}}{(1 - \omega)Q_{2t} + Q_{2M}} \leq R_U \qquad (10.26)$$

where $R_L = \min(Q_{1t}/Q_{2t})$ and $R_U = \max(Q_{1t}/Q_{2t})$ where the min and max are taken over all t. Together, firm A and firm B produce

$$Q_{1t} + 2Q_{1M} = Q_{1t}^*$$

$$Q_{2t} + 2Q_{2M} = Q_{2t}^* \qquad (10.27)$$

so that

$$Q_{1t} = Q_{1t}^* - 2Q_{1M}$$

$$Q_{2t} = Q_{2t}^* - 2Q_{2M} \qquad (10.28)$$

This allocation is possible only for $Q_i^* > 2Q_{iM}$. We restrict the test to output levels which satisfy this constraint.

Let

$$C_{At}(\phi, \omega) = C(Q_{At})$$

$$C_{Bt}(\phi, \omega) = C(Q_{Bt}) \qquad (10.29)$$

$$C_t^* = C(Q_{At} + Q_{Bt}) = C(Q_t^*)$$

We measure the degree of subadditivity by

$$\text{Sub}_t(\phi, \omega) = \frac{C_t^* - C_{At}(\phi, \omega) - C_{Bt}(\phi, \omega)}{C_t^*} \qquad (10.30)$$

If $\text{Sub}_t(\phi, \omega)$ is less than zero, the industry configuration given by (ϕ, ω) is less efficient than the monopoly configuration.

Our proposed statistical test is as follows. Calculate $Sub_t(\phi,\omega)$. If Max-$Sub_t(\phi,\omega)$ is negative and statistically significantly different from zero, then we can accept the hypothesis that the cost function is locally subadditive. If MaxSub$_t$ is not statistically significantly different from zero, we reject the hypothesis that the cost function is locally subadditive but we do not reject the hypothesis that the cost function locally additive.

We have applied this test to the Bell System. Table 10.9 reports cost and output data on the Bell System between 1947 and 1977. Between these years, costs increased more than fourteenfold, toll service increased almost fourteenfold, and local service increased more than fivefold. These data provide information on a reasonably large portion of the cost function. Both local and toll output doubled by 1958 making this year the first feasible year for our test. The ratio

Table 10.9 Local and Toll Output and Costs for the Bell System (1947–1977)

Year	Local	Toll	Cost
1947	.410	.346	2550.7
1948	.458	.372	2994.9
1949	.487	.383	3291.1
1950	.520	.416	3563.2
1951	.556	.466	4047.1
1952	.591	.501	4616.2
1953	.625	.522	4935.1
1954	.656	.550	5258.8
1955	.702	.619	5770.5
1956	.756	.683	6305.4
1957	.803	.740	6351.2
1958	.842	.776	6788.4
1959	.896	.862	7334.7
1960	.953	.935	7912.5
1961	1.000	1.000	8516.5
1962	1.054	1.082	9018.7
1963	1.110	1.174	9508.1
1964	1.159	1.317	10524.0
1965	1.228	1.474	11207.0
1966	1.306	1.684	11954.2
1967	1.383	1.842	12710.9
1968	1.465	2.055	13814.1
1969	1.558	2.334	14940.4
1970	1.638	2.536	16485.8
1971	1.709	2.697	17951.8
1972	1.804	2.969	20161.2
1973	1.912	3.316	21221.7
1974	2.007	3.605	23168.4
1975	2.075	3.864	27378.7
1976	2.173	4.244	31304.5
1977	2.291	4.684	36078.0

of local to toll has been between 0.5 and 1.3. For each year $t = 1958, \ldots,$ 1977 we calculated $Sub_t(\phi,\omega)$ for the unique combinations of (ϕ,ω) corresponding to[37]

$$\phi = 0, .1, .2, \ldots, .9, 1.0$$

$$\omega = 0, .1, .2, \ldots, .9, 1.0$$

(10.31)

from three estimated cost functions: (1) the general translog cost function with nonautoregressive errors; (2) the general translog cost function with first order autoregressive errors; and (3) the Box–Tidwell cost function with nonautoregressive errors. We calculated Sub_t for the third cost function because, although this function performed substantially more poorly than the other two cost function, it was the only cost function we estimated that exhibited cost complementarities between local and toll and is therefore the cost function most likely to exhibit subadditivity.

Table 10.10 Percentage of Gain or Loss from Multifirm Versus Single-Firm Production for Alternative Industry Configurations[a,b] Translog—AR(1) 1961

φ =	0.0	0.1	0.2	0.3	0.4	0.5	0.6	0.7	0.8	0.9	1.0	
ω =												
0.0	8											0.0
	(21)											
0.1	8	8										0.1
	(19)	(20)										
0.2	9	8	8									0.2
	(18)	(18)	(19)									
0.3	12	10	9	9								0.3
	(16)	(17)	(18)	(18)								
0.4	15	13	10	9	9							0.4
	(15)	(13)	(17)	(18)	(18)							
0.5	20	16	13	11	9	9						0.5
	(14)	(15)	(16)	(17)	(18)	(18)						
0.6	25	21	17	14	11	10	9					0.6
	(14)	(15)	(16)	(17)	(17)	(18)	(18)					
0.7			23	18	15	12	10	9				0.7
			(16)	(16)	(17)	(18)	(18)	(18)				
0.8				20	16	12	10	8				0.8
				(17)	(18)	(18)	(19)	(19)				
0.9						17	13	10	8			0.9
						(19)	(19)	(20)	(20)			
1.0									10	8		1.0
									(21)	(21)		
	0.0	0.1	0.2	0.3	0.4	0.5	0.6	0.7	0.8	0.9	1.0	

[a]Entries equal Sub_{1961} so that a positive number indicates that multifirm production is much more efficient than single-firm production.
[b]Standard errors are reported in parentheses.

Table 10.11 Percentage of Gain or Loss from Multifirm Versus Single-Firm Production for Alternative Industry Configurations[a,b] Translog—AR(0) 1961

$\phi =$	0.0	0.1	0.2	0.3	0.4	0.5	0.6	0.7	0.8	0.9	1.0	
$\omega =$												
0.0	−17											0.0
	(16)											
0.1	−16	−16										0.1
	(14)	(14)										
0.2	−13	−14	−15									0.2
	(12)	(13)	(13)									
0.3	−9	−12	−13	−14								0.3
	(11)	(12)	(12)	(12)								
0.4	−5	−9	−11	−13	−13							0.4
	(11)	(11)	(12)	(12)	(12)							
0.5	0	−5	−8	−11	−13	−13						0.5
	(11)	(11)	(11)	(12)	(12)	(12)						
0.6	5	0	−4	−8	−11	−13	−13					0.6
	(11)	(11)	(11)	(12)	(12)	(12)	(13)					
0.7		1	−4	−8	−11	−13	−14					0.7
		(13)	(12)	(12)	(12)	(12)	(12)					
0.8				−4	−8	−12	−14	−15				0.8
				(14)	(13)	(13)	(13)	(13)				
0.9						−8	−12	14	−16			0.9
						(15)	(15)	(14)	(14)			
1.0									−16	−17		1.0
									(16)	(16)		
	0.0	0.1	0.2	0.3	0.4	0.5	0.6	0.7	0.8	0.9	1.0	

[a]Entries equal Sub_{1961} so that a positive number indicates that multifirm production is more efficient than single-firm production.
[b]Standard errors are reported in parentheses.

We find that MaxSub_t is greater than zero for all three cost functions at the output configurations observed between 1958 and 1977. MaxSub_t, however, is never statistically significantly different from zero at conventional significance levels. Thus, we reject the hypothesis that the Bell System cost function is locally subadditive and that the Bell System is a natural monopoly at the output levels observed between 1958 and 1977. We do not reject the hypothesis that the Bell System cost function is locally additive. Tables 10.10 and 10.11, and 10.12 report the values of Sub_{1961} for the three cost functions we examined. Standard errors are reported in parentheses. The fact that Sub_{1961} is never significantly different from zero is consistent with the hypothesis that the cost function is locally additive at the level of demand observed in 1961.[38]

Table 10.12 Percentage of Gain or Loss from Multifirm Versus Single-Firm
Production for Alternative Industry Configurations[a,b]
Box–Tidwell AR(0) 1961

ϕ = 0.0	0.1	0.2	0.3	0.4	0.5	0.6	0.7	0.8	0.9	1.0
ω =										
0.0 10										
(14)										
0.1 10	11									
(14)	(14)									
0.2 11	11	11								
(14)	(14)	(14)								
0.3 11	11	11	11							
(14)	(14)	(14)	(14)							
0.4 11	11	11	11	11						
(14)	(14)	(14)	(14)	(14)						
0.5 11	11	11	11	11	11					
(14)	(14)	(14)	(14)	(14)	(14)					
0.6 10	11	11	11	11	11	11				
(14)	(14)	(14)	(14)	(14)	(14)	(14)				
0.7		11	11	11	11	11	11			
		(14)	(14)	(14)	(14)	(14)	(14)			
0.8				11	11	11	11	11		
				(14)	(14)	(14)	(14)	(14)		
0.9						11	11	11	11		
						(14)	(14)	(14)	(14)	
1.0										10	10
									(14)	(14)
	0.0	0.1	0.2	0.3	0.4	0.5	0.6	0.7	0.8	0.9	1.0

[a]Entries equal Sub_{1961} so that positive number indicates that multifirm production is more efficient than single firm production.
[b]Standard errors are reported in parentheses.

Summary

We estimated several alternative multiproduct cost functions for the Bell System. We rejected the modified translog cost function and the Box–Tidwell cost function in favor of the translog cost function. We tested and rejected the hypothesis that there exists an aggregate measure of local and long-distance telecommunications service. Therefore (a) the single-output cost functions estimated by previous researchers were rejected by the data and (b) scale economy estimates from these single-output cost functions provide little information about the optimal structure of the telecommunications industry.

We developed a test for natural monopoly that is more direct and more empirically tractable than the previous tests reported in the literature. For observed output configurations, our test compares the single-firm cost of production with the multifirm cost of production. We applied this test to the Bell System. We found that the Bell System did not have a natural monopoly over any of the output configurations which were realized between 1958 and 1977. Two firms were always able to produce these output configurations more cheaply than a single firm.

Appendix A
Alternative Formulations of the Cost Function

The Bell System provides numerous telecommunications services to consumers and businesses. Its operating companies provide local service through their exchange facilities; public telephone service; directory advertising; and interstate and intrastate long-distance services including various private line services. Its Long Lines Department provides some interstate long-distance and private line services and most wide-area toll services. These entities own different portions of the telecommunications network.

It is useful to view the telecommunications network as consisting of numerous nodes, which represent the local exchange facilities, interconnected by lines, which represent the long-distance facilities. This idealized network provides local service—including local telephone service, public telephone service, and local private line services—through the exchange facilities and long-distance service—including toll service, private line service, and wide-area toll service—through the long-distance facilities. AT&T and the Bell Operating Companies operate and maintain different chunks of this integrated network.

AT&T owns 100% of the stock of 17 of its operating companies, more than 85% of the stock of four others, and a minority interest in one. Its Long Lines Division and General Departments provide overall coordination and direction for the operating companies. It therefore ultimately owns and coordinates the numerous local exchanges as well as the long-distance facilities shared by the local exchanges. In examining whether there is a telecommunications natural monopoly, these considerations suggest the following cost function

$$C = C(l_1, \ldots, l_n, T) \tag{10.32}$$

where l_i represents the local service produced the ith local exchange and T is the long-distance (or toll) service produced by the shared long-distance facilities. If synergies arise from the joint production of local exchange services with each other and with the long-distance service, a natural monopoly may exist. Separation of the local exchange facilities from each other and from the shared long-distance service would decrease efficiency. With data on a cross section of telephone conglomerates having different numbers of local exchanges and not all having a shared long-distance facility, it is possible to estimate (10.32) and test whether there are economies of scope between local exchanges and long-distance facilities. Such data are not presently available. With time-series data on a conglomerate such as AT&T, it is possible to estimate (10.32) and test whether there are cost complementarities between local exchange facilities and long-distance facilities. For reasons discussed below, available data are not even sufficient for this purpose.

The Bell operating companies are distinct financial units although they are commonly owned by AT&T. They provide both local and long-distance services within their respective territories. In 1979, they earned about 21 billion dollars in local service revenue and 21 billion dollars in long-distance service revenue. AT&T Long Lines supplements the long-distance service provided by the operating companies. In 1979, it earned slightly more than three billion dollars in long-distance service revenues. These considerations suggest the following formulation of the cost function

$$C_i = C_i(l_i, t_i, L_R, T_R), \quad i = 1, \ldots, n \qquad (10.33)$$

where l_i is the local service output of the ith operating company, t_i is the long-distance service output of the ith operating company, L_R is the local service output of the other operating companies, T_R is the long-distance service output of the other operating companies and the Long Lines Division and C_i is the cost incurred by the ith operating company. This system of equations could be estimated with time series cross section data on the Bell operating companies and AT&T Long Lines. If increases in L_R and T_R decrease the costs of producing given levels of l_i and t_i and if there are product-specific scale economies in producing l_i and t_i, there may be a natural monopoly over the telecommunications system.

Unfortunately, reliable data are not available for estimating (10.33). The costs incurred and the revenues received by the various entities that provide telecommunications service are determined by the "separations and settlement" process. This process involves periodic negotiations between the Bell operating companies, AT&T, the independent telephone companies, the FCC, and the state regulatory commissions. These negotiations assign the costs of operating the network to the state rate bases examined by the state commissions and to the interstate rate base examined by the FCC. These negotiations also assign the revenues earned by the network to the assets residing in the various regulatory jurisdictions. The costs and revenues that appear on the financial record of AT&T Long Lines and the Bell operating companies are determined by a political rather than an economic process. The assignment formulas have changed over time as the political powers of the participants in the negotiations have ebbed and flowed. Therefore, the correspondence between reported costs and revenues and economic costs and revenues for the individual entities providing telecommunications services is tenuous and changeable.

These data limitations dictate the following formulation of the cost function

$$C = C(L, T) \qquad (10.34)$$

where L is the aggregate local service provided by the Bell operating companies,

T is the aggregate long-distance service provided by the Bell operating companies and the Long Lines Department, and C is the aggregate cost incurred by the Bell System. This formulation assumes that costs are independent of the allocation of a given level of output across the Bell operating companies and Long Lines. Because the levels of local and long-distance services have probably increased in tandem across the Bell operating companies, the aggregation of l_i and t_i into L and T and the estimation of an aggregate cost function is probably not too unreasonable.

Appendix B
Unrestricted Cost Function Estimates

When the homogeneity and symmetry restrictions implied by producer theory were relaxed, the parameter estimates changed dramatically. Table 10.13 reports the unrestricted parameter estimates for the nonautoregressive translog cost func-

Table 10.13 Parameter Estimates for the Translog Cost Function Unrestricted for Homogeneity and Symmetry with No Serial Correlation

Parameter*	Cost		Capital share		Labor share	
Constant	9.051	(.002)				
Capital	.197	(.050)	.539	(.004)		
Labor	.163	(.155)			.352	(.003)
Toll	−1.168	(.183)				
Local	.760	(.159)				
Technology	1.226	(.083)				
Capital^2	2.484	(.394)	.197	(.025)		
Labor^2	4.722	(3.487)			.203	(.062)
Materials^2	10.811	(4.057)				
Capital · Labor	−2.836	(.373)	−.218	(.063)	−.166	(.025)
Capital · Materials	−3.892	(1.042)	.208	(.115)		
Labor · Materials	−13.377	(3.542)			−.165	(.113)
Capital · Toll	12.723	(1.264)	−.499	(.103)		
Labor · Toll	4.210	(3.622)			.406	(.102)
Materials · Toll	−11.562	(6.477)				
Capital · Local	−11.814	(1.352)	.568	(.100)		
Labor · Local	−4.033	(3.448)			−.500	(.098)
Materials · Local	8.764	(5.783)				
Toll^2	36.462	(3.455)				
Local^2	45.660	(4.374)				
Toll · Local	−77.513	(7.387)				
Technology^2	16.041	(1.126)				
Capital · Tech.	−3.334	(.290)	.223	(.043)		
Labor · Tech.	.795	(.975)			−.192	(0.42)
Materials · Tech	14.942	(2.810)				
Toll · Tech.	−29.646	(2.435)				
Local · Tech.	20.436	(2.086)				

tion. The estimates are markedly different from the corresponding estimates in the first column of Table 10.2. The homogeneity and symmetry restrictions are rejected by the data at a high level of confidence. These restrictions require parameters to be identical across equations. As we see by comparing the columns of Table 10.13 parameter estimates are wildly different between equations.

Appendix C
Bell System Data

Our data on costs, input prices, and output quantities were obtained from Christensen.[39] Christensen calculated Tornqvist indices of output quantities and input quantities for the Bell system based on detailed yearly data for the period 1947–1977. The Tornqvist index can be written as

$$\ln\left(\frac{X_t}{X_{t-1}}\right) = \sum \overline{w}_{it} \ln\left(\frac{x_{it}}{x_{it-1}}\right)$$

where

$$\overline{w}_{it} = \frac{1}{2}\left(\frac{p_{it}x_{it}}{\Sigma p_{it}x_{it}} + \frac{p_{it-1}x_{it-1}}{\Sigma p_{it-1}x_{it-1}}\right)$$

and where X_t is an aggregate quantity index for period t, x_{it} is one of individual quantity indices, and p_{it} is the price of quantity i in period t. The quantity index X_t can be obtained from this formula by normalizing one of the X_t to one and calculating the X_t recursively.

Christensen collected operating revenue data for five categories: local, interstate toll, intrastate toll, directory advertising, and miscellaneous. In order to obtain quantity data from these revenue data he divided by price indices supplied by AT&T. He formed an aggregate measure of output from these quantity data and the price indices using the Tornqvist procedure described above. For our multiproduct study, we used the Tornqvist procedure to calculate aggregate output indices for local and long-distance service. Our local service output is based on local revenue, directory advertising revenue, and miscellaneous revenue.

Christensen collected data on hours worked by Bell System employees broken down by occupation and years of service. Using Bell System wage rates, he then calculated an index of labor output. He collected data on twenty different types of owned tangible assets. He says,

> For each of the twenty categories we obtained a time series of investment expenditures, which we then deflated by specific price indexes. The resulting real investment figures were used in conjunction with capital stock benchmarks and rates of replacement to obtain capital stock series via the perpetual inventory method. The benchmarks and replacement rates were based on surveys of Bell

System Capital Stock for 1958, 1965, and 1970. These capital stocks, their asset prices, and rates of replacement were used along with the Bell System's cost of capital and tax information to compute capital service price weights. These weights were constructed following the methodology originally proposed by Christensen and Jorgenson and modified for regulated firms by Caves, Christensen, and Swanson. We computed capital input for the Bell System as a Tornqvist index of the twenty types of owned capital, and one category of rented capital, using service price weights.[40]

He included data on seven categories of materials: electricity, accounting, machines, advertising, stationery and postage, services from Bell Labs, and mis-

Table 10.14 Bell System Data Used for Multiproduct Cost Function Estimates[a]

Year	Cost	Local output	Toll output	Capital price
1947	2550.68	.41014	.36642	.49948
1948	2994.94	.45783	.34642	.55879
1949	3291.06	.48703	.38296	.57440
1950	3563.20	.52004	.41592	.61810
1951	4047.07	.55560	.46552	.70031
1952	4616.23	.59149	.50116	.79500
1953	4935.13	.62452	.52271	.80853
1954	5258.76	.65669	.55000	.81269
1955	5770.47	.70289	.61941	.86056
1956	6305.44	.75645	.68394	.88033
1957	6351.19	.80355	.74006	.81997
1958	6788.40	.84224	.77663	.87304
1959	7334.71	.89657	.86274	.91051
1960	7912.48	.95314	.93512	.95733
1961	8516.46	1.00000	1.00000	1.00000
1962	9018.66	1.05411	1.08231	1.01457
1963	9508.12	1.11068	1.17451	1.00832
1964	10524.00	1.15909	1.31715	1.07804
1965	11207.00	1.22822	1.47436	1.06139
1966	11954.20	1.30609	1.68434	1.04475
1967	12710.90	1.38312	1.84266	1.04058
1968	13814.10	1.46568	2.05511	1.08325
1969	14940.40	1.55869	2.33437	1.04579
1970	16485.80	1.63899	2.53682	1.04891
1971	17951.80	1.70956	2.69772	1.04058
1972	20161.20	1.80454	2.96927	1.09157
1973	21221.70	1.91210	3.31628	1.00312
1974	23168.40	2.00785	3.60503	1.00104
1975	27376.70	2.07532	3.86421	1.18939
1976	31304.50	2.17307	4.24442	1.32778
1977	36078.00	2.29155	4.68449	1.53590

Source: See Appendix C.

[a]Arithmetic values of Tornqvist indices normalized to equal one in 1961.
[b]Tornqvist index of local and toll used for estimating single-product cost function.

cellaneous. Using Bell System price indices, he formed an index of materials inputs. He obtained price indices for capital, labor, and materials by dividing total expenditures on these items by the associated price index. We normalized input prices to equal one in 1961.

The research and development index was calculated

$$A_t = \left[\sum_{k=0}^{N_t} \left(\frac{\lambda^{k-1}}{\Gamma(k)} \right) \frac{Rand_{t-k}}{CPI_{t-k}} \right] \Big/ \sum_{k=0}^{N_t} \left(\frac{\lambda^{k-1}}{\Gamma(k)} \right)$$

Table 10.14 (continued)

Labor price	Materials price	R&D index	Capital share	Labor share	Aggregate[b] output
.53566	.66952	.57955	.39552	.49635	.37200
.58236	.75117	.55445	.40430	.48286	.41109
.60959	.74530	.55261	.41936	.47113	.43262
.63164	.76525	.56980	.44096	.45352	.46579
.66926	.81572	.59576	.45338	.44230	.50814
.70946	.82863	.62057	.46670	.43159	.54461
.73411	.84389	.63873	.46436	.43614	.57320
.76134	.85563	.65059	.46596	.42866	.60369
.80674	.87558	.66162	.47840	.41414	.66093
.81063	.90493	.68018	.47642	.41045	.72066
.84824	.93896	.71436	.47138	.41365	.77258
.85084	.95305	.76830	.50754	.38849	.81106
.91958	.97417	.83934	.52030	.37321	.88010
.95979	.99061	.91902	.53120	.36083	.94477
1.00000	1.00000	1.00000	.54381	.34605	1.00000
1.03632	1.01995	1.08533	.55077	.33966	1.06707
1.07393	1.03404	1.18984	.55139	.33353	1.13972
1.12970	1.08451	1.32815	.56240	.32693	1.22792
1.17121	1.10681	1.49998	.55286	.32925	1.33575
1.22827	1.14085	1.16877	.54302	.33698	1.47162
1.29702	1.17371	1.86844	.54079	.34058	1.58499
1.36057	1.21948	2.02744	.54614	.33406	1.72439
1.49416	1.28286	2.16342	.51402	.35802	1.89881
1.62387	1.35211	2.28416	.49799	.37133	2.03198
1.80415	1.42019	2.40026	.48313	.38304	2.14140
2.06226	1.47653	2.52124	.47953	.39061	2.31167
2.26329	1.56221	2.65447	.44558	.41442	2.52082
2.51621	1.74061	2.80468	.434068	.42485	2.69770
2.85473	1.91315	2.97195	.46178	.40606	2.84174
3.21920	2.01408	3.15081	.469773	.39508	3.05380
3.40726	2.12911	3.33422	.48680	.37808	3.30393

where $\Gamma(k)$ is the gamma function evaluated at k, $Rand_t$ is the research and development expenditure by Bell Labs charged to AT&T, CPI_t is the consumer price index, $\lambda = 6$ and

$$N_t = 22 - (1958\text{-}t)\ t = 1947\text{--}1957$$

$$22 \qquad\qquad t = 1958\text{--}1977$$

A_t is based on fewer than 22 lagged values prior to 1958 because we had data on $Rand_t$ only from 1936. We tried deflating by an R & D-specific deflator rather than the CPI but we obtained poor statistical results. We also tried different values of the lag parameter but found that our results were insensitive to this parameter. This index was initially proposed by Vinod.[41] Our long-distance service output is based on interstate and intrastate toll service. Table 10.14 reports the data we used.

NOTES

1. AT&T, *Defendants' Third Statement of Contentions and Proof*, in *US* v. *AT&T*, p. 35.

2. The major studies are Laurits Christensen, Diane Cummings, and Philip Schoech, "Econometric Estimation of Scale Economies in Telecommunications," SSRI Working Paper No. 8013 (Madison, WI: Social Systems Research Institute, University of Wisconsin at Madison August 1981) and M. Ishaq Nadiri and Mark A. Shankerman, "The Structure of Production, Technological Change, and the Rate of Growth of Total Factor Productivity in the US Bell System," in T. Cowing and R. Stevenson, eds., *Productivity Measurement in Regulated Industries* (New York: Academic Press, 1981). H.D. Vinod has published two studies that rely on ridge regression: "Applications of New Ridge Regression Methods to a Study of Bell System Scale Economies," *Journal of the American Statistical Association*, December 1976, pp. 835–841 and "Bell System Scale Economies and the Economics of Joint Production," FCC Docket 20003, Bell Exhibit No. 59. The latter paper is the only major study which estimates the parameters of a multiproduct technology for the Bell System. Unfortunately, this study is seriously flawed because it relies on canonical ridge regression which has no known optimality property, as Vinod admits in "Canonical Ridge and the Econometrics of Joint Production," *Journal of Econometrics*, August 1976, pp. 147–166. Several studies have estimated multiproduct cost functions for Bell Canada. See Melvyn Fuss and Leonard Waverman, *The Regulation of Telecommunications in Canada*, Technical Report No. 7, Economic Council of Canada, March 1981 and the references cited therein. See Chapter 6 of this volume for a critique of the existing studies on the cost characteristics of the telecommunications industry.

3. Christensen, Cummings, and Schoech, *ibid.*, p. 4.

4. W. Baumol, J. Panzar, and R. Willig, *Contestable Markets and the Theory of Industry Structure* (San Diego: Harcourt Brace Jovanovich, 1982).

5. This function is estimated under the assumption that AT&T has exogenously given output levels and exogenously given factor prices. The assumption that input prices are exogenously given is questionable. The Communications Workers of America represent most of the Bell System's nonmanagerial employees. The wage rates faced by the Bell System are therefore determined by negotiations between a powerful corporation and a powerful trade union rather than by competitive market forces. The Bell System purchases most of its capital, materials, and research and development from Western Electric and Bell Labs, both of which are owned by AT&T and both of which are large purchasers in their respective factor markets. The Bell System may have some control over the prices it faces for capital, materials, and research and development. The assumption that output levels are exogenously given is also questionable. The justification usually given for this assumption is that the prices charged for telephones service are regulated and that the Bell System is obligated to meet all demand. But the Bell System can file for tariff increases which may be approved and instituted quite quickly. Given the institutional evidence

against the hypothesis of output level and input price endogeneity, it is important to test this hypothesis. Using a Wu test, we reject the hypothesis that output and input prices are endogenous.

6. We have not modeled the regulatory process explicity because a sensible, empirically tractable model of regulation is presently lacking and we did not have the resources to develop one. Previous studies of regulated industries have used the Averch–Johnson theory of regulation to model the impact of regulation on profit maximization and input demand. These studies have had rather mixed success. See Fuss and Waverman, *op. cit.;* R. Spann, "Rate of Return Regulation and Efficiency in Production: An Empirical Test of the Averch–Johnson Thesis," *The Bell Journal of Economics and Management Science,* Spring 1974, pp. 38–52; and T. Cowing, "The Effectiveness of Rate-of-Return Regulation: An Empirical Test Using Profit Functions," in M. Fuss and D. McFadden, *Production Economics: A Dual Approach to Theory and Applications* (Amsterdam: North Holland, 1978). The Averch–Johnson model misconstrues several aspects of the regulatory process. In telecommunications, for example, regulators control prices directly through the tariff-setting mechanism and the rate of return only indirectly through these tariffs. See Paul Joskow, "Inflation and Environmental Concern: Structural Change in the Process of Public Utility Regulation," *Journal of Law and Economics,* October 1974, pp. 291–328 and Paul Joskow and Roger Noll, "Theory and Practice in Public Regulation: A Current Overview," in G. Fromm, ed., *Studies in Public Regulation* (Cambridge: MIT Press, 1981) for further discussion.

7. We combined local telephone service, directory advertising, and miscellaneous service into an aggregate called *local service* and interstate and intrastate long distance into an aggregate called *long distance.*

8. L.R. Christensen, D.W. Jorgenson, and L.J. Lau, "Transcendental Logarithmic Production Functions," *Review of Economics and Statistics,* August 1973, pp. 28–49.

9. These restrictions are termed the homogeneity and symmetry restrictions in what follows although they include several other restrictions as well. Symmetry refers to the independence of the second-order derivatives of cost with respect to price to the order of differentiation, that is,

$$\frac{\partial C}{\partial p_i \, \partial p_j} = \frac{\partial C}{\partial p_j \, \partial p_i} \text{ which implies } \gamma_{ij} = \gamma_{ji}.$$

Since only $(\gamma_{ij} + \gamma_{ji})$ is identified in the cost and share equations, symmetry is usually tested by determining whether the γ_{ij}'s are identical across equations.

10. See Ronald W. Shephard, *Theory of Cost and Production Functions* (Princeton: Princeton University Press, 1970), for definitions.

11. Michael Denny and Cheryl Pinto, "An Aggregate Model with Multiproduct Technologies," in M. Fuss and D. McFadden, *op. cit.,* p. 256. Note that the indices in their derivation are in an improper order.

12. The separability test tells us whether an aggregator function exists but not the form of the aggregator function. It is possible to accept separability but reject particular aggregator functions such as the translog aggregator used by Christensen, Cummings, and Schoech.

13. Denny and Pinto, *op cit.,* p. 258

14. Economies of scope are a necessary condition for natural monopoly. Economies of scope exist for the cost function $C(Q_1,Q_2)$ if $C(Q,0) + C(0,Q_2)$ exceeds $C(Q_1,Q_2)$ so that joint production is cheaper than separate production.

15. David F. Burgess, "Duality Theory and Pitfalls in the Specifications of Technologies," *Journal of Econometrics,* May 1975, pp. 105–123.

16. A.R. Gallant, "On the Bias in Flexible Functional Forms and an Essentially Unbiased Form," *Journal of Econometrics,* February 1981, pp. 211–246.

17. David K. Guilkey and C.A. Knox Lovell, "On the Flexibility of the Translog Approximation," *International Economic Review,* February 1980, pp. 137–147.

18. Melvyn Fuss and Leonard Waverman, *op. cit..*

19. Douglas W. Caves, Laurits R. Christensen, and Michael Trethway, "Flexible Cost Functions for Multiproduct Firms," *Review of Economics and Statistics,* August 1980, pp. 477–481.

20. Because the disturbances for the share equations must lie in the unit interval, the assumption that the ε_i in (10.16) and the u_i in (10.17) are multinormally distributed is clearly inappropriate. Woodland, however, found that parameter estimates obtained under the assumption that the error terms of the share equations are multinormally distributed were close to parameter estimates obtained under the assumption that the error terms are Dirichlet distributed. The Dirichlet distribution lies within the unit sphere. See A.D. Woodland, "Stochastic Specification of the Estimation of Share Equations." *Journal of Econometrics,* August 1979, pp. 361–384. We note, however, that the Dirichlet distribution is rather restrictive since it imposes negative covariances between the disturbances. Also, using the method proposed by H. White it is possible to correct for nonnormality of the error distribution terms in forming estimates of the standard error. H. White, "A Heteroskedasticity Consistent Covariance Matrix Estimator and a Direct Test of Heteroskedasticity," *Econometrica,* May 1980, 817–830.

21. See L. Christensen and W. Greene, "Economies of Scale in U.S. Electric Power Generation," *Journal of Political Economy,* October 1976, pp 655–676, for a discussion of the estimation of the translog cost function using the iterated Zellner method. The Zellner method for estimating a seemingly unrelated regression system involves two steps: (1) form a consistent estimate S_1 of the cross-equation covariance matrix Σ, and (2) using S_1 in place of Σ calculate the joint generalized least squares estimate of the parameter vector β. The iterated method forms a new estimate S_2 of Σ from the residuals of the second step and then forms a new estimate of the parameter vector β from S_2. This iteration continues until estimates of β and Σ converge. The Zellner method and the iterated Zellner method were developed for systems which are linear in their parameters but can be readily extended to systems which are nonlinear in their parameters. See A. Ronald Gallant, "Seemingly Unrelated Nonlinear Regressions," *Journal of Econometrics,* April 1975, pp. 35–50.

22. The likelihood function for the modified translog and Box–Tidwell cost functions exhibited considerable nonlinearities and plateaus. The likelihood optimization routines usually either failed to converge or converged to local but not global maxima. Seemingly unrelated nonlinear regression techniques were more successful for these specifications. Consequently, the estimates reported for the modified translog and Box–Tidwell cost functions were obtained from seemingly unrelated nonlinear regressions rather than maximum likelihood esimation. Asymptotically, both methods yield identical results.

23. A.P. Barten, "Maximum Likelihood Estimation of Complete Systems of Demand Equations," *European Economic Review,* May 1967, pp. 7–73.

24. E.R. Berndt and N.E. Savin, "Estimation and Hypothesis Testing in Singular Equation Systems with Autoregressive Disturbances," *Econometrica,* September–November 1975, pp. 937–957.

25. The specification $v_L = v_K$ successfully purged the autocorrelation from the residuals whereas the specification $v_L \neq v_K$ continued to exhibit autocorrelation.

26. H.D. Vinod, 1976, *op. cit.*

27. As alternative measures of technological changes, we used time, the percent of telephones with access to long-distance dialing facilities, and the percent of telephones connected to modern switching facilities, singly and in combination. Generally, these specifications gave less stable and less plausible estimates (e.g., negative marginal cost estimates and positive own-price factor demand elasticities) and higher generalized variances than the specifications reported below. Fuss and Waverman, *op. cit.,* assumed that technological change was output augmenting, with increases in direct distance dialing increasing the value of toll calls and increases in modern switching facilities increasing the value of local calls. This specification performed poorly with U.S. data. Christensen, Cummings, and Schoech's single-output study, *op. cit.,* estimated, *inter alia,* a factor augmenting technological change specification where increases in R & D lowered the effective prices for capital, labor, and materials, In the two-output case with autoregressive disturbances, this specification exhibited constant returns to scale. The generalized variance was, however, considerably higher than in a specification which included general technological change. This study also experimented with different lag parameters for the R & D index but found that estimates were relatively insensitive to the value of the lag parameter.

28. These comparisons are not entirely reliable since the translog specifications were estimated with maximum likelihood while the other specifications were estimated with seemingly unrelated nonlinear regression techniques. In order to compare generalized variances obtained from iden-

tical estimation procedures, we also estimated the translog specification with seemingly unrelated nonlinear regression. We found that the modified translog cost function with $\eta = .03$ gave the lowest generalized variance under the assumption that the disturbances are not temporally correlated. The modified translog had a generalized variance of 23.962 whereas the translog had a generalized variance of 24.378. Obviously, it is not possible to reject the hypothesis that $\eta = 0$. We found that the translog cost function continued to give the lowest generalized variance under the assumption that the disturbances are temporally correlated. Also, the translog cost function performed better than the Box–Tidwell cost function for both error specifications.

29. We usually rejected homogeneity and symmetry separately; rejected homogeneity with symmetry imposed; and rejected symmetry with homogeneity imposed. Table 10.13 in Appendix B reports estimates for the translog cost function with nonautoregressive disturbances when homogeneity and symmetry are not imposed.

30. When $\nu_L \neq \nu_K$ it was possible to accept separability. The generalized variance under the hypothesis of homogeneity and symmetry was 8.4177×10^{-15} and under the hypothesis of homogeneity, symmetry, and separability was 9.8276×10^{-15}. The likelihood ratio test statistic for separability was 4.79 compared with a critical value at the five percent level with two restrictions of 5.99. The acceptance of separability suggests that a consistent aggregate $A(L,T)$ exists but does not tell us whether the translog aggregate used by Christensen, Cummings, and Schoech and by Nadiri and Schankerman is appropriate. We tested the validity of the translog aggregation in the following fashion. Let $A(L,T) = \ln Q = W_L \ln L + W_T \ln T$ where $\ln Q$ is the aggregate formed by Christensen, Cummings, and Schoech. We chose W_L and W_T to satisfy this relationship. Under our general specification we have

$$\beta \ln Q = \beta_1 W_{\bar{L}} \ln L + \beta_2 W_{\bar{T}} \ln T \text{ and } \delta(\ln Q)^2 = \delta_{11}(W_{\bar{L}} \ln L)^2 + \delta_{22}(W_{\bar{T}} \ln T)^2 + 2\delta_{12} W_{\bar{T}} W_{\bar{L}} \ln L \ln T$$

with $z = 0$. Under the single-output specification, we have $\beta_1 = \beta_2 = \beta, \delta_{11} = \delta_{22} = 2\delta_{12}$, and $z = 1$. The generalized variance under the general multioutput specification is 9.827. The generalized variance under the single-output specification is 22.725. Separability, homogeneity, and symmetry were imposed on both specifications. Letting $z, \beta_1, \beta_2, \delta_{11}, \delta_{22}, \delta_{12}$, be unconstrained we obtained a generalized variance of 8.5281. The point estimate of z was -5.43 with a standard error of 1.92 leading to rejection of $z = 1$ at the .0001 level of significance.

31. Christensen, Cummings, and Schoech *op. cit.*, Table 1, column 12. At the sample mean and using their notation, $\varepsilon_{KK} = (\gamma_{KK} + \beta_K^2 - \beta_K)/\beta_K = .1494$, the elasticity of demand for labor is $\varepsilon_{LL} = (\gamma_{LL} + \beta_L^2 - \beta_L)/\beta_L = .0269$, and the elasticity of demand for materials is $\varepsilon_{MM} = (\gamma_{MM} + \beta_M^2 - \beta_M)/\beta_M = -.0979$. They claim that the cost function is better behaved when technological change is allowed to be factor augmenting in the following fashion; $p_i = p_i^{\lambda A}$, $i = L,K,M$ and where A denotes the technological change proxy. Their reported estimates (see column 10), however, show that the elasticity of demand for capital is positive: $\varepsilon_{KK} = (\gamma_{KK} + \beta_K^2 - \beta_K)/\beta_K = [.266 + (.518)^2 - (.518)]/.518 = .0315$ at the sample mean.

32. The estimated general translog cost functions were monotonic with respect to input prices in all years. They were also quasi-concave in all years. Quasi-concavity requires that the following conditions hold where

$$C_{ij} = \frac{\partial C}{\partial p_i \, \partial p_j}, \quad i,j = 1,2,3, = K,L,M \quad \text{for capital, labor, and materials, respectively}$$

$$(1) \; C_{KK} \leq 0 \qquad (2) \; C_{LL} \leq 0 \qquad (3) \; C_{MM} \leq 0$$

$$(4) \begin{vmatrix} C_{KK} & C_{KL} \\ C_{KL} & C_{LL} \end{vmatrix} \geq 0 \quad (5) \begin{vmatrix} C_{KK} & C_{KM} \\ C_{KM} & C_{MM} \end{vmatrix} \geq 0 \quad (6) \begin{vmatrix} C_{LL} & C_{LM} \\ C_{LM} & C_{MM} \end{vmatrix} \geq 0$$

$$(7) \begin{vmatrix} C_{KK} & C_{KL} & C_{KM} \\ C_{KL} & C_{LL} & C_{LM} \\ C_{KM} & C_{LM} & C_{MM} \end{vmatrix} \leq 0$$

where we have made use of the symmetry of the C_{ij}'s. The first three quantities were strictly less than zero in every year. The second three quantities were strictly greater than zero in every year. The last quantity was virtually zero in every year. The value of the determinant (7) was of the order $\pm \ 10^{-8}$ although typical elements were of the order 10^3. In the case of the nonautoregressive specification, the determinant (7) was exactly zero in the middle year of the sample. In the nonautoregressive specification the determinant (7) was slightly negative in 14 years, zero in two years, and positive in 15 years. In the autoregressive specification, the determinant (7) was slightly negative in 19 years, and slightly positive in 14 years. Thus, estimated cost functions are both quasi-concave. Monotonicity and quasi-concavity of the cost function are sufficient conditions for cost minimization.

33. Q_1 and Q_2 is feasible if there exist prices p_1 and p_2 such that $Q_1 = D(p_1)$, $Q_2 = D(p_2)$ and $R(p_1, p_2) = p_1 Q_1 + p_2 Q_2 - C[Q_1(p_1), Q_2(p_2)] \geq 0$ where $R(p_1, p_2)$ is the firm's profit function.

34. See Note 14 for a definition of economies of scope.

35. There are product-specific scale economies in product two if

$$\frac{C(Q_1, Q_2) - C(Q_1, 0)}{Q_2 \dfrac{\partial C}{\partial Q_2}} > 1$$

and similarly for product one.

36. See Fuss and Waverman, *op. cit.*, for a tentative test of economies of scope.

37. Excluding the complements of (ϕ, ω) and the values which do not satisfy the inequalities given by (10.26).

38. Of course, it is possible that the cost function is subadditive at output vectors other than those realized between 1958 and 1977. If introducing competition into the telecommunications industry lead to radical changes in prices, it is possible that an output vector could be realized at which the cost function is subadditive. If the introduction of competition leads to only minor changes in prices and if the demand is fairly stable over time, our results suggest that several firms could meet demand more cheaply than a single firm could. Our results are too imprecise to evaluate the relative costs of *particular* industry configurations although they show that *some* multifirm configurations would be more efficient than a single firm.

39. These data were submitted as an appendix to Christensen's written testimony in *US* v. *AT&T*.

40. Christensen, Cummings, and Schoech, *op cit.*, p. 9. In calculating the user cost of capital, he "used the embedded cost of capital used for capital budget planning in the Bell System." See Appendix 2 to Christensen's testimony in *US* v. *AT&T*.

41. Vinod (1976), *op. cit.*

Bibliography

Articles

Areeda, P., and Turner, D., "Predatory Pricing and Related Practices Under Section 2 of the Sherman Act," *Harvard Law Review*, February 1975, pp. 697–733.

Asbury, J., and Webb, S., "Decentralized Electric Power Generation: Some Probable Effects," *Public Utilities Fortnightly*, September 25, 1980, pp. 20–24.

Averch, H., and Johnson, L., "Behavior of the Firm under Regulatory Constraint," *American Economic Review*, December 1962, pp. 1052–1069.

Bailey, E., and Panzar, J., "The Contestability of Airline Markets during the Transition to Deregulation," *Journal of Law and Contemporary Problems*, Winter 1981, pp. 125–145.

———, and Willig, R., "Ramsey Optimal Pricing of Long-Distance Telephone Service," in J. Wenders, *Pricing in Regulated Industries*, Vol. I (Keystone, CO: Mountain States Telephone and Telegraph, 1977).

Barten, A., "Maximum Likelihood Estimation of Complete Systems of Demand Equations," *European Economic Review*, May 1967, pp. 7–73.

Baseman, K., "Open Entry and Cross-Subsidization in Regulated Markets," in G. Fromm, ed., *Studies in Public Regulation* (Cambridge: MIT Press, 1981).

Baumol, W., and Bradford, D., "Optimal Departures from Marginal Cost Pricing," *American Economic Review*, June 1970, pp. 265–283.

———, and Klevorick, A., "Input Choice and Rate-of-Return Regulation: An Overview of the Discussion," *Bell Journal of Economics and Management Science*, Autumn 1970, pp. 162–190.

———, Bailey, E., and Willig, R., "Weak Invisible Hand Theorems on the Sustainability of Natural Monopoly," *American Economic Review*, June 1977, pp. 350–365.

———, and Willig, R., "Fixed Costs, Sunk Costs, Public Goods, Entry Barriers, and the Sustainability of Natural Monopoly," undated, unpublished paper.

Beam, F., "Preface," in J. Ainsworth and G. Johnston, *A Discussion of Telephone Competition* (Chicago: International Independent Telephone Association, 1908).

Becker, G., "The Economics of Crime and Punishment," *Journal of Political Economy,* March/April 1968, pp. 169–237.

Bergson, A., "The Current Soviet Planning Reforms," in Alexander Bolinky, *et al., Planning and the Market in the USSR: The 1960s* (New Brunswick, NJ: Rutgers, 1967), pp. 43–64.

Berndt, E., and Savin, N., "Estimation and Hypothesis Testing in Singular Equation Systems with Autoregressive Disturbances," *Econometrica,* September/November 1975, pp. 937–957.

Bhandar, S., Soldofsky, L., and Boc, W., "Bond Quality Changes: A Multivariate Analysis," *Financial Management,* Spring 1979, pp. 74–81.

Bolter, W., "The FCC's Selection of a 'Proper' Costing Standard after Fifteen Years— What Can We Learn from Docket 18128?" in H. Trebing, ed., *Assessing New Pricing Concepts in Public Utilities* (East Lansing, MI: Institute of Public Utilities, Michigan State University, 1978).

Box, G., and Cox, D., "An Analysis of Transformations," *Journal of the Royal Statistical Society,* Series B, 26, 1964, pp. 211–243.

Breautigam, R., "Optimal Pricing with Intermodal Competition," *American Economic Review,* March 1979, pp. 219–240.

Brock, W., and Dechert, W., "Dynamic Ramsey Pricing," SSRI Working Paper No. 8253 (Madison, WI: Social Systems Research Institute, University of Wisconsin at Madison, June 1982).

————, and Magee, S., "The Economics of Special Interest Politics: The Case of a Tariff," *American Economic Review,* May 1978, pp. 246–250.

————, Miller, R., and Scheinkman, J., "Natural Monopoly and Regulation," unpublished paper.

————, and Scheinkman, J., "Free Entry and the Sustainability of Natural Monopoly: Bertrand Revisited by Cournot," SSRI Working Paper No. 8126 (Madison, WI: Social Systems Research Institute, University of Wisconsin at Madison, March 1981).

Burgess, D., "Duality Theory and Pitfalls in the Specifications of Technologies," *Journal of Econometrics,* May 1975, pp. 105–123.

Caves, D., Christensen, L., and Trethway, M., "Flexible Cost Functions for Multiproduct Firms," *Review of Economics and Statistics,* August 1980, pp. 477–481.

Christensen, L., Cummings, D., and Schoech, P., "Econometric Estimation of Scale Economies in Telecommunications," SSRI Working Paper No. 8013 (Madison, WI: Social Systems Research Institute, University of Wisconsin at Madison, August 1981).

————, Jorgenson, D., and Lau, L., "Transcendental Logarithmic Production Functions," *Review of Economics and Statistics,* February 1973, pp. 28–49.

Cowing, T., "The Effectiveness of Rate-of-Return Regulation: An Empirical Test Using Profit Functions," in M. Fuss and D. McFadden, eds., *Production Economics: A Dual Approach to Theory and Applications,* Vol. II (Amsterdam: North-Holland, 1978).

Debreu, G., "Excess Demand Functions," *Journal of Mathematical Economics,* March 1974, pp. 15–21.

Denny, M., and Pinto, C., "An Aggregate Model with Multiproduct Technologies," in M. Fuss and D. McFadden, eds., *Production Economics: A Dual Approach to Theory and Applications* (Amsterdam: North-Holland, 1978).

Dingee, A., Jr., Snollen, L., and Haslett, B., "Characteristics of a Successful Entrepreneur," in S. Rubel, ed., *A Guide to Venture Capital Sources* (Chicago: Capital Publishing Corporation, 1977).

Dixit, A., "The Role of Investment in Entry Deterrence," *Economic Journal,* March 1980, pp. 95–106.

———, "Recent Developments in Oligopoly Theory," *American Economic Review,* May 1982, pp. 12–17.

Easterbrook, F., "Predatory Strategies, and Counterstrategies," *University of Chicago Law Review,* Spring 1981, pp. 263–311.

Eaton, B., and Lipsey, R., "Capital, Commitment, and Entry Equilibrium," *Bell Journal of Economics.* August 1981, pp. 593–609.

Fisher, F., "The Use of Multiple Regression Analysis in Legal Proceedings," *Columbia Law Review,* May 1980, pp. 702–736.

Fisher, L., "Determinants of Risk Premiums on Corporate Bonds," *Journal of Political Economy,* June 1959, pp. 217–237.

Fuss, M., "A Survey of Recent Results in the Analysis of Production Conditions in Telecommunications" (Toronto: Institute for Policy Analysis, University of Toronto, June 1981).

Gabel, R., "The Early Competitive Era in Telephone Communications, 1893–1920," *Journal of Law and Contemporary Problems,* Spring 1969, pp. 340–359.

Gallant, A., "Seemingly Unrelated Nonlinear Regressions," *Journal of Econometrics,* April 1975, pp. 35–50.

———, "On the Bias in Flexible Functional Forms and an Essentially Unbiased Form," *Journal of Econometrics,* February 1981, pp. 211–246.

Gordon, R., and Malkiel, B., "Corporation Finance," in Henry J. Aaron and Joseph A. Pechman, eds., *The Effects of Taxation on Economic Behavior* (Washington, DC: Brookings, 1981).

Grossman, S., "Nash Equilibrium and the Industrial Organization of Markets with Large Fixed Costs," *Econometrica,* September 1981, pp. 1149–1172.

Guilkey, D., and Lovell, C., "On the Flexibility of the Translog Approximation," *International Economic Review,* February 1980, pp. 137–147.

Haslett, B., and Snollen, L., "Preparing a Business Plan," in S. Rubel, *A Guide to Venture Capital Sources* (Chicago: Capital Publishing Corporation, 1977).

Hayek, F., "The Use of Knowledge in Society," *American Economic Review,* September 1945, pp. 519–530.

Hoerl, A., and Kennard, R., "Ridge Regression: Biased Estimation for Nonorthogonal Problems," *Technometrics,* February 1970, pp. 55–67.

Horrigan, J., "The Determination of Long-Term Credit Standing with Financial Ratios," *Journal of Accounting Research,* supplementary issue entitled *Empirical Research in Accounting: Selected Studies,* 1966, pp. 44–62.

Joskow, P., "Inflation and Environmental Concern," *Journal of Law and Economics,* October 1974, pp. 291–328.

———, and Noll, R., "Theory and Practice in Public Regulation: A Current Overview," in G. Fromm, ed., *Studies in Public Regulation* (Cambridge: MIT Press, 1981).

Kaplan, R., and Urwitz, G., "Statistical Models of Bond Ratings: A Methodological Inquiry," *Journal of Business,* February 1974, pp. 231–291.

Klein, B., Crawford, W., and Alchian, A., "Vertical Integration, Appropriable Rents, and the Competitive Contracting Process," *Journal of Law and Economics*, October 1978, pp. 297–326.

Kiss, F., Karabadjian, and LeFabvre, B., "Economies of Scale and Scope in Bell Canada," presented at the Telecommunications in Canada Conference, March 4–6, 1981.

Kneips, G., and Vogelsang, K. "The Sustainability Concept Under Alternative Behavioral Assumptions," *Bell Journal of Economics*, Spring 1982, pp. 234–241.

Kreps, D., and Wilson, R., "On the Chain-Store Paradox and Predation: Reputation for Toughness," IMSSS Technical Report No. 319 (Palo Alto, CA: IMSSS, Stanford University, 1980).

———, "Reputation and Imperfect Information," *Journal of Economic Theory*, August 1982, pp. 280–312.

Leamer, E., "Multicollinearity: A Bayesian Interpretation," *Review of Economics and Statistics*, August 1973, pp. 371–380.

Mantel, R., "On the Characterization of Aggregate Excess Demand," *Journal of Economic Theory*, March 1974, pp. 248–353.

McGee, J., "Predatory Price Cutting: The Standard Oil (NJ) Case, *Journal of Law and Economics*, October 1958, pp. 133–167.

Miller, M., "Debt and Taxes," *Journal of Finance*, May 1977, pp. 261–275.

Milgrom, P., and Roberts, J., "Predation, Reputation, and Entry Deterrence," Research Paper No. 600 (Evanston, IL: Graduate School of Business, Northwestern University, 1981).

Modigliani, F., "New Developments on the Oligopoly Front," *Journal of Political Economy*, June 1958, pp. 215–232.

———, and Miller, M., "The Cost of Capital, Corporation Finance, and the Theory of Investment," *American Economic Review*, June 1958, pp. 261–297.

Nadiri, M., and Shankerman, M., "The Structure of Production, Technological Change, and the Rate of Growth of Total Factor Productivity in the U.S. Bell System," in T. Cowing and R. Stevenson, eds., *Productivity Measurement in Regulated Industries* (New York: Academic Press, 1981).

Nagel, T., "Interconnection and Reliability," in H. Trebing, ed., *Energy and Communications in Transition* (East Lansing, MI: Institute for Public Utilities, Michigan State University, 1981).

Panzar, J., "Regulation, Deregulation, and Economic Efficiency: The Case of the CAB," *American Economic Review*, May 1980, pp. 311–315.

———, "Sustainability, Efficiency, and Vertical Integration," D. P. No. 165 (Holmdel, NJ: Bell Laboratories, August 1980).

———, and Willig, R., "Free Entry and the Sustainability of Natural Monopoly," *Bell Journal of Economics*, Spring 1977, pp. 1–22.

Raj, B., and Vinod, H., "Bell System Scale Economies from a Randomly Varying Parameter Model," *Journal of Economics and Business*, February 1982, pp. 247–252.

Ramsey, F., "A Contribution to the Theory of Taxation," *Economic Journal*, March 1927, pp. 47–61.

Reece, J. and Cool, W., "Measuring Investment Center Performance," *Harvard Business Review*, May/June 1978, pp. 28–49.

Rothschild, M., and Stiglitz, J., "Increasing Risk I: A Definition," *Journal of Economic Theory*, September 1970, pp. 66–84.

————, "Increasing Risk II: Its Economic Consequences," *Journal of Economic Theory*, March 1971, pp. 66–84.

Rubel, S., "Guidelines for Dealing with Venture Capital Firms," in S. Rubel, ed., *A Guide to Venture Capital Sources* (Chicago: Capital Publishing Corporation, 1977).

Scherer, F., "Predatory Pricing and the Sherman Act: A Comment," *Harvard Law Review*, March 1976, pp. 869–900.

Scott, J., Jr., "On the Theory of Conglomerate Merger," *Journal of Finance*, September 1977, pp. 1235–1250.

Sharkey, W., and Telser, L., "Supportable Cost Functions and the Multiproduct Firm," *Journal of Economic Theory*, June 1978, pp. 23–37.

Smith, J. B., and Corbo, J., "Economies of Scale and Economies of Scope in Bell Canada," Working Paper (Montreal: Department of Economics, Concordia University, March 1979).

Sonnenschein, H., "Market Excess Demand Functions," *Econometrica*, April 1972, pp. 549–563.

Spann, R., "Rate of Return Regulation and the Efficiency of Production: An Empirical Test of the Averch–Johnson Thesis," *Bell Journal of Economics and Management Science*, Spring 1974, pp. 38–52.

Stiglitz, J., "A Re-Examination of the Modigliani and Miller Theorem," *American Economic Review*, December 1969, pp. 783–793.

Taplon, S., and Gerwitz, S., "The Effect of Regulated Competition on the Air Transport Industry," *Journal of Air Law and Commerce*, Vol. XXII, No. 2, 1955, pp. 527–562.

Tobin, J., "A General Equilibrium Approach to Monetary Theory," *Journal of Money Credit and Banking*, February 1969, pp. 15–29.

Turnovsky, S., and Brock, W., "Time Consistency and Optimal Government Policies in Perfect Foresight Equilibrium," *Journal of Public Economics*, April 1980, pp. 183–212.

Vinod, H., "Nonhomogenous Production Functions and Applications to Telecommunications," *Bell Journal of Economics and Management Science*, Autumn 1972, pp. 531–543.

————, "Bell System Economies and the Economics of Joint Production," Bell Exhibit No. 59, FCC Docket 20003.

————, "Canonical Ridge and the Econometrics of Joint Production," *Journal of Econometrics*, August 1976, pp. 147–166.

————, "Applications of New Ridge Regression Methods to a Study of Bell System Scale Economies," *Journal of the American Statistical Association*, December 1976, pp. 835–841.

————, "A Survey of Ridge Regression and Related Techniques for Improvements over Least Squares," *Review of Economics and Statistics*, February 1978, pp. 121–131.

von Weizsäcker, C., "A Welfare Analysis of Barriers to Entry," *Bell Journal of Economics*, Autumn 1980, pp. 399–420.

Weitzman, M., "Comment on Contestable Markets: An Uprising in the Theory of Industry Structure," Working Paper (Cambridge: Department of Economics, MIT, May 1982).

White, H., "A Heteroskedasticity Consistent Covariance Matrix Estimator and a Direct Test of Heteroskedasticity," *Econometrica*, May 1980, pp. 817–830.

Williamson, O., "The Vertical Integration of Production: Market Failure Considerations," *American Economic Review*, May 1971, pp. 114–118.

————, "Predatory Pricing: A Strategic and Welfare Analysis," *Yale Law Journal,* December 1977, pp. 284–340.

Woodland, A., "Stochastic Specification and Estimation of Share Equations," *Journal of Econometrics,* August 1979, pp. 361–384.

Books

Aaron, H., and Pechman, J., eds., *The Effects of Taxation on Economic Behavior* (Washington, DC: Brookings, 1981).

Ainsworth, J., and Johnston, G., *A Discussion of Telephone Competition* (Chicago: International Independent Telephone Association, 1908).

Anderson, G. W., *Telephone Competition in the Middle West and its Lesson for New England* (Boston: New England Telephone Company, September 1906).

AT&T, *Bell System Statistical Manual* (New York: AT&T, May 1980).

————, *Annual Report,* various years.

Bain, J., *Barriers to New Competition* (Cambridge: Harvard, 1956).

Baumol, W., Panzar, J., and Willig, R., *Contestable Markets and the Theory of Industry Structure* (San Diego: Harcourt, Brace, Jovanovich, 1982).

Bergson, A., *Planning and Productivity under Soviet Socialism* (Pittsburgh: Carnegie–Mellon University, 1968).

Bonbright, J., *Principles of Public Utility Rates* (New York: Columbia University Press, 1961).

Bork, R., *The Antitrust Paradox* (New York: Basic Books, 1978).

Breslaw, J., and Smith, J., *Efficiency, Equity, and Regulation: An Econometric Model of Bell Canada.* Report to the Department of Communications, Canada, Interim Report, December 1979, and Final Report, March 1980.

Brock, G., *The Telecommunications Industry* (Cambridge: Harvard University Press, 1981).

Brock, W., and Evans, D., *Federal Regulation of Small Business* (New York: Holmes and Meiers, forthcoming, 1984).

Brooks, J., *Telephone: The First Hundred Years* (New York: Harper and Row, 1975).

Chandler, A., *The Visible Hand: The Managerial Revolution in American Business* (Cambridge: Harvard University Press, 1977).

Cournot, A., *Researches into the Mathematical Principles of the Theory of Wealth,* Nathaniel Bacon, trans. (New York: Kelly, 1960).

Cowing, T., and Stevenson, R., eds., *Productivity Measurement in Regulated Industries* (New York: Academic Press, 1981).

Danielian, N., *AT&T: The Story of Industrial Conquest* (New York: Vanguard, 1974).

Diamond, P., and Rothschild, M., eds., *Uncertainty in Economics: Readings and Exercises* (New York: Academic Press, 1978).

Friedman, J., *Oligopoly and the Theory of Games* (Amsterdam: North Holland, 1977).

Fromm, G., ed., *Studies in Public Regulation* (Cambridge: MIT Press, 1981).

Fuss, M., and McFadden, D., *Production Economics: A Dual Approach to Theory and Applications,* 2 vols. (Amsterdam: North Holland, 1978).

————, and Waverman, L., *The Regulation of Telecommunications in Canada*. Technical Report No. 7, Economic Council of Canada, March 1981.

Gelfand, I., and Fomin, S., *Calculus of Variations* (Englewood Cliffs, N.J., Prentice-Hall, 1963).

Goldfeld, S., and Chandler, L., *The Economics of Money and Banking*, 8th ed. (New York: Harper and Row, 1981).

Hunt, C., Jr., *Competition in Telecommunications: A Surcharge as a Method to Promote Competition in Private Line Services*. Ph.D. Thesis, University of Colorado at Boulder, 1980.

International Monetary Fund, *International Monetary Fund Statistics* (Washington, DC: International Monetary Fund, October 1981).

Johnston, G., *Some Comments on the 1907 Annual Report of AT&T* (Chicago: International Independent Telephone Association, September 1908).

Judge, G., Griffiths, W., and Hill, R., *The Theory and Practice of Econometrics* (New York: Wiley, 1980).

Kahn, A., *The Economics of Regulation: Principles and Institutions*, 2 vols. (New York: John Wiley and Sons, 1970).

Kmenta, J., *Elements of Econometrics* (New York: MacMillan, 1971).

Knight, F., *Risk, Uncertainty and Profit* (Chicago: University of Chicago, Press, 1971).

Littlechild, S.C., *Elements of Telecommunications Economics* (London: Institute of Electrical Engineers, 1979).

MacMeal, H., *The Story of Independent Telephony* (Chicago: Independent Pioneer Telephone Association, 1934).

Meyer, J., Wilson, R., Baughcum, A., Burton, E., and Caouette, L., *The Economics of Competition in the Telecommunications Industry* (Cambridge: Oelgeschlager, Gunn, and Hain, 1980).

Moody's Financial Services, *Public Utility Manual*, 1979.

Owen, B., and Breautigam, R., *The Regulation Game* (Cambridge, Ballinger, 1978).

Posner, R., *Antitrust: Cases, Economic Notes, and Other Materials* (St. Paul: West Publishing Co., 1974).

————, *Economic Analysis of Law* (Boston: Little, Brown, 1972).

Rao, C., *Linear Statistical Inference and its Applications*, 2d ed. (New York: John Wiley and Sons, 1973).

Reed, O., and Welch, W., *From Tin Foil to Stereo* (New York: Bobbs-Merrill, 1959).

Rubel, S., ed., *A Guide to Venture Capital Sources* (Chicago: Capital Publishing Corporation, 1977).

Salop, S., *Strategic Predation and Antitrust Analysis* (Washington, DC: Federal Trade Commission, 1981).

Shephard, R., *Theory of Cost and Production Functions* (Princeton: Princeton University Press, 1970).

Skoog, R., *Design and Cost Characteristics of Telecommunications Networks* (Holmdel, NJ: Bell Laboratories, 1980).

Smith, A., *Wealth of Nations* (New York: The Modern Library, 1937).

Stehman, J., *The Financial History of the American Telephone and Telegraph Company* (Boston: Houghton Mifflin, 1925).

Stigler, G., *The Organization of Industry* (Homewood, IL: Irwin, 1968).

Sylos-Labini, P., *Oligopoly and Technical Progress* (Cambridge: Harvard University Press, 1962).

Takayama, A., *Mathematical Economics* (Hinsdale, IL: The Dryden Press, 1974).

Taylor, G., and Neu, I., *The American Railroad Network* (Cambridge, Harvard University Press, 1956).

Taylor, L., *Telecommunications Demand: A Survey and Critique* (Cambridge: Ballinger, 1980).

Vancil, R., *Decentralization: Managerial Ambiguity by Design* (Homewood, IL: Dow Jones-Irwin, 1978).

VanHorne, J., *Financial Management and Policy*, 5th ed. (Englewood Cliffs, NJ: Prentice Hall, 1980).

von Weizsäcker, C., *Barriers to Entry* (Berlin: Springer-Verlag, 1980).

Wenders, J., *Pricing in Regulated Industries*, 2 vols. (Keystone, CO: Mountain States Telephone and Telegraph, 1977).

Zajac, E., *Fairness or Efficiency: An Introduction to Public Utility Pricing* (Cambridge: Ballinger, 1978).

Court and Regulatory Decisions

Bluefield Water Works and Improvement Co. v. *Public Service Commission*, 263U.S.679 (1923).

Eastern Railroad Presidents' Conference v. *Noerr Motor Freight, Inc.*, 305U.S.127 (1961).

Federal Communications Commission, *Specialized Common Carriers*, 29FCC2d (1971).

Federal Power Commission v. *Hope Natural Gas*, 320U.S.591 (1944).

MCI Telecommunications Corp. v. *FCC*, 580F.2d.590 (1978).

Permian Basin Area Rate Cases, 390U.S.747 (1968).

United Mine Workers v. *Pennington*, 318U.S.657 (1965).

U.S. Telephone Co. v. *Central Union Telephone Co.*, 202Fed66, 6thCir (1913).

Documents and Testimony in *US* v. *AT&T*

AT&T, *Defendants' Third Statement of Contentions and Proof*, 3 vols, March 10, 1980.

―――, *Defendants' Memorandum in Support of Defendants' Motion for Involuntary Dismissal Under Rule 41(b)*, July 31, 1981.

Greene, Judge Harold, *Opinion on Defendants' Motion for Involuntary Dismissal Under Rule 41(b)*, September 11, 1981.

―――, *Opinion*, August 11, 1982.

U.S. Department of Justice, *Plaintiff's Third Statement of Contentions and Proof*, 2 vols., January 10, 1982.

―――, *Plaintiff's Memorandum in Opposition to Defendants' Motion for Involuntary Dismissal Under Rule 41(b)*, August 16, 1981.

See also oral and written testimony by Kenneth Arrow, William Baumol, Laurits Christensen, William Nordhaus, Almarin Phillips, James Rosse, Malcom Schwartz, Ronald Skoog, David Teece, and H. D. Vinod for AT&T and Nina Cornell, Walter Hinchman, Aaron Kerschenbaum, and Bruce Owen for the U.S. Department of Justice.

Government Publications

Bureau of the Census, *Historical Statistics of the United States: Colonial Times to 1970* (Washington, DC: Government Printing Office, 1975).

———, *Telephones and Telegrams 1912* (Washington, DC: Government Printing Office, 1915).

Federal Communications Commission, *Statistics of Communications Common Carriers* (Washington, DC: Government Printing Office, 1947–1979).

———, *Telephone Investigation: Proposed Report* (Washington, DC: Government Printing Office, 1938).

Securities Exchange Commission, *Prospectus*, filed by MCI Communications Corporation, University Computer Corporation, and Wyly Corporation, 1969–1980.

Magazine and Newspaper Articles

"American Bell Telephone and Proposed Competition," *Commercial and Financial Chronicle*, December 16, 1899, pp. 1222–1224.

Beam, F., "Address," *Telephony*, April 1908, p. 253.

Clausen, H., "Standardization Possibilities," *Telephony*, January 1908, pp. 32–34.

"Credit Quality of Some Units Questioned," *The New York Times*, January 12, 1982, p. 27.

Dickson, F., "Independent Finances," *The Telephone Magazine*, January 1905, p. 29.

"Fortune 500 Industrials," *Fortune Magazine*, May 5, 1980.

Hoge, J., "National Inter-State Telephone Association," *The Telephone Magazine*, July 1905, p. 374.

———, "Necessity of State and National Organizations," *The Telephone Magazine*, June 1905, pp. 373–375.

Holt, D., "The Tax Break that Turned into a Nightmare," *Fortune Magazine*, September 10, 1981, pp. 110–115.

Houck, Z., "Long Distance Telephone Lines," *The Telephone Magazine*, March 1905, pp. 181–183.

Lauback, W., "Standard Toll Signs," *The Telephone Magazine*, July 1905, pp. 60–62.

Lindermuth, A., "Address Before the West Virginia and Indiana Telephone Conventions," *Telephony*, June 1908, p. 383.

"S&P Debates Maintaining AT&T Rating," *The Washington Post*, January 13, 1982, p. D-8.

Special Tariff Committee, "Interchanged Toll Business," *The Telephone Magazine*, July 1905, pp. 103–108.

Stanton, W., "Standardization of Toll Line Equipment," *Telephony*, May 1907, p. 306.

Thomas, J., "Address," *Telephony*, July 1900, p. 126.

Tiffany, F., "The Chicago Bell Franchise," *Telephony*, December 1907, p. 323.

———, "The Chicago Situation," *Telephony*, October 1907, p. 202–204.

Author Index

Subject Index